Annals of Theoretical Psychology

Volume 2

A Continuation Order Plan is available for this series. A continuation order will bring delivery of each new volume immediately upon publication. Volumes are billed only upon actual shipment. For further information please contact the publisher.

Annals of Theoretical Psychology

Volume 2

Edited by
JOSEPH R. ROYCE
and
LEENDERT P. MOS

Center for Advanced Study in Theoretical Psychology
University of Alberta
Edmonton, Alberta, Canada

Springer Science+Business Media, LLC

ISBN 978-1-4757-9193-8 **ISBN 978-1-4757-9191-4 (eBook)**
DOI 10.1007/978-1-4757-9191-4

Originally published by Plenum Press, New York in 1984.
Softcover reprint of the hardcover 1st edition 1984

Contributors

William J. Baker, Center for Advanced Study in Theoretical Psychology and Department of Linguistics, University of Alberta, Edmonton, Alberta, Canada

Lewis Wolfgang Brandt, Department of Psychology, University of Regina, Regina, Saskatchewan, Canada

Gordon F. Derner, Late of the Institute of Advanced Psychological Studies, Adelphi University, Garden City, New York

Rudolf Ekstein, 536 South Westgate Avenue, Los Angeles, California

Mark F. Ettin, Department of Psychiatry, UMDNJ-Rutgers Medical School, Piscataway, New Jersey

David A. Freedman, Department of Psychiatry, Baylor College of Medicine, Texas Medical Center, Houston, Texas

Michael E. Hyland, Department of Psychology, Plymouth Polytechnic, Drake Circus, Plymouth, Devon, England

Arthur R. Jensen, Institute of Human Learning, University of California at Berkeley, Berkeley, California

K. B. Madsen, Royal Danish School of Educational Studies, Copenhagen N.V., Denmark

Philip K. Peake, Department of Psychology, Smith College, Northampton, Massachusetts

Lawrence A. Pervin, Department of Psychology, Rutgers University, New Brunswick, New Jersey

William T. Powers, 1138 Whitfield Road, Northbrook, Illinois

Joel O. Raynor, Department of Psychology, State University of New York at Buffalo, Buffalo, New York

Daniel N. Robinson, Department of Psychology, Georgetown University, Washington, D.C.

J. Philippe Rushton, Department of Psychology, University of Western Ontario, London, Ontario, Canada

Robin J. H. Russell, Department of Psychology, Goldsmiths' College, University of London, London, England

Joseph F. Rychlak, Department of Psychology, Loyola University of Chicago, Chicago, Illinois

Dirk L. Schaeffer, Psychological Assessment Consultation Evaluation (PACE), 10322A–121 Street, Edmonton, Alberta, Canada

Jan Smedslund, Institute of Psychology, University of Oslo, Blindern, Oslo, Norway

Herman Tennessen, Department of Philosophy and Center for Advanced Study in Theoretical Psychology, University of Alberta, Edmonton, Alberta, Canada

Philip E. Vernon, Department of Educational Psychology, University of Calgary, Calgary, Alberta, Canada

Fred Vollmer, Department of Psychology, University of Bergen, Bergen, Norway

Walter B. Weimer, Department of Psychology, Pennsylvania State University, University Park, Pennsylvania

Pamela A. Wells, Department of Psychology, Goldsmiths' College, University of London, London, England

K. V. Wilkes, St. Hilda's College, Oxford University, Oxford, England

Benjamin B. Wolman, Suite 6D, 10 West 66th Street, New York, New York

Call for Papers and Commentary

Although the space for Volumes 3 and 4 is committed, readers are invited to submit papers and comments for subsequent volumes in this annual series.

Papers (up to 15,000 words) concerned with substantive theory, metatheory, or a mixture thereof are eligible. We are also soliciting comments (up to 2,500 words) on previously published papers and commentaries.

Send the original and four carbon copies, following APA guidelines, to the editors, Center for Advanced Study in Theoretical Psychology, University of Alberta, Edmonton, Alberta, Canada, T6G 2E9.

Preface

As such things happen, several manuscripts in the present volume were under review prior to the ones that appeared in Volume I of the *Annals*. A major difficulty encountered in the preparation of these volumes—apart from working up to three years in advance of publication—is eliciting appropriate commentary. If this format is to succeed, the commentary must be both engaging to the reader and satisfying to the author. It is not yet clear how successful we have been in this regard and, indeed, we do not feel bound to publish commentary with each manuscript that is accepted for publication. Nevertheless, we do invite readers' commentaries on published materials.

The contributions by Jan Smedslund and Benjamin Wolman in this volume have been through an inordinately long publication lag. We have been in receipt of both manuscripts since early in 1981 and Dr. Smedslund, especially, has since clarified and advanced his views elsewhere in print. K. B. Madsen and Joseph Rychlak submitted their manuscripts in the fall of 1981 while Michael Hyland and J. Philippe Rushton had first drafts of their manuscripts accepted for publication in the fall of 1982. We are grateful to our contributors for their expressed commitment to the *Annals* and assure potential contributors that the delay in publication is a mere matter of getting the series off the ground.

We thank Mrs. E. Murison and Mrs. F. Rowe for their secretarial assistance; and The University of Alberta for the opportunity to be editorially engaged.

LEENDERT P. MOS

Contents

1

Sociobiology

Toward a Theory of Individual and Group Differences in Personality and Social Behavior

J. Philippe Rushton

Abstract. Sociobiology, the latest synthesis of Darwinian theory, has many implications for the psychology of individual differences. Six issues are reviewed within a general context of sociobiological considerations: (a) the notion of genetic variance; (b) the fundamental postulate of sociobiology, that is, that individuals behave so as to maximize their inclusive fitness; (c) an application of the sociobiological perspective to possible universals in human behavior; (d) the inheritance of individual differences in activity level, aggression, altruism, chronogenetics, criminality, dominance, emotionality, intelligence, locus of control, political attitudes, sexuality, sociability, values, and vocational interest; (e) group differences (e.g., sex, socioeconomic, and ethnic) in inherited behavior; and (f) genetic trait × social learning interactions. It is concluded that a significant proportion of human personality is inherited and that this has important implications for the behavioral sciences.

Developmental, personality, and social psychologists have focused much attention in recent years on how human behavior is acquired and modified through socialization. Particular attention has been placed on such processes of social learning as classical conditioning, instrumental and operant learning, observational learning, and learning through verbal

---

J. Philippe Rushton • Department of Psychology, University of Western Ontario, London, Ontario, Canada N6A 5C2. Portions of this paper were written while I was a Visiting Scholar at the Institute of Human Development, University of California, Berkeley, January to June 1981. Other portions were completed while on sabbatical leave (1982–83) at the University of London Institute of Psychiatry supported by Social Sciences and Humanities Research Council of Canada Leave Fellowship 451-82-0603.

instruction. Modern social learning theories also emphasize the role played by cognition, as in symbolic and self-regulatory processes (e.g., Bandura, 1977). Cognitive social learning theory amounts to an integrative paradigm in developmental-social-personality psychology providing systematized knowledge of human behavior in areas such as aggression (Bandura, 1973), altruism (Rushton, 1980), cognition (Rosenthal & Zimmerman, 1978), deviancy (Akers, 1977), personality (Mischel, 1981), and psychopathology (Wilson & O'Leary, 1980).

Despite recent advances in knowledge, large areas of uncertainty remain in understanding human behavior. Some investigators believe that additional research similar to the type that has produced successful results to date is the optimal strategy for completing this undertaking. An alternative view, adopted here, is to broaden current theory by taking evolutionary biology into account. This paper explores several ways in which sociobiology—the latest synthesis of Darwinian theory—may illuminate human social behavior and, in particular, individual differences. Perhaps one reason why evolutionary biology has had so little impact on current theorizing in psychology is its traditional focus on morphology rather than behavior. However, *sociobiology,* defined as "the systematic study of the biological basis of all social behavior" (Wilson, 1975, p. 4), makes explicit the attempt to unify "all aspects of social evolution, including that of man" (Wilson, 1975, p. 4).

The "new synthesis" of sociobiology has at its roots the view that "the organism is only DNA's way of making more DNA" (Wilson, 1975, p. 3). This represents a conceptual advance over Darwin's idea of the survival of the "fittest" individual, for it is now DNA, not the individual, that is "fit." According to this view, an individual organism is only a vehicle, part of an elaborate device that ensures the survival and replication of genes with the least possible biochemical alteration. Thus an appropriate unit of analysis for understanding natural selection and a variety of behavior patterns is the gene. Any means by which a pool of genes, in a group of individuals, can be transmitted more effectively to the next generation will be adopted (Hamilton, 1964). Here, it is suggested, are the origins of maternal behavior, sterility in castes of worker ants, aggression, cooperation, and self-sacrificial altruism. All these phenomena are means by which genes can be more readily transmitted. Dawkins (1976) captures this idea perfectly in the title of his book: *The Selfish Gene.*

The general framework of sociobiology has ordered an immense amount of disparate data, provided a theoretical framework for unrelated disciplines, and offered insights into the human condition (Alexander, 1979; Barash, 1982; Chagnon & Irons, 1979; Daly & Wilson, 1983; Dawkins, 1976, 1982; Freedman, 1979; Lumsden & Wilson, 1981, 1983; Wilson,

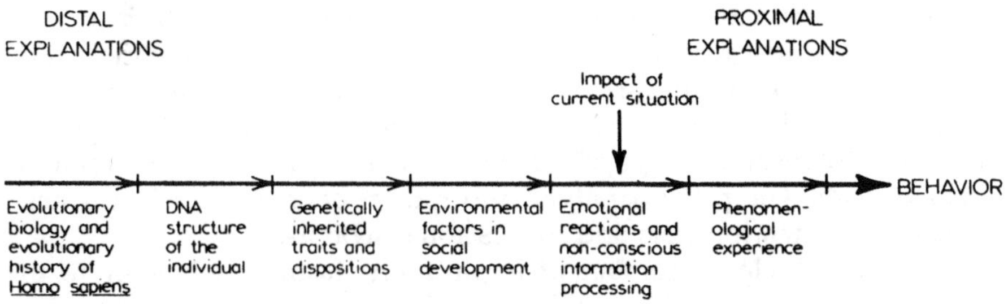

Figure 1. The distal-proximal dimension and levels of explanation in developmental, personality, and social psychology.

1975, 1978). There have been many criticisms of sociobiology, however, from several different perspectives. These will not be addressed here, although for a full exposition of the debate surrounding sociobiology, see Barlow and Silverberg (1980), Campbell (1975), Gould (1981), Gregory, Silvers, and Sutch (1978), Lewontin (1979), Montagu (1980), Ruse (1979), Wispe and Thompson (1976), Wyers *et al.* (1980), and the commentaries on Lumsden and Wilson's (1981) *Genes, Mind and Culture* (see Lumsden & Wilson, 1982).

Although several issues are involved in the controversy over sociobiology, many result from a confusion between distal and proximal levels of explanation (see Figure 1). When explanations move from distal to proximal levels controversy does not normally ensue. Evolutionary biologists do not usually find the heritability of traits problematic, and most trait theorists accept that behavioral dispositions are modified by later learning. In addition, learning theorists believe that the products of early experiences interact with subsequent situations to produce emotional arousal and cognitive information processing which in turn give rise to the person's phenomenology just prior to his or her behavior. Disagreement and uncertainty are more likely, however, when explanation moves from proximal to distal levels. Thus some phenomenologists, situationists, and cognitivists, who focus attention on processes just prior to the behavior, mistrust the view that these processes themselves are partly determined by previous learning. Learning theorists, in turn, often do not readily accept the view that a person's previous learning history is partly a function of inherited traits. Often even behavior geneticists ignore the broader context of the evolutionary history of the animal from which they are attempting to breed selected traits.

Proximal wariness of distal explanation may be due in part to concern about extreme reductionism, for example, that phenomenology is entirely reducible to learning, or that learning is only secondary to genetics. Unfortunately, another reason for dispute arises from lack of knowledge. Most researchers seem devoted to an exclusive orientation (Royce, 1982). It is rare for cognitive social learning theorists to expose themselves to behavior genetics, or for humanistic phenomenologists to immerse themselves in psychometrics, or for trait theorists to pursue behaviorism. The psychoanalytic and radical behaviorist schisms even create their own journals and professional schools.

In this article, a number of implications from sociobiology to the psychology of personality will be considered. First, a general introduction to the notion of genetic variance will be presented. From the perspective of evolutionary biology, the genes provide the initial structure of the personality. Since all humans belong to the same species, there are universals in the structure of the personality. Since, however, individuals differ from one another in their genetic makeup, people inherit variations on the basic structure which result in genetically based individual differences in behavior. Following this general introduction, five issues will be reviewed: (a) the fundamental postulate of sociobiology and some of the evidence that has compelled biologists to take it seriously, (b) an application of the sociobiological perspective to possible universals in human social behavior, (c) the inheritance of individual differences in behavior traits, (d) group differences (e.g., sex, socioeconomic, and ethnic) in inherited behavior, and (e) genetic trait × social learning interactions. Although many of these issues are not new, their repetition may well be worthwhile in the new context of sociobiological considerations.

1. The Variability of Genetic Material

The first premise of the modern synthesis of Darwin's (1859) theory of natural selection is that individuals of the same species are not identical and that their differences are capable of being inherited by their offspring. Such differences may have arisen previously from the natural mutations and recombinations that occur within genetic material. The second premise of Darwin's theory is that some individuals are more successful than others in producing offspring that grow to reproductive maturity. This differential success results in some genetic characteristics increasing in frequency and others decreasing, in the next generation. This is the defining feature of evolution, and natural selection is the process by which evolution occurs.

Natural selection is perhaps simplest to understand in terms of a readily observable physical dimension such as skin coloration. In hot climates, for example, humans are more likely to survive when there is more pigment in their skin. This is because pigment absorbs the excessive ultraviolet radiation that occurs in these climates before it can reach and harm the sensitive layers of the skin. Conversely, in regions of high latitude and seasonal cloudiness, a white skin is advantageous, at least in winter, for it permits more vitamin D formation. As a result, people indigenous to hot climates are darker skinned, and people indigenous to cold and cloudy climates are lighter skinned (Coon, 1962).

Behavioral capacities and dispositions are comparable to skin color. Obvious examples of the inheritance of behavior include horses that run fast, dogs that point or round up sheep, and cats that like the company of human beings. Such animal traits have been selectively bred by humans for centuries, and experimental studies in laboratories have extended these to include such exotic traits as alcohol preference in mice, courtship and mating speed in fruit flies, dispersal tendency in milkweed bugs, and aggressiveness in domestic fowl (Plomin, DeFries, & McClearn, 1980).

Herding, which occurs in a number of species, provides an example of a social behavior that is under genetic control. Animals that herd typically display signs of discomfort if removed from conspecifics. This is to some extent naturally selected, since predators are better able to kill those individuals that do not stay with the herd. Any genes that dispose an animal to stray are thus selected out, while genes disposing the animal to remain with the herd increase in frequency.

2. Inclusive Fitness Strategies

The fundamental postulate of sociobiology is that individual organisms behave so as to maximize their inclusive fitness by propagating as many of their genes as possible into the next generations. By analyzing social behaviors in the way biologists have previously approached physical structures, that is, as adaptations that contribute to genetic fitness, sociobiologists have had some notable successes. Some of the most illuminating insights of the new approach involve altruistic behavior, sexuality, and notions of parental investment.

2.1. The Paradox of Altruism

Wilson (1975) describes altruism as constituting the "central theoretical problem of sociobiology" (p. 3). (By *altruism* sociobiologists mean behavior that benefits another.) The existence of altruism in animals presents a major problem for theories of evolution as was recognized early by Darwin (1871, p. 130). How could altruism evolve through his hypothesized "survival of the fittest individual" when such behavior would appear to diminish personal fitness? If the most altruistic members of a group sacrificed themselves for others, they would run the risk of leaving fewer offspring to carry forward their genes for altruistic behavior. Hence altruism would be selected out and, indeed, selfishness would be selected in. Many naturalistic studies, however, have demonstrated that altruistic behaviors are pervasive in animal species as disparate as social insects, birds, rabbits, deer, elephants, porpoises, and chimpanzees. The observed altruistic behaviors include parental behavior, mutual defense, rescue behavior, and food sharing (Wilson, 1975). Some species are altruistic to the point of self-sacrifice. For example, honey bees die when they sting in the process of protecting their nests. How could such behavior possibly evolve through Darwinian selection?

The solution to the paradox of altruism is one of the triumphs that led to the new synthesis of sociobiology. The answer lies in kin-selection. The central tenet of sociobiology is that individuals behave so as to maximize their *inclusive fitness* rather than only their *individual fitness;* they maximize the production of successful offspring by both themselves *and their relatives* (Hamilton, 1964). This is because it is genes that survive and are passed on. Some of the same genes will be found in siblings, nephews and nieces, grandchildren, and cousins as well as offspring. If an animal sacrifices its life for its siblings' offspring, it ensures the survival of *common genes,* for, on average, it shares 50% of its genes with each of its siblings and 25% with these siblings' offspring. It could be predicted, then, that the percentage of genes shared would be an important determinant of the amount of altruism displayed, and this is borne out in a number of species. Social ants, for example, are one of the most altruistic species so far discovered. The self-sacrificing, sterile worker and soldier ants do little else than serve their colony. However, they also share 75% of their genes with their sisters and so by devoting their entire existence to the needs of others and sacrificing their lives if necessary they help to propagate their own genes. A similarly extreme form of altruism occurs in clones (e.g., aphids), where individuals are 100% related (Ridley & Dawkins, 1981). Altruism and degree of genetic similarity are closely related, and the unit of conceptual analysis has been redirected from the individual organism to his or her genes.

An additional mechanism has been proposed by Trivers (1971) to account for the natural selection of altruism: *reciprocity.* In this case, there is no requirement for kinship. All that is posited is that the performance of an altruistic act will result in a return of altruistic behavior. An excellent example of this has been provided by Packer (1977). When a male olive baboon (*Papio anubis*) is consorting with a female, it is hard for a single male to supplant him but relatively easy for two in coalition to do so. Packer observed that pairs of unrelated males would often join forces to achieve this. Then one of the two males would copulate while the other, the "altruist," did not. On a later occasion when another female was in oestrus, the same two males were likely to get together again, but this time their roles would be reversed, the former beneficiary now assuming the role of the altruist. Axelrod and Hamilton (1981) have proposed a model of cooperative reciprocity that can even be extended to bacteria.

2.2. The Nature of Sex and Parental Care

One important question is: What differentiates a male from a female? The answer lies not in terms of external appearance. Rather, it is that the male is categorized as the organism with the smaller and more numerous sex cells or gametes. This is true of plants as well as of animals and is of use in organizing disparate data. Stemming from this basic difference, it has been suggested, there are two rather different strategies for maximizing genetic fitness. The optimal strategy for a male is to spread his numerous (and therefore cheap) sperm as often and as widely as possible by being relatively indiscriminate with whom he mates. Each sperm is not a major consideration for a male, who can usually remove himself from the consequences of copulation. For females, the consequences are more serious. The best female strategy, therefore, is often the opposite of the male, that is, to be very discriminating and sure that the male has desirable characteristics, including, perhaps, the ability to help raise the offspring. These differential strategies can be observed in many species.

Males often compete for females, sometimes through the establishment of territoriality and/or dominance hierarchies. In this competition, males are differentially successful, with some males impregnating more females than others, a phenomenon known as the Bateman effect (Bateman, 1948). Thus, in terms of propagating genes into the future, dominant healthy males are the most successful. In turn, it is more desirable, from the animal's, or the gene's, "point of view," to produce a dominant healthy male offspring than an unhealthy male offspring, whereas both healthy and unhealthy female offspring fall in between these two male types. The reason for this patterning is that healthy male offspring always have the greatest opportunity to propagate their genes into future generations, whereas unhealthy males always have the least. This accounts for the fact that, in times of drought and ecological hardship, a higher percentage of female offspring are born. The optimal strategy appears to be: Produce males when nutrition is good and you are healthy; produce females when nutrition is poor and you are less healthy (Trivers & Willard, 1973). That this occurs is supported by data from pigs, sheep, mink, seals, deer, and humans (Barash, 1982; Freedman, 1979). In addition, it has been reported that under conditions of stress during pregnancy both deer and humans are more likely spontaneously to abort male rather than female fetuses (Barash, 1982). Moreover, Freedman (1979) reported that sex ratios favoring males increase with estimates of a country's health standards and, within countries (e.g., the United States), with increasing socioeconomic status.

A good example of this sex difference in reproductive potential is found in the sex changes of tropical fish, colonies of which sometimes consist of harems of one male and many females. The females are prevented from becoming male by the dominant male. When the male dies, the largest female changes sex and keeps the others female by dominating them. The dominant male is more fit than any female because he can mate with each female in his harem, whereas each female mates only once. Individuals that change from female to male enjoy the best of both worlds: guaranteed reproduction while small as a female and a chance to be maximally productive as a large male.

Finally, a sociobiological perspective offers explanations for why it is the female who usually provides the most parental care for offspring. First, each child is more potentially valuable to the female as a vehicle for the replication of her genes than it is to the male. Males of most species may have offspring from several females. In cases of strict monogamy, however, as in eagles, geese, and foxes, male and female investment and reproductive performance are the same. Significantly, male and female parental care is then distributed about equally. A second explanation for greater female care of the young involves knowledge of relatedness. Whereas females can be relatively certain that offspring are theirs, males have much less assurance. A striking confirmation of the hypothesis that parental certainty relates to parental investment is found in those species which practice external fertilization (such as many fish), where it is often the *male* who provides the care. In these cases, the male fertilizes the eggs of many females. Since he is related to all the offspring, whereas each female is related to only some, it is more appropriate that he should care for them (Barash, 1982).

3. Human Sociobiology

Sociobiological enquiries into how the patterning of animal social behaviors such as altruism and sexuality maximize inclusive fitness has illuminated some behavior otherwise difficult to explain, including anomalies for Darwin's (1859, 1871) theory. By examining distal causation and ultimate effects, an organizational order has been created that adds power to the conceptual model. Because it is assumed that *Homo sapiens* is subject to the laws of evolution no less than other species, it is natural to apply sociobiological theorizing to humans. This is where sociobiology has become controversial. However, the consequences of sociobiology for psychology are too important to ignore (Buss, 1983; Cunningham, 1981). This section will briefly consider ways in which sociobiology may alert us to important considerations about such human behaviors as aggression, altruism, dominance, emotionality, intelligence, and sexuality.

3.1. Aggression

Aggression is a pervasive characteristic of most human societies. In Western Europe alone, between the years A.D. 275 and A.D. 1025, there was a war every two years on average (Wilson, 1975). Recent history shows that there has been little change. In World War II, 18 million people were killed. War has often directly and substantially affected the gene pool, as when genocide was practiced (a not uncommon occurrence during the history of *Homo sapiens*).

In searching for the causes of aggression, sociobiologists might look for its historical adaptive significance. Lorenz (1966) offered the view that humans branched off from other primates because they were hunters. He viewed war and interpersonal aggression as partially the result of behavior patterns that evolved due to hunting. The veridicality of this perspective, however, has been disputed, partly because it is now believed that different mechanisms underly predatory behavior and intraspecific aggression. An alternative hypothesis in regard to aggression is that it evolved fairly directly from male–male competition for access to females (Barash, 1982). Both these views may have some truth, for aggression has now been categorized into several types, including territorial, dominance, sexual, predatory, antipredatory, imitative, and moralistic (Wilson, 1975 pp. 242–243). Durham (1976) has proposed a model for the prevalence of warfare, based on the adaptive advantage of aggressive intergroup behavior under conditions of resource competition.

In regard to proximal causes of human aggression, frustration has been of particular interest for the psychology of individual differences. Often the relationship between frustration and aggression has been viewed as a consequence of a maladaptive personality (Adorno, Frenkel-Brunswik, Levison, & Sanford, 1950). However, there may be sound biological reasons for the relationship: For example, the aggression may lead to the enhancement of the aggressor's DNA at the expense of others. In accord with this hypothesis is the finding, over a 50-year period in the American South, that as resources became scarce, as indexed by a fall in cotton prices, the lynching of blacks increased (Hovland & Sears, 1940). Other evidence suggests that crowding, which often leads to a reduction of resources, leads to aggressive behavior in other species (Calhoun, 1962). Colinvaux (1980) has argued that crowding is one of the main reasons for war—thus relating war to population biology, one of the main disciplines contributing to the new synthesis that is sociobiology.

3.2. Altruism

In regard to kin-selected altruism, the prediction is that we are most altruistic toward those who are genetically similar to ourselves, that is, family rather than friends and friends rather than strangers. Some research bears this out. Freedman (1979) cited several studies in which respondents reported that their intention would be to help close kin over distant kin and distant kin over strangers. Other studies have found that people are more likely to help members of their own race or country than members of other races or foreigners (Brigham & Richardson, 1979; Feldman, 1968). People are also more likely to help people they perceive as similar to themselves (Einswiller, Deaux, & Willits, 1971).

In regard to reciprocal altruism (Trivers, 1971), there is much evidence that human societies have very strong reciprocity rules prescribing that people should help those who have helped them in the past. It is certainly a very widespread human behavior. On the basis of much comparative anthropological data, Mauss (1954) concluded that three types of obligation are widely distributed in human societies in both time and space: (a) the obligation to give, (b) the obligation to receive, and (c) the obligation to repay. Reciprocal exchanges breed cooperation and good feelings. A failure (or inability) to reciprocate, on the other hand, breeds bitterness and dislike (Fisher, DePaulo, & Nadler, 1981). Numerous studies have demonstrated the tendency of individuals to reciprocate favors (Rushton, 1980). The tendency appears to be there even among preschoolers (Strayer, Wareing, & Rushton, 1979).

3.3. Dominance

Rather than having individuals continually competing with one another in an open and scrambling fashion, most social species "solve" the problem of competition by establishing dominance hierarchies. Once these are established, individuals "know their place" and relative peace ensues. Accordingly, and perhaps paradoxically, dominance hierarchies *decrease* aggression. The advantage of being closer to the top of the hierarchy is the greater access to important resources, and particularly to females. Generally, those at the top of the hierarchy will, for several reasons, be expected to leave more offspring behind than those lower in the hierarchy.

In regard to dominance hierarchies and the human species, the evidence does appear to favor the hypothesis that we organize ourselves into stratification systems in many different types of group—from preschool (Strayer, 1980) to academic science (Cole & Cole, 1973). As many ethological studies have shown, even in naturally occurring preschool groups, the members can readily identify, by a variety of independent, objective techniques such as visual gaze, physical displacement, and peer nominations, who the top, medium and low ranking persons are. Furthermore, the evidence is that these hierarchies are linear and stable over time (Strayer, 1980). Among preschool males, the hierarchies are based on "toughness," and those developed by age 6 still hold at age 14 (Freedman, 1979, p. 71). In academic science, the status hierarchies are often based on publication success (Rushton & Meltzer, 1981).

3.4. Emotionality

Psychologists have been limited in the range of emotional expressions they have studied, often concentrating on the emotions of fear and anger. Sociobiologists have been even more limited in their analyses (neither the word *anxiety* nor the word *fear* appears in the subject indices of the books by Barash or Wilson, for example). Nonetheless, there is reason to expect emotionality to be central to the evolutionary perspective (Gray, 1971, 1982; Plutchik, 1980).

Gray (1971) discussed the origins of fear in humans and proposed a model of the way in which the central nervous system organizes avoidance behavior. He suggests that many fearful stimuli have one of four general characteristics: intensity, novelty, special evolutionary dangers, or development from social interaction. This last, he suggests, arises from the dominance and submission behaviors that occur among conspecifics during social encounters. The interesting aspect of this view is that it predicts that when both males and females of the species belong to overlapping dominance hierarchies (as in many primates, including humans, and when there is sexual dimorphism favoring the male in size), females will show more fearfulness than males. If there is no overlap in the dominance hierarchy (as in other primates and mammals such as rodents), this sex difference will disappear or be *reversed*. There are no doubt other reasons why anxiety and fearfulness have evolved as adaptive emotional responses. In animals that are capable of learning, stimuli associated with aversive experiences can be avoided in the future.

3.5. Intelligence

Comparative psychologists have long been interested in relating the intelligence of animals to their place on the phylogenetic scale, and physical anthropologists in the evolution of brain size in the evolutionary line leading to *Homo sapiens*. Among humans, crude brain size does have some relation to intelligence. *Home habilis*, who evolved two million years ago from the *Australopithecus afarensis* line, had a a brain size of 800 cc; *Home erectus* emerged one and a half million years ago with a brain size of 1000 cc; and *Homo sapiens*, emerging perhaps only 500,000 years ago, has a brain size, on average, of 1,300 to 1,500 cc (Johanson & White, 1979). Among present day humans, Passingham (1979) has demonstrated that a positive correlation exists between cranial capacity and IQ, even when body height and weight are controlled.

Brain size also provides a rough index of the "intelligence" of other animals, although problems arise with those like the dolphin and elephant which have even larger brains than *Homo sapiens*. The size of the brain is related to the size of the body, as is that of any other bodily organ such as the heart. To overcome this problem, Passingham (1975) proposed a measure of brain development—the neocortex–medulla volume—and found that for a number of primate species this measure correlates with responsiveness to novel objects and with performance on visual discrimination learning, the latter of which has been shown to relate to measures of intelligence in human children. Furthermore, since the neocortex–medulla volume was closely related to indices of cranial capacity, Passingham suggested it was possible to relate the measure to the fossil evidence.

3.6. Sexuality

One of the most comprehensive accounts of human sexuality from a sociobiological perspective has been provided by Symons (1979). As we mentioned earlier, many genetically based sex differences derive from the numerosity and size of the male and female gametes (sperm and ova). Male gametes are usually tiny and produced in the millions, whereas female gametes are large and are produced in small numbers. Numerous predictions follow from this. First, there is a possible genetic basis for the sexual double standard. Young males, on average, will maximize their inclusive fitness by being more active, approach-oriented, and vigorous in their pursuit of sex, wanting to engage in sexual activity fairly speedily with a variety of females. In terms of promulgating their DNA, there is maximal gain and little cost from adopting this strategy. Young females, on the other hand, should be relatively selective as to whom they allow to have intercourse with them, for each impregnation represents a major genetic investment. Females should be inclined to delay intercourse until they ascertain that the male has sufficiently desirable characteristics (e.g., is healthy, is high in the status hierarchy, and is likely to stay around to help raise the child). This is one of the reasons why females may be more nurturant and sexually more conservative than males. Some of the strongest support for these expectations comes from the study of male and female homosexuality, where male and female subcultures can develop, unconstrained by compromise with the opposite sex. Homosexual males are typically found to be promiscuous and not to maintain long-term relationships. The opposite is usually true of homosexual females (Symons, 1979).

Three additional implications of the sociobiological perspective will be offered to explain aspects of human sexual behavior. First, males should be more jealous and object more to females' having casual sex than vice versa. This follows from the male fear of being cuckolded and thereby tricked into investing his time and energy to raising another male's offspring as his own. Daly, Wilson, and Weghorst (1982) found supporting data for this sociobiological perspective from cross-cultural and historical reviews of both adultery laws and of homicides, as well as from analysis of motives for current homicides in Detroit. Second, males should generally have a strong preference for mating with young females, whereas females may be relatively more likely to find older males attractive. This is because males are primarily concerned with finding mates who will produce healthy offspring, whereas females are concerned with mating as high up in the status hierarchy as possible (where older males tend to predominate). This process of "marrying up" is known as hypergamy and is advantageous to the female if it leads her (a) to become impregnated by a male with the good genes to become high in the hierarchy and (b) to gain access to the greater resources usually available at the top of hierarchies (see van den Berghe & Barash, 1977, for a discussion of human family structure from a sociobiological perspective). Finally, perhaps related to hypergamy, Freedman (1979) suggested that a male, in order to mate successfully, must feel superior to the female—or at least be unafraid of her—and he speculated that this is the reason why males tend to demean women, belittle their accomplishments, and, in the vernacular (clearly laden with symbolism), "put them down" (Freedman, 1979, p. 74).

4. The Inheritance of Individual Differences in Behavior Traits

Most of the work in sociobiology has focused on differences between species in social behavior or on universals in human behavior. Yet the theory of evolution requires that there be genetic differences *within* species. Indeed, the first premise of evolutionary theory, as we stressed above, is that individuals of the same species are not identical. To date, sociobiologists have not seriously addressed the implications of genetic variability within *Homo sapiens*. There is, however, a growing body of research from the behavior genetic and psychometric traditions which is of direct relevance to the sociobiological enterprise. This is the study of genetically based individual (and group) differences in personality and social behavior.

4.1. The Existence of Stable Individual Differences in Behavior

The sociobiological perspective is quite compatible with the traditional trait approach to personality. This approach consists of a search for general laws in which consistent patterns of individual differences in behavior play a central role. Basic assumptions of this approach include substantial consistencies of people's behavior when it has been reliably assessed and considerable predictive power of measures of traits in accounting for behavior (Rushton, Jackson, & Paunonen, 1981). Numerous dimensions of personality have been investigated over the last few decades and assessment techniques created for their measurement (Anastasi, 1982).

In recent years, the traditional wisdom of the trait approach has been challenged (Kenrick & Stringfield, 1980; Mischel, 1968). Critics propose that (a) consistencies are so low as to be unimportant and (b) whatever consistency exists is primarily in the eye of the beholder. It is now realized, however, that such criticisms are largely due to a major error of interpretation—that is, they are based on the low correlations of .2 or .3 found between single items of behavior. When behaviors are more reliably assessed, by aggregating over items to remove error variance, substantial consistencies are readily found (Epstein, 1979, 1980; Eysenck, 1981; Rushton, Brainerd, & Pressley, 1983). Moreover, such behavior traits appear to be longitudinally stable (Block, 1971, 1981; Conley, in press; Olweus, 1979). An interesting question then becomes: Where do such traits originate? One answer lies in evolutionary history.

4.2. Estimating the Heritability of Behavior Traits

Several procedures are available for estimating the proportion of variance in a set of measurements that is attributable both to the genes and to the environment (Eaves, Last, Young, & Martin, 1978; Falconer, 1981; Fulker, 1981; Plomin *et al.*, 1980). The basic assumption is that phenotypic (observed) variance in measurements can be partitioned into environmental (E) and genetic (G) components, which combine in an additive manner. The model usually also allows for a nonadditive, or interaction, term (G × E) to deal with possible nonadditive combinations of genetic and environmental effects. Symbolically:

$$\text{Phenotypic variance} = G + E + [G \times E]$$

The estimate of the genetic contribution to phenotypic variance is often referred to as a heritability coefficient and represented as h^2. The heritability of individual differences in behavior may be assessed by several methods. For example, selective breeding studies of animals may be undertaken, using cross-fostering to control for upbringing. In humans, correlations may be calculated between scores on the trait in question and the degree of relatedness within the family, the best known example being twin studies. Adoption studies also permit the investigator to separate the effects of environment and heredity. Finally, the trait in question may be studied in infancy to ascertain whether individual differences emerge early and remain stable over time. When studies such as these have been carried out, a degree of genetic influence has been detected (Loehlin & Nichols, 1976; Plomin, 1983; Plomin *et al.*, 1980). In short, the evidence from converging methods confirms the role of heredity in human personality.

Adoption studies and the comparison of twins are the most widely used procedures for estimating h^2. In twin studies, monozygotic (MZ) twins are assumed to share 100% of their genes and dizygotic (DZ) twins are assumed to share, on average, 50% of their genes. By comparing such twins on a set of measures, one can derive estimates of h^2. If the correlation between scores on a trait is higher for the MZ twins than for DZ twins, the difference can be attributed to genetic effects if it is assumed that the differential environment of each type of twin is roughly equal. Doubling the difference between the MZ and DZ twin correlations is one widely used estimate of h^2 (Falconer, 1981; Plomin *et al.*, 1980). Some have argued that the equal environment assumption is not valid since MZ twins are said to be treated more similarly than DZ twins. Much evidence, however, suggests that it is a valid assumption. For example, when zygosity is wrongly defined by the parents, degree of twin similarity is better predicted by true zygosity (defined by blood and fingerprint analyses) than by social definition (Scarr & Carter-Saltzman, 1979).

Mittler (1971) reviewed available twin data using the concordance method. This involves finding twins with a clearly established disorder (e.g., in psychiatric hospitals) and then determining whether the cotwin displays the same disorder. To the degree that monozygotic twins are more similar to each other than dizygotic twins, the influence of heredity is established. Table 1 presents the weighted averages of the concordance rates from the studies reviewed by Mittler. There appears to be a significant heritable component to most of these behavioral categories. Subsequent reviews of concordance data by Plomin *et al.* (1980) and Willerman (1979) provided further support for this conclusion.

The typical strategy for calculating heritabilities is to use questionnaire data to compare MZ and DZ twins reared together. Loehlin and Nichols (1976) carried out one of the most extensive of this type of twin study by comparing 514 pairs of MZ twins with 336 pairs of DZ twins who, as high school students, had taken the National Merit Scholarship test. Each participant took a wide variety of personality, attitude, and interest questionnaires. The results showed the MZ twins to be roughly twice as much alike as the DZ twins over a wide range of personality measures—exactly as would be predicted by genetic theory.

Dramatic examples of identical twin similarity and the heritability of personality are currently being found at the University of Minnesota by Thomas Bouchard and his colleagues (Bouchard, Heston, Eckert, Keyes, & Resnick, 1981). The focus of their study is on identical twins separated at birth and raised apart. Bouchard (1983) reported that the 34 pairs of identical twins studied to date demonstrated almost as much similarity on such objective personality scales as the Differential Personality Questionnaire, the Minnesota Multiphasic Personality Inventory, and the California Psychological Inventory, as did identical twins raised together. Although individual cases must be interpreted with great caution, many remarkable similarities of life-style, personal preferences and idiosyncrasies between members of these twin pairs have also been documented. One pair is the "Jim twins" (Holden, 1980). Both

Table 1. The Percentage of MZ and DZ Twins Falling into the Same Category as Their Co-twins (Adapted from Mittler, 1971)

			Number of pairs		Percentage concordant	
Behavioral category	Number of studies	Total	Monozygotic	Dizygotic (same sex)	Monozygotic	Dizygotic (same sex)
Adult crime	6	225	107	118	71%	34%
Alcoholism	1	82	26	56	65%	30%
Childhood behavior disorder	2	107	47	60	87%	43%
Juvenile delinquency	2	67	42	25	85%	75%
Male homosexuality	1	63	37	26	100%	12%
Manic depressive psychosis	5	518	168	350	74%	12%
Mental subnormality	2	586	197	389	96%	56%
Neurosis	10	1267	560	707	22%	11%
Schizophrenia	13	1251	503	748	53%	11%

were adopted as infants into separate working-class Ohio families. Incredibly, their lives have been marked by a trail of similar names. Both had childhood pets named Toy. Both married and divorced women named Linda and had second marriages with women named Betty. They named their sons James Allen and James Alan. In addition, their personality profiles are extremely alike. Another pair is 47-year-old Oskar and Jock (Holden, 1980). While one was raised as a German Catholic and Nazi youth, the other lived as a Jew in Trinidad, Israel, and the United States. Similarities between the two were apparent from the outset. Both arrived at the research center wearing wire-rimmed glasses and mustaches. They share many idiosyncrasies: they like spicy foods and sweet liqueurs, are absentminded, flush the toilet before using it, store rubber bands on their wrists, and have domineering relationships with women. They also have extremely similar profiles on objectively measured personality tests.

In the remainder of this section, a brief review is offered on the heritability of individual differences in several areas: activity level, aggression, altruism, chronogenetics, criminality, dominance, emotionality, intelligence, locus of control, political attitudes, sexuality, sociability, values, and vocational interests.

4.3. Activity Level

Several investigations have found evidence that individual differences in activity level are in part inherited. These include studies by Buss, Plomin, and Willerman (1973), Owen and Sines (1970), Scarr (1966), and Willerman (1973). In one of these, Scarr (1966) assessed activity using a cluster of related measures including ratings, experimental tasks, and interviews. The subjects were 61 pairs of MZ and DZ girls between the ages of 6 and 10. Although the particular heritabilities differed from measure to measure, the average heritability was found to be .31. An interesting aspect of Scarr's work was her analysis of data from those twins mistakenly identified by the mothers as being DZ when they were actually MZ, and MZ when they were actually DZ (as correctly determined by blood grouping). She found that mothers' incorrect beliefs about their children's zygosity did not affect ratings of their children on activity level. That is, MZ twins mistakenly identified by their mothers as DZ were similar in scores to correctly identified MZ twins, and DZ twins mistakenly identified as MZ were similar in scores to other DZ twins. In another study, Willerman (1973) tested 93 sets of same-sexed twins and found the heritability of activity-level to be close to .70. Additional studies, reviewed by Buss and Plomin (1975), also suggest that there is substantial heritability to activity level.

4.4. Aggression

Several studies have been conducted on the heritability of individual differences in aggressiveness (Eysenck & Eysenck, 1976; Loehlin & Nichols, 1976; Owen & Sines, 1970; Rushton, Fulker, Neale, Nias, & Eysenck, 1984; Scarr, 1966). In Scarr's study, parents completed the Adjective Check List to describe their children. On this measure aggressiveness had a heritability of .40. In Loehlin and Nichols' investigation with 850 twin pairs, cluster analyses were carried out of self-ratings on various traits. Two clusters that Loehlin and Nichols labelled "argumentative" and "family quarrel" showed the MZ twins to be about twice as alike as the DZ twins. Rushton *et al.*, (1984) gave a 47-item questionnaire measuring both aggressiveness and assertiveness to 573 adult twin pairs and found about 50% of the variance on each scale to be associated with genetic effects. Finally, psychoticism, a dimension correlated with hostility, has a reported heritability of .80 (Eysenck & Eysenck, 1976; Fulker, 1981).

4.5. Altruism

At least three studies have been carried out to test for the existence of genetically based individual differences in human altruism (Loehlin & Nichols, 1976; Matthews, Batson, Horn, & Rosenman, 1981; Rushton, Fulker, Neale, Blizard, & Eysenck, in press). Loehlin and Nichols carried out cluster analyses of self-ratings made by 850 twin pairs on various traits. One cluster that Loehlin and Nichols labelled "kind" demonstrated a heritability of .44. Matthews *et al.* (1981) analyzed twin responses to a self-report measure of empathy and estimated a heritability of .72. In the Rushton *et al.* study, three separate questionnaires measuring altruistic tendencies were completed by 573 twin pairs. Approximately 50% of the variance on each scale was found to be associated with additive genetic influences.

4.6. Chronogenetics

Genetic mechanisms turn on and off over the course of a lifetime. Common phenomena that reflect such genetic clockworks are the age of onset of puberty and menopause. Identical twins are highly concordant for both events, whether reared apart or together (Bouchard, 1982). Comparisons of MZ and DZ twins have demonstrated that the genes also influence the age of first sexual intercourse (Martin, Eaves, & Eysenck, 1977). Another example is Huntington's chorea, a degenerative disorder of the central nervous system caused by a dominant gene. Age of onset varies from 5 to over 75, but family studies show that it is under genetic control. Chronogenetics also affects cognitive development. Wilson (1983) examined genetic influence on the developmental spurts and lags so characteristic of young children. He compared a large sample of MZ and DZ twins from 3 months to 6 years of age, with measures made of height and mental development. The synchronies in developmental lags and spurts averaged about .9 for MZ twins but only about .5 for DZ twins, demonstrating the high heritability of these developmental trajectories.

4.7. Criminality

Historically there has been a belief that criminals are born as well as made (Eysenck, 1977). Studies of the concordance rates of MZ and DZ twins provide evidence in favor of this hypothesis (see Table 1). Additional support derives from adoption studies. Plomin *et al.* (1980) reviewed four of these carried out in Denmark and the United States by Crowe (1972, 1974), Hutchings and Mednick (1975), and Schulsinger (1972). These studies included 321 first-degree biological relatives of adopted criminal or psychopathic probands and 316 controls (biological relatives of adoptees who had shown no criminality). Twenty-five percent of the biological relatives of criminal probands either had criminal records or were diagnosed as psychopathic. In the control group, only 13% of the biological relatives were similarly diagnosed. Plomin *et al.* (1980) concluded: "These studies thus provide significant evidence for the involvement of heredity in criminal behavior" (p. 352). Ellis (1982) reviewed the evidence from four classes of research design bearing on the genetics of criminality: general pedigree (or family) studies, twin studies, karyotype studies, and adoption studies. He concluded that "most of the evidence is extremely supportive of the proposition that human variation in tendencies to commit criminal behavior is significantly affected by some genetic factors" (p. 43).

Conversely, support for the inheritance of law-abiding behavior comes from studies assessing the heritability of such scales on the California Psychological Inventory as Responsibility, Socialization, and Self-control. A review of several studies using these dimensions demonstrates heritabilities ranging from .30 to .40 (Carey, Goldsmith, Tellegan, & Gottesman, 1978).

4.8. Dominance

Using a variety of assessment techniques, several studies have found individual differences in interpersonal dominance to be largely inherited (e.g., Gottesman, 1963, 1966; Loehlin & Nichols, 1976). In a longitudinal study of 42 twin pairs, Dworkin, Burke, Maher, and Gottesman (1976) found that individual differences in dominance, as assessed on the California Psychological Inventory, remained stable over a 12-year time period, as did the heritability estimate. Carey *et al.* (1978), in a review of the literature, reported that, of all traits, dominance is one of those most reliably found to be heritable, with a weighted mean heritability coefficient, over several samples, of .56.

4.9. Emotionality

Individual differences in emotional reactivity have long been thought to be partly inherited, and several studies have reported substantial heritability coefficients (e.g., Buss, Plomin, & Willerman, 1973; Cattell, Blewett, & Beloff, 1955; Dworkin *et al.*, 1976; Fulker, 1981; Scarr, 1966; Vandenberg, 1962). All of these focused on emotionality as anxiousness and "neuroticism." The largest heritability study of this trait was carried out by Floderus-Myrhed, Pedersen, and Rasmuson (1980). They administered the Eysenck Personality Inventory to 12,898 unselected twin pairs of the Swedish Twin Registry. The heritability index for neuroticism was 0.50 for men and 0.58 for women. The opposite side of the coin, emotional stability (measured by the California Psychological Inventory's Sense of Well-Being Scale), has also been found to have significant heritabilities, both in adolescence and 12 years later in adulthood, as in the previously mentioned study by Dworkin *et al.* (1976).

4.10. Intelligence

Ever since Galton (1869), more heritability estimates of intelligence have been computed than of any other trait. The data published prior to 1963 were reviewed by Erlenmeyer-Kimling and Jarvik (1963) and were compatible with an estimated heritability as high as .80. Many of these studies were subsequently criticized by Kamin (1974), who argued that flaws in them required an estimation of the heritability of intelligence to be closer to zero. Newer data and reviews (e.g., Cattell, 1980, 1982; Loehlin & Nichols, 1976; Plomin & DeFries, 1980), however, have confirmed the high heritability of intelligence. The most extensive review is that by Bouchard and McGue (1981), based on 111 studies identified

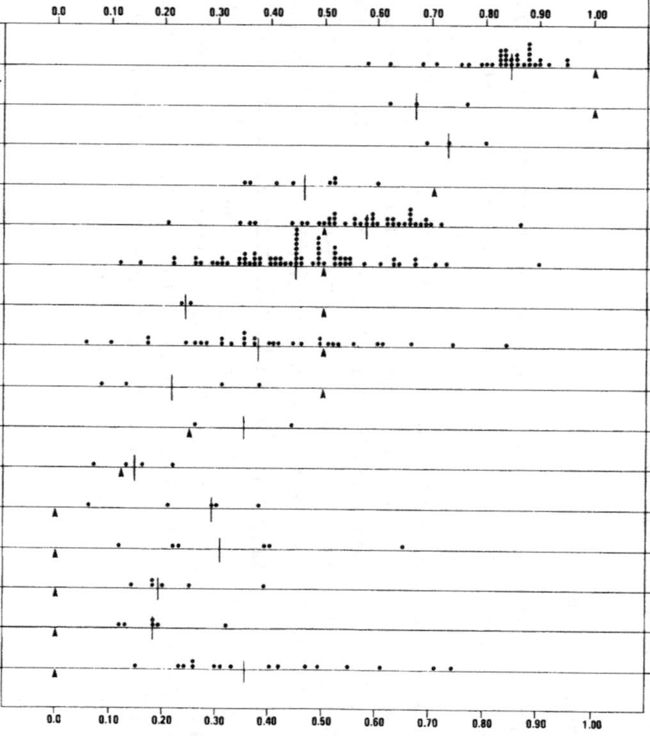

	No. of correlations	No. of pairings	Median correlation	Weighted average
MONOZYGOTIC TWINS REARED TOGETHER	34	4672	.85	.86
MONOZYGOTIC TWINS REARED APART	3	65	.67	.72
MIDPARENT–MIDOFFSPRING REARED TOGETHER	3	410	.73	.72
MIDPARENT–OFFSPRING REARED TOGETHER	8	992	.475	.50
DIZYGOTIC TWINS REARED TOGETHER	41	5546	.58	.60
SIBLINGS REARED TOGETHER	69	26,473	.45	.47
SIBLINGS REARED APART	2	203	.24	.24
SINGLE PARENT–OFFSPRING REARED TOGETHER	32	8433	.385	.42
SINGLE PARENT–OFFSPRING REARED APART	4	814	.22	.22
HALF-SIBLINGS	2	200	.35	.31
COUSINS	4	1,176	.145	.15
NON-BIOLOGICAL SIBLING PAIRS (ADOPTED/NATURAL PAIRINGS)	5	345	.29	.29
NON-BIOLOGICAL SIBLING PAIRS (ADOPTED/ADOPTED PAIRINGS)	6	369	.31	.34
ADOPTING MIDPARENT–OFFSPRING	6	758	.19	.24
ADOPTING PARENT–OFFSPRING	6	1397	.18	.19
ASSORTATIVE MATING	16	3817	.365	.33

Figure 2. Familial correlations for IQ. The vertical bar in each distribution indicates the median correlation; the arrow, the correlation predicted by a simple genetic model. From T. J. Bouchard, Jr., and M. McGue "Familial Studies of Intelligence: A Review," *Science*, 1981, *212*, 1055–1059. Copyright 1981 by the American Association for the Advancement of Science. Reprinted by permission.

in a survey of the world literature. Altogether there were 652 familial correlations, based on 113,942 pairings. The results were in accord with a polygenic model of the inheritance of IQ. Figure 2 displays the correlations between relatives, biological and adoptive, in the 111 studies.

4.11. Locus of Control

The Internal-External Locus of Control Scale (I-E scale) was developed as a continuous measure of the attitude with which individuals relate their own behavior to its contingent reward or punishment (Rotter, 1966). That one's own actions are largely affected by luck or chance or some more powerful force was labeled a belief in external control. The converse attitude, that outcomes are contingent on one's own behavior, was termed internal control. A recent study by Miller and Rose (1982) reported a twin family study of variation in locus of control. The I-E scale was administered to a total of 598 individuals; pair-wise resemblance was assessed in 109 twin-siblings, 106 spousal pairs, and 54–62 pairings of single parents and their offspring. The results revealed heritability estimates > .50. In the above study, the heritability estimates based on the comparison of MZ and DZ twins were corroborated by also estimating heritability through the regression of offspring on parent and the correlation between nontwin siblings.

4.12. Political Attitudes

It has generally been assumed that political attitudes are for the most part environmentally determined. However, in a large-scale twin study of social and political attitudes, Eaves and Eysenck (1974) found that a dimension of Radicalism–Conservatism had a heritability of .65; Tough-mindedness, a factor identifiable with ideological commitment, had a heritability of .54; and the tendency to voice extreme views, irrespective of right- or left-wing bias, had a heritability of .37.

4.13. Sexuality

This can be viewed in at least two ways as a personality trait. One is a continuum of masculine-feminine attitude, the other is strength of sex drive. Studies carried out with the Minnesota Multiphasic Personality Inventory Masculinity-Femininity (Attitude) Scale show no apparent heritability for this measure (Dworkin *et al.*, 1976; Gottesman, 1963, 1966). A large study of twins, using questionnaire measures of strength of sex drive, found direct evidence that inheritance plays a substantial role in accounting for individual differences in strength of sex drive

(Eysenck, 1976). Differences in sex drive were found to be predictive of many phenomena, including age of first sexual intercourse, which itself has been shown to be under genetic influence (Martin, Eaves, & Eysenck, 1977).

4.14. Sociability

Sociability is another well-researched trait, and again the evidence favors the hypothesis of a large genetic component. Using different paper and pencil indices of the trait, some studies have found greater than 50% of the variance in individual differences in sociability to be inherited (Carey *et al.*, 1978; Cattell, 1981; Dworkin *et al.*, 1976; Eaves & Eysenck, 1975; Floderus-Myrhed *et al.*, 1980; Fulker, 1981; Gottesman, 1963, 1966; Loehlin & Nichols, 1976; Owen & Sines, 1970; Scarr, 1969). In the largest of these studies, Floderus-Myrhed *et al.* gave the Eysenck Personality Inventory to 12,898 unselected twin pairs of the Swedish Twin Registry. The heritability index for extraversion, highly related to measures of sociability, was .54 (men) and .66 (women).

4.15. Values and Vocational Interests

Loehlin and Nichols' (1976) study of 850 twin pairs raised together provides evidence for the heritability of both values and vocational interest. Values such as the desire to be well-adjusted, popular and kind were found to have a significant genetic component. Having scientific, artistic, and leadership goals were similarly found to be genetically influenced as were a range of career preferences, including those for sales, bluecollar management, teaching, banking, literary, military, social service, and sports. Bouchard (1983) reported that, on measures of vocational interest, his 34 MZ twins raised apart were just as alike as MZ twins raised together. Moreover, both types of MZ twins were twice as similar as related individuals who share half their genes and live together (e.g., parents compared with offspring, or siblings, including DZ twins, compared). Adoption studies also confirm the heritability of vocational interests. Grotevant, Scarr, and Weinberg (1977) contrasted 194 adopted with 237 biological siblings, all of whom had spent an average of 18 years in their families. While biological siblings shared modestly similar interests, adoptive siblings did not.

4.16. A Summary of the Evidence

The evidence from comparisons of MZ and DZ twins demonstrates significant heritabilities for individual differences in such areas as activity level, aggression, altruism, criminality, dominance, emotionality, intelligence, locus of control, political attitudes, sexuality, sociability, values, and vocational interest. Additional evidence in support of the hypothesis of heritability of human personality was available from sources other than twin studies. Adoption studies, for example, demonstrated the heritability of individual susceptibility to criminality (Ellis, 1982; Plomin *et al.*, 1980). Another research procedure was to calculate correlations between scores on the trait in question and degree of genetic relatedness within the extended family. When this was done for intelligence, for example, the results favor a genetic model (Figure 2). In short, on the basis of the findings from several lines of investigation, we may conclude that the evidence favors the hypothesis that a large and significant component of the individual difference variance in human personality is inherited.

A cautionary note is essential: Despite the increasing number of studies using increasingly sophisticated techniques (e.g., Cattell, 1982; Eaves *et al.*, 1978; Fulker, 1981) which point to the role of the genes in shaping personality, many uncertainties remain. The number of studies on the heritability of personality lags far behind equivalent research efforts on environmental determinants. Also, unlike the studies with intelligence, heritability studies of personality have rested primarily on the comparison of MZ and DZ twins. Although some of the criticisms of this approach (e.g., that MZ and DZ twins have very different environments) do not seem to be true (see the discussion of Scarr's (1966) study on activity level, above, or her further discussion in Scarr and Carter-Saltzman, 1979), nonetheless, confidence would accrue if corroborative findings were obtained using different procedures. When other procedures have been used, the heritability estimates for personality have sometimes been in the region of .20 to .30, compared to the .50 often found with MZ–DZ comparisons (Ahern, Johnson, Wilson, McClearn, & Vandenberg, 1982; Cattell, Vaughan, Schuerger, & Rao, 1982; Loehlin, Horn, & Willerman, 1981; Scarr, Webber, Weinberg, & Wittig, 1981). Ahern *et al.* (1982), for example, measured 54 personality traits with such psychometric tests as the Adjective Check List, the Eysenck Personality Inventory, the Comrey Personality Scales, and the Cattell Sixteen Personality Factor Scales, in 118 families (n = 456 individuals). They then computed regression coefficients and correlation coefficients between all possible kinships, for example, offspring on midparent, sib–sib. The mean value over all 54 midparent–offspring regressions was .21 and the average sib–sib correlation was 0.10. Both these figures yield heritability estimates of 20%. Regardless of the exact figure, however, it seems clear that a significant proportion of variance in human personality is inherited.

5. Group Differences in Inherited Behavior Traits

One aspect of human sociobiology that has been relatively unexplored is the question of inherited differences between groups. If groups become susceptible to different environmental selection pressures, this will lead to distributions of traits about different means. This is readily understandable and accepted when it is applied to group differences in skin color (as in our example above) or, say, tolerance to heat. Baker (1977), for example, discussed how Peruvian Indians evolved body systems that allow them to survive in the cold mountain tops of the Andes. At the other extreme, black Africans have evolved physiological systems that enable them to survive the heat of tropical climates. As it is with morphology, so it is with behavior. If hunting is adaptive in one ecological setting, then any genetically based traits that enhance that ability (e.g., agility, endurance) will increase in that group. All that is required is that individuals who are high on those traits produce more kin that reach reproductive maturity than those lower on the same trait. If the ecological pressures derive from an agricultural setting, then behavior traits that enhance agriculture will increase. Campbell (1965) conjectured that traits such as the ability to delay gratification, industriousness, and the ability to save might be selected for in agricultural communities. *A priori*, then, it is to be expected that groups that have been subjected to different selection pressures will exhibit differences in behavior attributable to different genotypes.

Before beginning this section it is worthwhile to repeat what many others have said in this context: That variations in personality *within* groups are greater than those *between* groups. In other words, despite mean differences, there is significant overlap in the group distributions being compared. Three sets of group differences will be reviewed: sex, socioeconomic status, and ethnic.

5.1. Sex Differences

Differences in the behavior of males and females in regard to sexual activity have already been discussed. In a review of the sex difference literature, Maccoby and Jacklin (1974) concluded that males had higher mean scores than females in aggressiveness, dominance, social exhibitionism, and spatial IQ, whereas females had higher mean scores than males in verbal IQ and possibly in social responsivity.

This review was criticized by Block (1976), who argued that it was, in fact, biased *against* detecting sex differences. Block provided an alternative retabulation demonstrating that sex differences occur on an even greater variety of traits. In a discussion of this and subsequent studies, Rushton, Brainerd, and Pressley (1983) retabulated Block's analyses. This retabulation can be seen in Table 2.

Block's meta-analysis led her to rather different conclusions from Maccoby and Jacklin's. Block (1976) concluded that males not only are higher on spatial and quantitative abilities and aggressiveness, but also are

> better on insight problems requiring restructuring, and more dominant and have a stronger, more potent, self-concept, are more curious and exploring, more active, and more impulsive. (p. 307)

In addition, she suggested that females not only score higher on tests of verbal ability but also

> express more fear, are more susceptible to anxiety, are more lacking in task confidence, seek more help and reassurance, maintain greater proximity to friends, score higher on social desirability, and, at the younger ages at which compliance has been studied, are more compliant with adults. (p. 307)

That males are more aggressive than females appears to be due, at least in part, to heredity, for the difference appears in most other mammalian species and is strongly influenced by the amount of prenatal gonadal hormones (Hines, 1982). Unless protecting their young, females will usually not fight, despite severe provocation. In contrast, males in many species fight readily, even in the absence of external provocation. Moreover, males of many species, given injections of testosterone in infancy, exhibit an increase in fighting behavior when adults, whereas males castrated before puberty rarely fight. Opposites of aggression, for example, empathy and altruism, also exhibit evidence of sex linkage. In reviews of the literature, Eisenberg and Lennon (1983) and Rushton (1980) found that on average females were more empathic and concerned about others than were males.

Freedman (1979) summarized evidence that these two basic sex differences, active aggression and social responsiveness, begin to emerge at infancy. Male babies cry more, kick more, and respond less to vocal cajoling. Female babies kick less and allow more cuddling. By 9 months, female babies smile more than males (a sex difference that lasts a lifetime) and are more socially oriented (they can better discriminate male from female voices, attend to faces more, and babble responsively more). By one year of age, males are relatively more mechanical and more given to problem solving; they pull dolls apart and try to put them back together, whereas females are more inclined to cuddle them. These mechanical differences also show up in children blind from birth. Females, on the other hand, are on average more alert than males to vocal sounds

Table 2. Proportions of Studies Demonstrating Sex Differences Based on Block's (1976) Reanalysis of Maccoby and Jacklin's (1974) Literature Review[a]

Behavior assessed	Ratio of significant comparisons to total number of comparisons			
	Girls and women (significantly higher)		Boys and men (significantly higher)	
	Ratio	Proportion	Ratio	Proportion
Cognitive dimensions				
Verbal abilities	45/160	.28	18/160	.09
Spatial abilities	5/100	.05	35/100	.35
Quantitative abilities	6/35	.17	14/35	.40
Analytic impulsivity	6/80	.08	22/80	.28
Breaking set-responses to "insight" problems	0/14	.00	12/14	.86
Anagrams—breaking up words to form new words	4/10	.40	0/10	.00
Descriptive, analytic sorting style	0/6	.00	1/6	.17
Auditorially oriented	6/26	.23	2/26	.08
Social dimensions				
Aggressiveness	5/94	.05	52/94	.55
Empathy; sensitivity to social cues	7/31	.23	3/31	.10
Fear, timidity, anxiety	36/79	.46	0/79	.00
Activity level	6/109	.06	39/109	.36
Competitiveness	6/50	.12	14/50	.28
Dominance	4/89	.05	35/89	.39
Compliance and rule following	26/51	.51	1/51	.02
Nurturance, maternal behavior, helping, donating, and sharing	10/58	.17	7/58	.12
Sociability	60/215	.28	36/215	.17
Suggestibility	36/125	.29	8/125	.06
Achievement orientation	5/23	.22	4/23	.17
Dependency	28/88	.32	10/88	.11
Curiosity and exploration	8/50	.16	20/50	.40
Social desirability	7/9	.78	0/9	.00
Self-concept				
Strength and potency of self-concept	0/8	.00	7/8	.88
Low self-esteem	20/84	.24	13/84	.16
Confidence on task performance	0/33	.00	25/33	.76
Other				
Tactile sensitivity	5/13	.38	0/13	.00

[a] After Rushton, Brainerd, and Pressley, 1983.

from the first few weeks of life. Throughout later life females are better on average at all verbal and linguistic tasks, including learning new languages.

5.2. Socioeconomic Status (SES) Differences

Sociobiological theorizing might well lead to the expectation that terrestrial primates such as *Homo sapiens* would form themselves into dominance hierarchies in which those at the top exhibit high levels of whatever traits make for success in that culture and in turn get a greater than equal share of whatever scarce resources are available. In hunting societies those at the top will be the best hunters; in warrior societies those at the top will be the best warriors, etc. Furthermore, it would perhaps be expected that those traits which led to mobility up or down the status hierarchy would have an inherited, genetic basis.

The socioeconomic status dominance hierarchies of our own industrial-technological societies in the late twentieth century are partly built on intelligence, as measured, for example, by standard IQ tests. Several reviews of this literature have appeared (e.g., Eysenck, 1979; Herrnstein, 1973; Jensen, 1980, 1981a; Scarr, 1981). The basic finding is that there is a difference of nearly 3 standard deviations (40 IQ points) between *average* members of the professional and the unskilled classes. These are group-mean differences, with considerable overlap of distributions. Nonetheless, the overall correlation between IQ and social class appears to range from +.50 to +.90, depending on how the correlations are computed. Evidence of this relationship between IQ and SES comes from studies in the countries of continental Europe, the United Kingdom, and the United States. (For a recent study carried out in Poland, see Firkowska-Mankiewicz & Czarkowski, 1982.) Furthermore, it appears that the relationship is partly based on genetically inherited IQ.

The evidence for the overall inheritance of IQ has already been discussed (see Figure 2). The particular evidence for the inheritance of these socioeconomic status differences arises from at least three sources. First, causal modeling studies demonstrate the capacity of IQ to relate to occupational level and performance even when the effects of education, parental income, and the like are controlled (see Jensen, 1980, pp. 339–353). Second, there is the phenomenon of "regression to the mean." Compared to their parents, children of high IQ parents have lower average IQs and children of low IQ parents have higher average IQs. These data are predicted by genetic theory through the mechanism of dominant and recessive gene combinations (see Eysenck, 1979, pp. 120–122). Third, there is evidence from studies of intergenerational social mobility. In one study, Waller (1971) obtained the IQ scores of 130 fathers and their 172 adult sons, all of whom had been routinely tested during their high school years in Minnesota. The IQs ranged from below 80 to above 130 and were highly related to social class. The most interesting finding was in regard to social mobility: Children with lower IQs than their fathers went down in social class as adults, and those with higher IQs went up ($r = .37$ between difference in father–son social class and difference in father–son IQ).

5.3. Ethnic-Group Differences

In this section some of the most consistently found ethnic-group differences will be discussed.

5.3.1. Activity Level

There appear to be replicable ethnic-group differences in activity level. Freedman (1979) provided Afro-, Chinese-, and Euro-American one-day-olds with a variety of tests measuring how active or passive they were. Consistently, babies of Chinese ancestry were quieter and more readily soothed than the more easily aroused, more active, and harder to soothe Euro-American babies. Afro-American babies were in turn more active than Euro-American ones. One measure involved pressing the baby's nose with a cloth, forcing it to breathe with its mouth. Most Euro-American and Afro-American babies fought this immediately whereas the average baby of Chinese ancestry continued to lie on its back and breathe through its mouth. Subsequent infant studies replicated these findings in other countries and with different samples (Freedman, 1979). Among the most quiescent were the Navajo Indians of the southwestern United States. These infants stoically spend much of their first six months of life wrapped to a cradleboard. For many years anthropologists interpreted this as an environmental cause of later Indian impassiveness. Freedman (1979), however, believed the cause to be genetic. Attempts to get Euro-American children to accept the cradleboard have apparently met with no success. (The Navajo, like the Chinese, are classified as being of mongoloid ancestry.) Japanese babies seem to have temperaments similar to the Chinese and Navajo, thus providing further evidence for a genetic basis.

5.3.2. Intelligence

The previous section on passivity–activity found that Europeans scored between Asians and Africans. This ordering may also be true with intelligence. Evidence is accumulating, from international as well as intranational investigations, that, on Euro-American-originated, standardized intelligence tests, some Asian peoples score from $^1/_3$ to $^2/_3$ of a standard deviation higher than Europeans. Europeans, in turn, score about one standard deviation higher than African-descended peoples. Jensen's (1969) monograph is an often cited starting point for discussion. Jensen pointed to the 40 IQ point difference, just discussed, between adults in the unskilled working class and those in the professional class and considered the evidence that such differences were partly genetic in origin. He then pointed to the difference of 15 IQ points (one standard deviation), established over several decades, between Afro-Americans and Euro-Americans. While acknowledging the problem of extrapolating from within-group heritability to between-group, he conjectured that some of this ethnic-group difference in intelligence might be inherited. Since that time, more data have come to light.

First, it would seem that, despite social changes since the 1960s and attempts to ameliorate the situation (desegregation, busing, affirmative-action programs, head-start schooling, etc.) the ethnic-group difference betwen Afro- and Euro-Americans in mean IQ has not disappeared (e.g., Hall & Kaye, 1980; Scarr, 1981); the differences are about as large today as they were at the time of the First World War (Loehlin, Lindzey, & Spuhler, 1975). From an environmental perspective it can be argued that social action has not gone far enough or been implemented long enough to counteract historical inequalities. This may be the case; as yet, though, no social changes have succeeded in eliminating the differences.

Second, Jensen has addressed criticisms which argue against his hypothesis of a genetic basis to the difference in ethnic-group IQ. The most common argument used to discount reported IQ differences among ethnic groups is that the IQ tests themselves are culturally biased. Jensen (1980) provided analyses of the difference scores between Afro- and Euro-Americans on the types of items typically found on IQ tests. Those items judged to be most culturally biased showed the smallest differences between Afro-American and Euro-American children. Those items, on the other hand, that were least culturally biased (and loaded most highly on "g") showed the greatest differences between these ethnic groups (Jensen, 1980, 1983). ("g" is the label given to the general factor of intelligence that emerges when factor analysis is carried out on different measures of complex mental ability.) Moreover, Jensen (1977) and Vernon and Jensen (in press) have found that reaction time measures, which are positively related to intelligence tests and which assess the speed with which individuals perform basic cognitive processes, likewise demonstrate the ethnic group difference. On a manifest level at least, these tests lack cultural bias. Thus, the cultural bias argument has been considerably weakened.

Third, Lynn (1978) reviewed studies of ethnic and national differences in intelligence from around the world. Although sampling procedures of some of the studies may be questionable, they do show a consistent difference of one standard deviation between Europeans and sub-Saharan Africans—including the children of middle-class Africans in such postcolonial countries as Nigeria, Tanzania, and Uganda. More recent data support this. For example, Buj (1981) tested the IQ of 10,737 Europeans in 21 different countries using Cattell's CFT3 Scale, a nonverbal culture-fair test. Although the mean IQs varied somewhat from country to country, the overall European mean was 102.2 with a standard deviation of 18.7. The same test was given to 225 Ghanians in Akkra, who obtained a mean IQ score of 82.2. Furthermore, similar lower scores are found among African-descended children in Jamaica and the United Kingdom (Lynn, 1978; Scarr, Caparulo, Ferdman, Tower, & Caplan, 1983).

There is a welter of additional data regarding the genetics of these ethnic-group differences which limited space does not allow us to pursue. These are concerned with such issues as (a) whether heritabilities calculated on IQs of Afro-Americans are lower than they are for Euro-Americans (if lower, the environment might well be having a suppressant effect on Afro-American IQ); (b) whether Afro-American IQs increase to the level of Euro-Americans if these children are adopted at birth by families of European descent and raised in upper-middle-class environments; and (c) whether Afro-American IQs vary with the amount of European genetic admixture. These issues are currently debated (Jensen, 1981b; Kamin, 1981; Osborne, 1978; Scarr, 1981). Continuing research should ultimately allow resolution.

Recently, Lynn (1977, 1978, 1982; Lynn & Dziobon, 1980) and Vernon (1982) have extended the ethnic-group IQ literature to include Asians. In one study, Lynn (1977) reported that when the WAIS was standardized on 1,070 children and 1,682 adults in Japan in the early 1950s, the average Japanese IQ was 106.6 (compared with the mean IQ of 100 for Euro-Americans and British peoples, i.e., $^1/_3$ of a standard deviation higher). In a second study, Lynn and Dziobon (1980) showed that the higher Japanese IQ was maintained when Japanese and Euro-American scores were recalibrated against a British sample, which made relative comparisons more meaningful. More recently still, Lynn (1982) reported an analysis of results from the standardization in Japan in 1975 of the new revised version of the American Wechsler Intelligence Scale for Children. This showed that the Japanese-American disparity in mean IQ has increased since the early 1950s. Among the younger generation the mean Japanese IQ is approximately 111, some 11 points (i.e., $^2/_3$ of a standard deviation) above the mean IQ of the United States and other

Western populations. Lynn pointed out that with this difference in mean IQ, 10% of the Japanese younger generation had IQs above 130 (gifted). Among the population as a whole, 77% of the Japanese had a higher IQ than the average American or European.

Vernon (1982) has documented a great deal of research concerned with the abilities and achievements of Chinese and Japanese immigrants in Canada and the United States. His findings demonstrate that, despite discrimination and deprivation, on the average, they appear to have reached higher educational and occupational levels than Euro-Americans, as well as having scored higher on intelligence tests. Of interest is the fact that the initial Chinese immigrants came from poor and uneducated peasant backgrounds and yet even their first-generation children were making their way up the educational and socioeconomic ladder. Vernon (1982) allowed that genetic factors may be involved in these mental differences between Asian and European peoples.

5.3.3. Physical Coordination

Ethnic-group differences in physical coordination have been found from birth onwards and again we find the interesting rank ordering in which Europeans fall midway between Asian and African people. Freedman (1979) summarized the results of 15 independent studies, including some of his own. African babies, tested in various parts of East and West Africa, were more advanced in physical coordination compared with those of Asian and European descent. African newborns, for example, were often found to hold their heads erect. These results are unlikely to be due to current cultural differences, for the same ethnic-group differences emerge when Afro-Americans are tested. Afro-American children also walk at an average age of 11 months, compared with 12 months in Euro-Americans and 13 months in American Indians.

Do these relationships remain in adulthood? Some evidence suggests that they might. Relative to their overall percentage in the general population, Afro-Americans are "overrepresented" in United States professional sporting events (*Time Magazine*, 1977). Moreover, African-descended people living in Britain are increasingly "overrepresented" in British sports (Cashmere, 1982). These data are compatible with genetically based group differences in physical coordination.

5.3.4. Other Personality Traits

A surprisingly large number of studies have been carried out to test the personality of the Chinese and Japanese, both in their homelands and in North America (Vernon, 1982). Many investigators gave university students standardized personality tests such as Cattell's Sixteen Personality Factor Questionnaire, the Eysenck Personality Questionnaire, and the Edwards Personal Preference Schedule. Other studies relied on naturalistic observation and interviews. The evidence consistently favored the hypothesis that on average Asians were both more introverted and more anxious than Euro-Americans and less dominant and less aggressive. These differences also manifested themselves in play behavior, with Oriental children being quieter, more cautious, and less competitive and aggressive than Euro-Americans (see, also, Freedman, 1979, pp. 155–156). Interestingly enough, Eskimos, who are also of mongoloid origin, were also seen as behaviorally restrained (LeVine, 1975, p.19). To Eskimos, Euro-Americans seemed "emotionally volatile" (LeVine, 1975, p. 19), as they also did when contrasted with Chinese-Americans (Freedman, 1979, p.156).

If the framework advocated here is correct, then open-ended but exciting empirical questions can be raised. Are there other group differences in personality that might stem from genetically based traits: for example, in aggression, altruism, criminality, dominance, emotionality, locus of control, political attitudes, sexuality, sociability, values, and vocational interest? Englishmen are said to be reserved and circumspect, and Americans are said to be open and direct. Do these and similar stereotypes reflect real psychological differences among human populations? Do these differences subsequently lead to the particular social and cultural institutions which people generate and participate in? At the moment most of our information stems from stereotypes. The study of cross-cultural differences in (partly inherited) personality and their relation to culture could be an empirical gold mine.

6. Sociobiology and Social Learning

Social learning is particularly important for a species such as our own. It is a characteristic of *Homo sapiens* that there is a great deal of plasticity in our nervous systems. We are genetically programmed to learn from our environments. We even have our own species-specific ways of learning, such as verbal instruction.

Moving into the realm of social learning, however, does not leave sociobiology behind. From a sociobiological perspective, social learning is an additional mechanism affecting the transmission of DNA into the next and subsequent generations. One consequence of social learning is to increase enormously the range of phenotypic variation that is displayed. This increases the range of ecological niches that humans can fill and provides more material on which natural selection can operate. Individual differences among people (and groups) are to a large extent also a result of social learning. Such procedures as classical conditioning and operant and observational learning have major effects on the development and maintenance of individual differences. Indeed, effective therapeutic programs have come into being based on these principles (Wilson & O'Leary, 1980).

One question that arises is whether genetic differences between people in personality *interact* with learning processes. Significant advances in understanding people are particularly likely to grow out of such study, for information is gained simultaneously about procedures of social learning, about the core structure of personality, and of the very heart of *interactionism* which constitutes a consensual framework for personality psychology (Bandura, 1978; Endler & Magnusson, 1976).

Two genetic trait × learning process interactions will be described to demonstrate the possibilities in this relatively untapped area of psychological research: These are: (1) IQ × learning procedure and (2) extraversion × conditioning.

6.1. IQ × Learning Procedure Interactions

Jensen (1973) suggested that factor analyses of tests of IQ, scholastic achievement, and information-processing ability reveal two types of cognitive ability, which he calls Level I and Level II. Level I ability appears to be more dependent on associative and memory processes, where Level II ability appears to be more dependent on abstract and conceptual processes. Intelligence tests typically assess Level II abilities to a greater extent than Level I. Although the two ability types are themselves correlated, the correlations are low enough to allow some children of poor Level II ability to do very well on Level I ability tests. Jensen found that, although ethnic group and socioeconomic status differences in Level II are substantial, they are only slight, or nonexistent, on Level I.

Jensen (1973) has proposed that these differences of type in ability level have implications for education. Whereas those with Level I ability will learn most readily through rote memory training, those with Level II ability will learn most readily if the information is presented more abstractly. There is an increasing amount of research evidence favoring this hypothesis (Hall & Kaye, 1980; Vernon, 1981). From this perspective, then, acknowledgment of genetic diversity and employment of corresponding learning environments which maximize ultimate performance behavior is an example of a genetic trait × learning environment interaction that, if applied, has consequences of potential benefit to society.

6.2. Extraversion × Social Learning Interactions

Both Eysenck (1967, 1981) and Gray (1970, 1981) have proposed theories of personality functioning in which genetically based individual differences in conditionability play a part. Both theories are concerned with the dimensions of Extraversion–Introversion, and Emotionality–Stability. Eysenck's view is that extraverts should condition less well than introverts due to their low cortical arousal. Gray proposed that extraverts condition less well only under punishment learning due to their relative insensitivity to punishment. Gray's theory predicts extraverts to condition *better* than introverts under reward learning. Both Eysenck and Gray expect those high on measures of emotionality to be more conditionable under both reward and punishment learning than those low on emotionality.

Both Eysenck's and Gray's theories order disparate data and make testable predictions. For example, they explain why "clusters" develop in types of neurotic disorder. One group, comprising the "dysthymic" disorders, includes generalized anxiety, depression, excessive guilt, obsessive-compulsive behaviors, and phobias. Another group, the "character" disorders, includes criminality and delinquency. Both groups are different from "normal" in being high on anxiety. They differ from each other in introversion and concomitant conditionability. Those with character disorders tend to be extraverted and therefore more difficult to condition. The dysthymic disordered group tends to be introverted and particularly susceptible to conditioning. A different scheme that fits these data and helps create order is the dimension of "over control–under control" (Block, 1971; Block & Block, 1980). Regardless of how the relationship is conceptualized, evidence does exist for the relationship between extraversion and character disorder (Eysenck, 1977; Rushton & Chrisjohn, 1981). There is also direct evidence for the theory of differential conditionability. Three studies will be described.

In the first, Gupta (1976) carried out an operant verbal conditioning experiment in which rewards or punishments were made contingent on the subject's choice of personal pronouns. The results showed that punishment decreased responding significantly more among introverts than extraverts, thus supporting Gray's theorizing. In the second, Nagpal and Gupta (1979) again made rewards and punishments contingent on the use of personal pronouns. In accord with predictions, extraverts conditioned best under reward, introverts best under punishment. Finally, in a test of Gray's hypothesis of differential sensitivity to stimuli previously associated with punishment, Harvey and Hirschmann (1980) found that introverts were the most reactive. Heart rate acceleration, a physiological index of a defensive reaction, was the measure of sensitivity, and slides of people who had met violent deaths were the aversive stimuli. Interestingly, extraverts showed heart rate deceleration, indicative of the orienting response.

The processes of learning discussed so far have been the elemental ones of conditioning. Much human learning, however, is observational in nature (Bandura, 1977). Observational learning is so powerful that many governments have instituted investigations to examine whether there is inadvertent observational learning from watching television (U.S. Department of Health and Human Services, 1982). The evidence suggests that there is, from both aggressive portrayals (Murray & Kippax, 1979) and prosocial ones (Rushton, 1979). It is interesting to speculate whether genetic trait × learning interactions also occur in observational learning. Are introverts more susceptible to vicarious punishment and extraverts to vicarious reward, for example, as might be expected from Gray's theory (Rushton & Campbell, 1977)? Are those low in dominance or high on sociability more likely to learn from the observation of others? And are dispositionally aggressive individuals predisposed to acquire aggressive patterns of behavior or dispositionally altruistic ones to acquire prosocial patterns?

7. Challenging Issues That Remain

This review has brought together some related issues in the psychology of personality and social development under the umbrella of sociobiology. Sociobiology, a new science, is defined as "the systematic study of the biological basis of all social behaviors" (Wilson, 1975, p. 4). It aims to unify "all aspects of social evolution, including that of man" (p. 5). Its central tenet is that the purpose of life is the propagation of DNA into future generations with the least possible biochemical alteration. Here, it is suggested, are the origins and mainstays of consistent patterns of individual differences (traits) and their manifestation in such phenomena as maternal behaviors; altruism; aggression; anxiety and fear; sociability; dominance hierarchies; class, ethnic group, and sex differences; human social learning; and many other phenomena pertinent to social, personality, and developmental psychology.

Needless to say, many challenging issues remain. One objection to the account given above is that little evidence has yet been provided of a relationship between variation in personality and differential reproductive success. This, after all, is the core of the theoretical structure of sociobiology. The objection is well taken. Relatively little investigation has been carried out on the relationship between inherited traits and reproductive success. The research that does exist, however, is in line with expectations. Epidemiological and demographic studies of abnormal personality demonstrate that those who suffer from extreme anxiety, depression, and low IQ have fewer children than those with more moderate behavior patterns (Rosenthal, 1970). A related issue is whether the genetic basis to individual differences in social behavior simply reflects imcomplete stabilizing selection or whether, rather, directional selection is involved. For traits such as high intelligence there is likely to be directional dominance. For others, however, the information currently available is too limited for us to know.

A separate issue concerns the nature of the structures that are inherited that relate to the individual differences found in altruism, aggression, criminality, intelligence, etc. In many cases the genes may determine specific neural and chemical substrates that directly underlie particular traits. For example, Gray (1982) has described the cytoarchitecture of the "brain inhibition system" and linked activity in these fiber tracts to personality differences in anxiety level. The work on the evoked potential and other physiological correlates of IQ constitutes another prime example of matching individual differences in behavior with those in neurophysiological systems (Eysenck, 1982; Hendrickson & Hendrickson, 1981). In other cases, however, inherited individual differences in social behavior may be byproducts of other traits. For example, criminality may arise from individual differences in aggressiveness, extraversion, anxiety, and intelligence (Eysenck, 1977; Rushton & Chrisjohn, 1981).

Some may object that there is nothing new here, that behavior genetics and the psychology of individual differences were progressing quite well long before sociobiology appeared on the scene. This complaint of "cannibalization" would not be limited to behavior geneticists and personality psychologists. In his book, *Sociobiology*, Wilson (1975) subsumes disciplines as wide-ranging as cellular biology, neurophysiology, ethology, physiological psychology, population biology, anthropology, sociology, and ethics. In Wilson's more recent work, with Lumsden (1981, 1982), the attempt to unify the social sciences with biology is taken even further. Indeed, profound interactions between inherited differences in personality and the environment are derivable from their theory of gene–culture reciprocal coevolution. The central tenet of this theory is that genes causally affect culture and that culture, in turn, causally affects relative gene frequency. What is being suggested is that genes influence the structure and neurochemical functioning of an individual's brain. This influences emotions and cognitions and hence behavior. Thus the likelihood of assimilating or producing a particular culturgen (unit of culture) is affected. In this fashion the relative gene frequencies present in a particular society will influence the character of that society. On the other hand, the nature of the society in which people find themselves will affect their chance of survival and reproduction. A particular behavioral quality may in some societies be advantageous and in others disadvantageous. Just as gene frequencies affect the culture, so the culture affects the gene frequencies present in the next generation. The term *coevolution* is used instead of evolution to describe this reciprocal influence. From an individual difference perspective, it is possible schematically to present this as a feedback loop such that: individual differences in genes → individual differences in neural and chemical substrates → individual differences in minds → individual differences in the assimilation and production of culturgens (units of culture) → individual differences in genes, with the environment exerting influence at each link (Rushton & Russell, in press).

The above formulation leads to interesting lines of inquiry. Thus, it follows that variance in (partly inherited) measurable personality traits will be correlated with (a) variance in the physiological systems underlying those traits, (b) variance in the culturgens produced and assimilated, and (c) variance in genetic fitness. Preliminary evidence can be gathered in support of each of these predictions. The most important of these from the present perspective is (b), different personality types producing or assimilating different culturgens. Consider, for example, the studies examining the role that personality plays in scientific creativity. Many studies have found successful scientists to be more socially introverted than average (e.g., Cattell, 1962; Terman, 1955), whereas other studies have found them to be also more intellectually curious, needful of cognitive structure, aggressive, dominant, and independent (Rushton, Murray, & Paunonen, 1983). Thus individual differences in scientific creativity are in part inherited (see also Karlsson, 1978).

The synthesis of gene–culture coevolution with the psychology of personality has only just begun. The implications, however, may be far-reaching. One might conjecture, for example, that some personality types would thrive more in some cultures than others. To take some speculative examples: (a) genetically similar personality types may detect and seek each other out in order to provide mutually supportive cultures (there is, for example, some evidence of assortative mating for personality traits—Jensen, 1978; Vandenberg, 1972); (b) there may be natural antipathies toward others who have genetically dissimilar personalities; (c) cross-cultural differences in behavior may be partly genetic in origin; and (d) religious, political, and other ideological battles may become as heated as they do partly because they have implications for genetic fitness, that is, certain genotypes will thrive more in some ideological cultures than others.

Irrespective of the above, clearly Darwin's (1859, 1871) revolution has implications for the study of human personality, as indeed was recognized from the beginning (Galton, 1869). Inherited individual and group differences in personality are potentially enormously important, involving human happiness, marital adjustment, medicine, psychopathology, crime and delinquency, education, ethnic relations, politics, social disorder, war, and the very direction human history will take. Surely it is now time for the Darwinian perspective to be adopted, or at least to give it the close attention it undoubtedly deserves.

ACKNOWLEDGMENTS

I would like to thank J. Block, C. J. Brainerd, D. M. Buss, H. J. Eysenck, M. A. Goodale, J. Higgins, D. N. Jackson, A. R. Jensen, C. P. Lawrence, C. H. Littlefield, P. H. Mussen, H. L. Roediger III, R. J. H. Russell, C. H. Vanderwolf, P. A. Vernon, and P. A. Wells for comments and discussion. This must not be taken to imply, of course, that these individuals necessarily agree with the contents of this paper.

8. References

Adorno, T., Frenkel-Brunswik, E., Levison, D., & Sanford, R. *The authoritarian personality.* New York: Harper & Row, 1950.

Ahern, F. M., Johnson, R. C., Wilson, J. R., McClearn, G. E., & Vandenberg, S. G. Family resemblances in personality. *Behavior Genetics,* 1982, *12,* 261–280.

Akers, R. L. *Deviant behavior: A social learning approach.* Belmont, Ca.: Wadsworth, 1977.

Alexander, R. D. *Darwinism and human affairs.* Seattle, Wash.: University of Washington Press, 1979.

Anastasi, A. *Psychological testing* (5th ed.). New York: Macmillan, 1982.

Axelrod, R., & Hamilton, W. D. The evolution of cooperation. *Science,* 1981, *211,* 1390–1396.

Baker, P. T. *The biology of high altitude peoples.* New York: Cambridge University Press, 1977.

Bandura, A. *Aggression: A social learning analysis.* Englewood Cliffs, N.J.: Prentice-Hall, 1973.

Bandura, A. *Social learning theory.* Englewood Cliffs, N.J.: Prentice Hall, 1977.

Bandura, A. The self-system in reciprocal determinism. *American Psychologist,* 1978, *33,* 344–358.

Barash, D. P. *Sociobiology and behavior* (2nd ed.). New York: Elsevier, 1982.

Barlow, G. W., & Silverberg, J. (Eds.). *Sociobiology: Beyond nature/nurture? Reports, definitions and debate. AAAS Selected Symposium 35.* Boulder, Colo: Westview Press, 1980.

Bateman, A. J. Intra-sexual selection in *Drosophila. Heredity,* 1948, *2,* 349–368.

Block, J. *Lives through time.* Berkeley, Ca.: Bancroft Books, 1971.

Block, J. Some enduring and consequential structures of personality. In A. I. Rabin, J. Aronoff, A. M. Barclay, & R. A. Zucker (Eds.), *Further explorations in personality*. New York: Wiley, 1981.

Block, J. H. Issues, problems, and pitfalls in assessing sex differences: A critical review of "The Psychology of Sex Differences." *Merrill-Palmer Quarterly*, 1976, *22*, 283–309.

Block, J. H., & Block, J. The role of ego-control and ego-resiliency in the organization of behavior. In W. A. Collins (Ed.), *Minnesota Symposium on Child Psychology: Vol. 13: Development of Cognition, Affect and Social Relations*. Hillsdale, N.J.: Lawrence Erlbaum, 1980.

Bouchard, T. J., Jr. Twins nurture's twice-told tale. In *1983 Yearbook of Science and the Future*. New York: Encyclopedia Britanica, 1982.

Bouchard, T. J., Jr. *Traits and the concepts of convergence and divergence in the development of human personality*. Paper presented to the Fourth International Congress on Twin Studies, London, England, June 28 to July 1, 1983.

Bouchard, T. J., Jr., & McGue, M. Familial studies of intelligence: A review. *Science*, 1981, *212*, 1055–1059.

Bouchard, T. J., Jr., Heston, L., Eckert, E., Keyes, M., & Resnick, S. The Minnesota study of twins reared apart: Project description and sample results in a developmental domain. In L. Gedda, P. Parisi, & W. E. Nance (Eds.), *Twin Research 3: Part B. Intelligence, personality, and development*. New York: Liss, 1981.

Brigham, J. C., & Richardson, C. B. Race, sex and helping in the marketplace. *Journal of Applied Social Psychology*, 1979, *9*, 314–322.

Buj, V. Average IQ values in various European countries. *Personality and Individual Differences*, 1981, *2*, 168–169.

Buss, A. H., & Plomin, R. *A temperament theory of personality development*. New York: Wiley, 1975.

Buss, A. H., Plomin, R., & Willerman, L. The inheritance of temperaments. *Journal of Personality*, 1973, *41*, 513–524.

Buss, D. M. Evolutionary biology and personality psychology: Implications of genetic variability. *Personality and Individual Differences*, 1983, *4*, 51–63.

Calhoun, J. B. Population density and social pathology. *Scientific American*, 1962, *206*, (February), 139–148.

Campbell, D. T. Ethnocentric and other altruistic motives. In D. Levine (Ed.), *Nebraska symposium on motivation*. Lincoln: University of Nebraska Press, 1965.

Campbell, D. T. On the conflicts between biological and social evolution and between psychology and moral tradition. *American Psychologist*, 1975, *30*, 1103–1112.

Carey, G., Goldsmith, H. H., Tellegan, A., & Gottesman, I. I. Genetics and personality inventories: The limits of replication with twin data. *Behavior Genetics*, 1978, *8*, 299–313.

Cashmere, E. *Black sportsmen*. London: Routledge & Kegan Paul, 1982.

Cattell, R. B. The personality and motivation of the researcher from measurements of contemporaries and from biography. In C. W. Taylor & F. Barron (Eds.), *Scientific creativity: Its recognition and development*. New York: Wiley, 1962.

Cattell, R. B. The heritability of fluid (fg) and crystallized (gc) intelligence estimated by a least squares use of the MAVA method. *British Journal of Educational Psychology*, 1980, *50*, 253–265.

Cattell, R. B. *The inheritance of ability and personality*. New York: Academic Press, 1982.

Cattell, R. B., Blewett, D. B., & Beloff, J. R. The inheritance of personality: A multiple variance analysis of approximate nature–nurture ratios for primary personality factors in Q-data. *American Journal of Human Genetics*, 1955, *7*, 122–146.

Cattell, R. B., Vaughan, D. S., Schuerger, J. M., & Rao, D. C. Heritabilities, by the multiple abstract variance analysis (MAVA) model and objective test measures, of personality traits U.I. 23, capacity to mobilize, U.I. 24, anxiety, U.I.26, narcistic ego, and U.I. 28, aesthenia, by maximum-likelihood methods. *Behavior Genetics*, 1982, *12*, 361–378.

Chagnon, N. A., & Irons, W. *Evolutionary biology and human social behavior: An anthropological perspective*. North Scituate, Mass.: Duxbury Press, 1979.

Cole, J., & Cole, S. *Social stratification in science*. Chicago: University of Chicago Press, 1973.

Colinvaux, P. *The fates of nations: A biological theory of history*. New York: Simon & Schuster, 1980.

Conley, J. J. The longitudinal stability of personality traits: A multitrait-multimethod-multioccasion analysis. *Journal of Personality and Social Psychology*, in press.

Crowe, R. R. The adopted offspring of women criminal offenders: A study of their arrest records. *Archives of General Psychiatry*, 1972, *27*, 600–603.

Crowe, R. R. An adoption study of antisocial personality. *Archives of General Psychiatry*, 1974, *31*, 785–791.

Cunningham, M. R. Sociobiology as a supplementary paradigm for social psychological research. In L. Wheeler (Ed.), *Review of personality and social psychology* (Vol. 2). Beverly Hills, Ca.: Sage, 1981.

Daly, M. & Wilson, M. *Sex, evolution, and behavior* (2nd ed.). Boston, Mass.: Willard Grant Press, 1983.

Daly, M., Wilson, M., & Weghorst, S. J. Male sexual jealousy. *Ethology and Sociobiology*, 1982, *3*, 11–27.

Darwin, C. R. *On the origin of species by means of natural selection, or, the preservation of favored races in the struggle for life*. London: Murray, 1859.

Darwin, C. R. *The descent of man and selection in relation to sex*. London, Murray, 1871.

Dawkins, R. *The selfish gene*. Oxford: Oxford University Press, 1976.

Dawkins, R. *The extended phenotype*. San Francisco, Ca.: Freeman, 1982.

Durham, W. H. Resource competition and human aggression. Part 1: A review of primitive war. *Quarterly Review of Biology*, 1976, *51*, 385–415.

Dworkin, R. H., Burke, B. W., Maher, B. A., & Gottesman, I. I. A longitudinal study of the genetics of personality. *Journal of Personality and Social Psychology*, 1976, *34*, 510–518.

Eaves, L. J., & Eysenck, H. J. Genetics and the development of social attitudes. *Nature*, 1974, *249*, 288–289.

Eaves, L. J., & Eysenck, H. J. The nature of extraversion: A genetical analysis. *Journal of Personality and Social Psychology*, 1975, *32*, 102–112.

Eaves, L. J., Last, K. A., Young, P. A., & Martin, N. G. Model fitting approaches to the analysis of human behavior. *Heredity*, 1978, *41*, 249–320.

Einswiller, T., Deaux, K., & Willits, J. E. Similarity, sex and requests for small favors. *Journal of Applied Social Psychology*, 1971, *1*, 284–291.

Eisenberg, N., & Lennon, R. Sex differences in empathy and related capacities. *Psychological Bulletin*, 1983, *94*, 100–131.

Ellis, L. Genetics and criminal behavior. *Criminology*, 1982, *20*, 43–66.

Endler, N. S., & Magnusson, D. Toward an interactional psychology of personality. *Psychological Bulletin*, 1976, *83*, 956–976.

Epstein, S. The stability of behavior: 1. On predicting most of the people much of the time. *Journal of Personality and Social Psychology*, 1979, *37*, 1097–1126.

Epstein, S. The stability of behavior: II. Implications for psychological research. *American Psychologist*, 1980, *35*, 790–806.

Erlenmeyer-Kimling, L., & Jarvik, L. R. Genetics and intelligence: A review. *Science*, 1963, *142*, 1477–1479.
Eysenck, H. J. *The biological basis of personality*. Springfield: Thomas, 1967.
Eysenck, H. J. *Sex and personality*. London: Open Books, 1976.
Eysenck, H. J. *Crime and personality*. (3rd ed.). London: Routledge & Kegan Paul, 1977.
Eysenck, H. J. *The structure and measurement of intelligence*. New York: Springer-Verlag, 1977.
Eysenck, H. J. (Ed.). *A model for personality*. New York: Springer-Verlag, 1981.
Eysenck, H. J. (Ed.). *A model for intelligence*. New York: Springer-Verlag, 1982.
Eysenck, H. J., & Eysenck, S. B. G. *Manual of the Eysenck Personality Questionnaire*. San Diego, Cal.: Educational and Industrial Testing Service, 1975.
Eysenck, H. J., & Eysenck, S. B. G. *Psychoticism as a dimension of personality*. London: Hodder & Stoughton, 1976.
Falconer, D. S. *Introduction to quantitative genetics*. (2nd ed.). London: Longman, 1981.
Feldman, R. E. Response to compatriots and foreigners who seek assistance. *Journal of Personality and Social Psychology*, 1968, *10*, 202–214.
Firkowska-Mankiewicz, A., & Czarkowski, M. P. Social status and mental test performance in Warsaw children. *Personality and Individual Differences*, 1982, *3*, 237–247.
Fisher, J. D., DePaulo, B. M., & Nadler, A. Extending altruism beyond the altruistic act: The mixed effects of aid on the help recipient. In J. P. Rushton & R. M. Sorrentino (Eds.), *Altruism and helping behavior: Social personality and developmental perspectives*. Hillsdale, N.J.: Lawrence Erlbaum, 1981.
Floderus-Myrhed, B., Pedersen, N., & Rasmuson, I. Assessment of heritability for personality, based on a short form of the Eysenck Personality Inventory: A study of 12,898 twin pairs. *Behavior Genetics*, 1980, *10*, 153–162.
Freedman, D. G. *Human sociobiology: A holistic approach*. New York: Free Press, 1979.
Fulker, D. W. The genetic and evnvironmental architecture of psychoticism, extraversion and neuroticism. In H. J. Eysenck (Ed.), *A model for personality*. New York: Springer-Verlag, 1981.
Galton, F. *Hereditary genius: An inquiry into its laws and consequences*. London: Macmillan, 1869.
Gottesman, I. I. Heritability of personality: A demonstration. *Psychological Monographs*, 1963, *77*, No. 9 (Whole No. 572).
Gottesman, I. I. Genetic variance in adaptive personality traits. *Journal of Child Psychology and Psychiatry and Allied Disciples*, 1966, *7*, 199–208.
Gould, S. J. *The mismeasure of man*. New York: Norton, 1981.
Gray, J. A. The psychophysiological basis of introversion–extraversion. *Behavior Research and Therapy*, 1970, *8*, 249–266.
Gray, J. A. *The psychology of fear and stress*. London: Weidenfeld and Nicholson, 1971.
Gray, J. A. A critique of Eysenck's theory of personality. In H. J. Eysenck (Ed.), *A model for personality*. New York: Springer-Verlag, 1981.
Gray, J. A. *The neuropsychology of anxiety: An enquiry into the functions of the septo-hippocampal system*. New York: Oxford University Press, 1982.
Gregory, M. S., Silvers, A., & Sutch, D. *Sociobiology and human nature: An interdisciplinary critique and defense*. San Francisco: Jossey-Bass, 1978.
Grotevant, H. D., Scarr, S., & Weinberg, R. A. Patterns of interest similarity in adoptive and biological families. *Journal of Personality and Social Psychology*, 1977, *35*, 667–676.
Gupta, B. S. Extraversion and reinforcement in verbal operant conditioning. *British Journal of Psychology*, 1976, *67*, 47–52.

Hall, V. C., & Kaye, D. B. Early patterns of cognitive development. *Monographs of the Society for Research in Child Development*, 1980, *45*, (2, Serial No. 184).
Hamilton, W. D. The genetical evolution of social behavior: I and II. *Journal of Theoretical Biology*, 1964, *7*, 1–52.
Harvey, F., & Hirschman, R. The influence of extraversion and neuroticism on heart rate responses to aversive visual stimuli. *Personality and Individual Differences*, 1980, *1*, 97–100.
Hendrickson, D. E., & Hendrickson, A. E. The biological basis of individual differences in intelligence. *Personality and Individual Differences*, 1981, *1*, 3–34.
Herrnstein, R. *IQ in the meritocracy*. New York: Allen Lane, 1973.
Hines, M. Prenatal gonadal hormones and sex differences in human behavior. *Psychological Bulletin*, 1982, *92*, 56–80.
Holden, C. News and comment: Identical twins reared apart. *Science*, 1980, *207*, 1323–1328.
Hovland, C. E., & Sears, R. R. Minor studies of aggression. VI. Correlation of lynchings with economic indices. *Journal of Psychology*, 1940, *9*, 301–310.
Hutchings, B., & Mednick, S. A. Registered criminality in the adoptive and biological parents of registered male criminal adoptees. In R. R. Fieve, D. Rosenthal, & H. Brill (Eds.), *Genetic research in psychiatry*. Baltimore: Johns Hopkins University Press, 1975.
Jensen, A. R. How much can we boost IQ and scholastic achievement? *Harvard Educational Review*, 1969, *39*, 1–123.
Jensen, A. R. *Educability and group differences*. Hew York: Harper & Row, 1973.
Jensen, A. R. Race and mental ability. In A. H. Halsey (Ed.), *Heredity and environment*. London: Methuen, 1977.
Jensen, A. R. Genetic and behavioral effects of nonrandom mating. In R. T. Osborne, C. E. Noble, & N. Weyl (Eds.), *Human variation: The biopsychology of age, race, and sex*. New York: Academic Press, 1978.
Jensen, A. R. *Bias in mental testing*. New York: The Free Press, 1980.
Jensen, A. R. *Straight talk about mental tests*. New York: Free Press, 1981.(a)
Jensen, A. R. Commentary. In S. Scarr (Ed.), *Race, social class, and individual differences in IQ: New studies on old issues*. Hillsdale, N.J.: Lawrence Erlbaum, 1981.(b)
Jensen, A. R. *The nature of the white-black difference on various psychometric tests*. Invited address, Division 16, Annual Meeting of the American Psychological Association, Anaheim, California, August, 1983.
Johanson, D. C., & White, T. D. A systematic assessment of early African hominids. *Science*, 1979, *203*, 321–330.
Kamin, L. J. *The science and politics of IQ*. Hillsdale, N.J.: Lawrence Erlbaum, 1974.
Kamin, L. J. Commentary. In S. Scarr (Ed.), *Race, social class, and individual differences in IQ: New studies on old issues*. Hillsdale, N.J.: Lawrence Erlbaum, 1981.
Karlsson, J. L. *Inheritance of creative intelligence*. Chicago: Nelson-Hall, 1978.
Kenrick, D. T., & Stringfield, D. O. Personality traits and the eye of the beholder: Crossing some traditional philosophical boundaries in the search for consistency in all of the people. *Psychological Review*, 1980, *87*, 88–104.
LeVine, R. A. *Culture, behavior, and personality*. Chicago: Aldine, 1975.
Lewontin, R. C. Sociobiology as an adaptationist paradigm. *Behavioral Science*, 1979, *24*, 5–14.
Loehlin, J. C., & Nichols, R. C. *Heredity, environment, and personality*. Austin: University of Texas Press, 1976.
Loehlin, J. C., Lindzey, G., & Spuhler, J. N. *Race differences in intelligence*. San Francisco: W. H. Freeman, 1975.

Loehlin, J. C., Horn, J. M., & Willerman, L. Personality resemblance in adoptive families. *Behavior Genetics*, 1981, *11*, 309–330.
Lorenz, K. *On aggression*. New York: Harcourt Brace Jovanovich, 1966.
Lumsden, C. J., & Wilson, E. O. *Genes, mind and culture: The coevolutionary process*. Cambridge: Harvard University Press, 1981.
Lumsden, C. J., & Wilson, E. O. Précis of *Genes, mind, and culture*, with commentary. *The Behavioral and Brain Sciences*, 1982, *5*, 1–38.
Lumsden, C. J. , & Wilson, E. O. *Promethean fire*. Cambridge: Harvard University Press, 1983.
Lynn, R. The intelligence of the Japanese. *Bulletin of the British Psychological Society*, 1977, *30*, 69–72.
Lynn, R. Ethnic and racial differences in intelligence: International comparisons. In R. T. Osborne, C. E. Noble, & N. Weyl (Eds.), *Human variation: The biopsychology of age, race, and sex*. New York: Academic Press, 1978.
Lynn, R. IQ in Japan and the United States shows a growing disparity. *Nature*, 1982, *297*, 222–223.
Lynn, R., & Dziobon, J. On the intelligence of the Japanese and other Mongoloid peoples. *Personality and Individual Differences*, 1980, *1*, 95–96.
Maccoby, E. E., & Jacklin, C. N. *The psychology of sex differences*. Stanford. Ca: Stanford University Press, 1974.
Martin, N. G., Eaves, L. J., & Eysenck, H. J. Genetical, environmental and personality factors influencing the age of first sexual intercourse in twins. *Journal of Biosocial Science*, 1977, *9*, 91–97.
Matthews, K. A., Batson, C. D., Horn, J., & Rosenman, R. H. "Principles in his nature which interest him in the fortune of others . . ." The heritability of empathic concern for others. *Journal of Personality*, 1981, *49*, 237–247.
Mauss, M. *The gift: Forms and function of exchange in archaic societies*. Glencoe, Ill.: Free Press, 1954.
Miller, J. Z., & Rose, R. J. Familial resemblance in locus of control: A twin-family study of the Internal-External Scale. *Journal of Personality and Social Psychology*, 1982, *42*, 535–540.
Mischel, W. *Personality and assessment*. New York: Wiley, 1968.
Mischel, W. *Introduction to personality*. New York: Holt, Rinehart and Winston, 1981.
Mittler, P. *The study of twins*. Harmondsworth, Middlesex: Penguin, 1971.
Mantagu, M. F. A. (Ed.). *Sociobiology examined*. New York: Oxford University Press, 1980.
Murray, J. P., & Kippax, S. From the early window to the late night show: A cross-national review of television's impact on children and adults. In L. Berkowitz (Ed.), *Advances in experimental social psychology* (Vol. 12). New York: Academic Press, 1979.
Nagpal, M., & Gupta, B. S. Personality, reinforcement and verbal operant conditioning. *British Journal of Psychology*, 1979, *70*, 471–476.
Olweus, D. Stability of aggressive reaction patterns in males: A review. *Psychological Bulletin*, 1979, *86*, 852–875.
Osborne, R. T. Race and sex differences in heritability of mental test performance: A study of Negroid and Caucasoid Twins. In R. T. Osborne, C. E. Noble, & N. Weyl (Eds.), *Human variation: The biopsychology of age, race, and sex*. New York: Academic Press, 1978.
Owen, D. R., & Sines, J. O. Heritability of personality in children. *Behavior Genetics*, 1970, *1*, 235–248.
Packer, C. Reciprocal altruism in *Papio anubis*. *Nature*, 1977, *265*, 441–443.
Passingham, R. E. The brain and intelligence. *Brain, Behavior and Evolution*, 1975, *11*, 1–15.

Passingham, R. E. Brain size and intelligence in man. *Brain, Behavior and Evolution*, 1979, *16*, 253–270.
Plomin, R. (Ed.), Special section on developmental behavioral genetics. *Child Development*, 1983, *54*, 253–435.
Plomin, R., & DeFries, J. C. Genetics and intelligence: Recent data. *Intelligence*, 1980, *4*, 15–24.
Plomin, R., DeFries, J. C., & McClearn, G. E. *Behavioral Genetics: A Primer*. San Francisco, Ca.: Freeman, 1980.
Plutchik, R. *Emotion: A psychoevolutionary synthesis*. New York: Harper & Row, 1980.
Ridley, M., & Dawkins, R. The natural selection of altruism. In J. P. Rushton & R. M Sorrentino (Eds.), *Altruism and helping behavior. Social, personality, and developmental perspectives*. Hillsdale, N.J.: Lawrence Erlbaum, 1981.
Rosenthal, D. *Genetic theory and abnormal behavior*. New York: McGraw-Hill, 1970.
Rosenthal, T. L., & Zimmerman, B. J. *Social learning and cognition*. New York: Academic Press, 1978.
Rotter, J. B. Generalized expectancies of internal versus external control of reinforcement. *Psychological Monographs*, 1966, *80*, (1, Whole No. 609).
Rotter, J. B. Some problems and misconceptions related to the construct of internal versus external control of reinforcement. *Journal of Consulting and Clinical Psychology*, 1975, *43*, 56–67.
Royce, J. R. Philosophic issues, division 24, and the future. *American Psychologist*, 1982 *37*, 258–266.
Ruse, M. *Sociobiology: Sense or nonsense?* Dordrecht, Holland: D. Reidel, 1979.
Rushton, J. P. The effects of prosocial televison and film material on the behavior of viewers. In L. Berkowitz (Ed.), *Advances in experimental social psychology* (Vol. 12). New York: Academic Press, 1979.
Rushton J. P. *Altruism, socialization, and society*. Englewood Cliffs, N.J.: Prentice-Hall, 1980.
Rushton, J. P., & Campbell, A. C. Modeling, vicarious reinforcement, and extraversion on blood donating in adults: Immediate and long term effects. *European Journal of Social Psychology*, 1977, *7*, 297–306.
Rushton, J. P., & Chrisjohn, R. D. Extraversion, neuroticism, psychoticism, and self-reported delinquency: Evidence from eight separate samples. *Personality and Individual Differences*, 1981, *2*, 11–20.
Rushton, J. P., & Meltzer, S. Research productivity, university revenue, and scholarly impact (citations) of 169 British, Canadian, and United States universities. *Scientometrics*, 1981, *3*, 275–303.
Rushton, J. P., & Russell, R. J. H. Gene–culture theory and inherited individual differences in personaltiy. *The Behavioral and Brain Sciences*, in press.
Rushton, J. P., Jackson, D. N., & Paunonen, S. V. Personality: Nomothetic or idiographic? A response to Kenrick and Stringfield. *Psychological Review*, 1981, *88*, 582–589.
Rushton, J. P., Brainerd, C. J., & Pressley, M. Behavioral development and construct validity: The principle of aggregation. *Psychological Bulletin*, 1983, *94*, 18–38.
Rushton, J. P., Fulker, D. W., Neale, M. C., Nias, D. K. B., & Eysenck, H. J. *Altruism and aggression: Individual differences are inherited*. Unpublished manuscript: University of Western Ontario, 1984.
Rushton, J. P., Murray, H. G., & Paunonen, S. V. Personality, research creativity and teaching effectiveness in university professors. *Scientometrics*, 1983, *5*, 93–116.
Rushton, J. P., Fulker, D. W., Neale, M. C., Blizard, R. A., & Eysenck, H. J. Altruism and genetics, *Acta Geneticae Medicae et Gemellogiae: Twin Research*, in press.

Scarr, S. Genetic factors in activity motivation. *Child Development*, 1966, *37*, 663–673.
Scarr, S. Socal introversion–extraversion as a heritable response. *Child Development*, 1969, *40*, 823–832.
Scarr, S. (Ed.), *Race, social class, and individual differences in IQ*. Hillsdale, N.J.: Lawrence Erlbaum, 1981.
Scarr, S., & Carter-Saltzman, L. Twin method: Defense of a critical assumption. *Behavior Genetics*, 1979, *9*, 527–542.
Scarr, S., Webber, P.L., Weinberg, R. A., & Wittig, M. A. Personality resemblance among adolescents and their parents in biologically related and adoptive families. *Journal of Personality and Social Psychology*, 1981, *40*, 885–898.
Scarr, S., Caparulo, B. K., Ferdman, B. M., Tower, R. B., & Caplan, J. Developmental status and school achievements of minority and non-minority children from birth to 18 years in a British Midlands town. *British Journal of Developmental Psychology*, 1983, *1*, 31–48.
Schulsinger, F. Psychopathy: Heredity and environment. *International Journal of Mental Health*, 1972, *1*, 190–206.
Strayer, F. F. Social ecology of the preschool peer-group. In W. C. Collins (Ed.), *Minnesota Symposium on Child Psychology: Volume 13: Development of cognition, affect, and social relations*. Hillsdale, N.J.: Lawrence Erlbaum, 1980.
Strayer, F. F., Wareing, S., & Rushton, J. P. Social constraints on naturally occurring preschool altruism. *Ethology and Sociobiology*, 1979, *1*, 3–11.
Symons, D. *The evolution of human sexuality*. New York: Oxford University Press, 1979.
Terman, L. M. Are scientists different? *Scientific American*, 1955, No. 437.
Time Magazine (1977).
Trivers, R. L. The evolution of reciprocal altruism. *Quarterly Review of Biology*, 1971, *46*, 35–57.
Trivers, R. L., & Willard, D. E. Natural selection of parental ability to vary the sex ratio of offspring. *Science*, 1973, *179*, 90–92.
U. S. Department of Health and Human Services. *Television and behavior: Ten years of progress and implications for the eighties*. Vol. 1: *Summary Report;* Vol. 2: *Technical Reviews*. Washington, D. C.: U. S. Government Printing Office, 1982.
van den Berghe, P. L., & Barash, D. P. Inclusive fitness and human family structure. *American Anthropologist*, 1979, *79*, 809–823.
Vandenberg, S. G. The hereditary abilities study: Hereditary components in a psychological test battery. *American Journal of Human Genetics*, 1962, *14*, 220–237.
Vandenberg, S. G. Assortative mating, or who marries whom? *Behavior Genetics*, 1972, *2*, 127–157.
Vernon, P. A. Level I and Level II: A review. *Educational Psychologist*, 1981, *16*, 45–64.
Vernon, P. A., & Jensen, A. R. Individual and group differences in intelligence and speed of information processing. *Personality and Individual Differences*, in press.
Vernon, P. E. *The abilities and achievements of Orientals in North America*. New York: Academic Press, 1982.
Waller, J. H. Achievement and social mobility: Relationships among IQ score, education, and occupation in two generations. *Social Biology*, 1971, *18*, 252–259.
Willerman, L. Activity level and hyperactivity in twins. *Child Development*, 1973, *44*, 288–293.
Willerman, L. *The psychology of individual and group differences*. San Francisco, Ca.: Freeman, 1979.
Wilson, E. O. *Sociobiology: The new synthesis*. Cambridge: Harvard University Press, 1975.
Wilson, E. O. *On human nature*. Cambridge: Harvard University Press, 1978.

Wilson, G. T., & O'Leary, K. D. *Principles of behavior theory*. Englewood Cliffs, N.J.: Prentice-Hall, 1980.

Wilson, R. S. The Louisville Twin Study: Developmental synchronies in behavior. *Child Development*, 1983, *54*, 298–316.

Wispé, L. G., & Thompson, J. W. The war between the words: Biological vs. social evolution and some related issues. *American Psychologist*, 1976, *31*, 341–384.

Wyers, E. Y., Adler, H. E., Carpen, K., Chiszar, D., Demarest, J., Flannagan, O. J., Jr., Von Glasersfled, E., Glickman, S. E., Mason, W. A., Menzel, E. W., & Tobach, E. The sociobiological challenge to psychology: On the proposal to "cannibalize" comparative psychology. *American Psychologist*, 1980, *35*, 955–979.

1

Sociobiology and Differential Psychology

The Arduous Climb from Plausibility to Proof

Arthur R. Jensen

Rushton has performed a most necessary service to the advancement of behavioral science. He has indicated, succinctly yet quite comprehensively, how differential psychology can fruitfully be brought under the purview of the newly developing science of sociobiology. We may rather safely predict that his effort will meet at least temporary resistance. In the long history of differential psychology (i.e., the study of individual and group differences in behavioral traits) and in the comparatively short history of sociobiology (i.e., the study of the biological basis of social behavior) we have seen tides of opposition beyond the usual technical criticism and analysis which are a normal accompaniment to all important scientific endeavor. The resistance has evinced more of the character of the resistance which has been seen historically in connection, not with normal science, but with true scientific revolutions in the Kuhnian sense. In such cases, science has invaded areas of deep human concern and seemingly threatened entrenched theories of man's nature and place in the universe. The apposition of evolutionary biology and human individual and group differences and social behavior, as proposed in Rushton's essay, will be perceived by some as a similar threat to humans' welfare and self-esteem.

This will be especially true, of course, when the subjects of race

Arthur R. Jensen • Institute of Human Learning, University of California at Berkeley, Berkeley, California 94720.

and behavioral differences are juxtaposed in an evolutionary and genetic perspective. Rushton has not in the least soft-pedalled this topic, although many may wish he had, because race is undoubtedly the bugbear of both differential psychology and sociobiology. Many scholars justifiably fear that some critics still foster popular confusion between the scientific study of human variation and the anathema of racism (Jensen, 1982). Yet, a positive value of our openly viewing the controversial questions about race and behavioral differences from a sociobiological perspective is that we thereby confront head-on what is probably the chief focus of anxiety about sociobiology. Perhaps if that one source of resistance is overcome, the rest will dwindle automatically. Such a thrust, however, stands the best chance of advancing the theoretic integration of differential psychology and sociobiology only by our exercising the most scrupulous and explicit concern with evidence and inference.

Many reasonable scientists would argue that we should shun, as subjects of scientific investigation, those topics which are socially sensitive and which do not, for technical or ethical reasons, allow the kind of research that directly yields definitive answers. Hence, "speculation," "conjecture," "hypothesis," "plausibility," and the like, although quite acceptable as theoretical scaffolding in socially noncontroversial fields of scientific enterprise, are all terms quite severely frowned upon when it comes to research on race differences. In this realm, research results and theoretical interpretations that fall short of rigorous proof, some would say, should be disdained.

I disagree with this position for two main reasons. First, what we mean by objective scientific proof, as contrasted with mathematical or logical proof, is in fact a continuum rather than a dichotomous classification—"proved" versus "not proved." Some kinds of evidence add more, and some less, to the plausibility of a conjecture or hypothesis. Certainty always varies by degrees, and progress in any science would be virtually impossible if it were forced to treat all items of evidence as either 0 or 1. By trying to integrate theoretically an enlarging network of correlated items of evidence, it is possible to advance a theory up the continuum of plausibility. (The point on this continuum at which "proof" is established is purely a matter of consensus among qualified scientists.) My second reason is that by enlarging the network of correlated items of evidence we stand a better chance of finding more feasibly and rigorously testable hypotheses than those we had entertained at the outset. Many items of evidence, each of which, when viewed singly, allows only quite limited inference, when viewed all together may permit much broader inferences. What may seem puzzling when standing alone may be theoretically explainable as a part of the network of interrelated items of evidence. Much of sociobiology, like Darwin's theory of evolution, will depend on this kind of building of a network of interrelated items of evidence. The credibility of the effort depends in part upon the logical rigor with which any given network of correlated items of evidence is interpreted. Most correlated facts make scientific sense only within some theoretical framework.

This position should not be confused with the fagot fallacy, which is the mistaken belief that a theory can be proven by amassing a large number of items of evidence, each of which is in some way too defective to stand on its own. For a network of correlated facts to be scientifically useful, each of the separate elements of the network and each of its correlations with other elements must be firmly established beyond dispute. Arguments would then concern which theoretical interpretation best comprehends the largest number of indisputably established observations in a given correlational network.

Many of the questions addressed by sociobiology, as by evolutionary theory and by differential psychology, cannot be tackled with the methodological power and rigor which the experimental method ideally affords. In these fields, practical or ethical limitations generally restrict investigators to observation of the correlations among natural variations. The problem is how to make the most of them. Many important questions, such as the question of whether the observed statistical differences in IQ between racial groups have a genetic basis, could probably be answered quite directly and definitively if the methods of *experimental* genetics could be brought to bear on them. But they simply cannot be brought to bear, given the moral constraints which we value above scientific knowledge. In the case of the the race–IQ question just mentioned, for example, it would require the randomized cross-racial mating (in every possible sex × race combination) of truly random samples of the two racial populations in question and randomized cross-fostering of all the progeny. Such an experiment would be out of the question. Therefore we are left with only various *correlations* among observable natural phenomena.

Correlation coefficients, of course, have always been the primary grist for the analytic mill of differential psychology. The more that one works with correlations (and their extension to multivariate methods such as multiple regression, canonical correlation, and factor analysis), as all differential psychologists such as I must do, the more one comes to realize how treacherous they can be when we come to theoretical interpretation. Yet, with the preclusion of experimental methods, investigation and hypothesis generation must begin somewhere. And in this field the starting point is generally a correlation, or a pattern of intercorrelations among three or more variables. And the cardinal rule here is that more information than the correlations *per se* is always needed for a scientifically justifiable interpretation. The study of racial or ethnic differences in mental ability illustrates this rule nicely.

1. Correlates of Race and IQ

Rushton has reviewed some of the best known findings about racial or ethnic differences in behavioral traits and suggests that many such observable or phenotypic differences may be related to genetic differences arising from evolutionary divergence. A central variable in discussions of psychological racial variations has been general intelligence. Operationally, it is roughly measured by present-day standard IQ tests. Theoretically, at present, it is conceived of as a construct best represented as the general factor common to virtually all complex cognitive tests. Spearman referred to it as *g*, and it is this construct which is of greater interest than a score on any single test. However, the best modern IQ tests are quite valid indicators of individual differences in *g* for the vast majority of people with a common language and a common culture. These indicators of *g* also show distributional differences between races. Because much more data are available with respect to Euro-American and Afro-American (henceforth termed white and black) groups, I will focus the discussion on these.

It seems to me that the eventual achievement of some sociobiologic understanding of the observed relationship between race and *g* can best begin by examining the known correlates of both variables (i.e., race and *g*) which would seem to have some biological implications, and by also seeking evidence of the correlations among the other variables that are correlated with race and *g*. That correlation *per se* does not necessarily imply causation hardly bears repeating. In Figure 1 are shown a number of variables (22 in all) among which reliable correlations have been reported in the scientific literature. A more thorough search of the literature would very likely turn up more variables which could be added to the picture, but these 22 will do for illustration. The lines between the variables represent reported correlations, most of which are well substantiated by a number of studies. The sizes of the correlations, of course, vary widely. In any system of multiple causation, it should not be surprising to find some quite small (albeit reliable) correlations in the network. If a given variable, say, IQ, is multiply determined, we should not expect to find high correlations between IQ and each of its causal factors viewed singly.

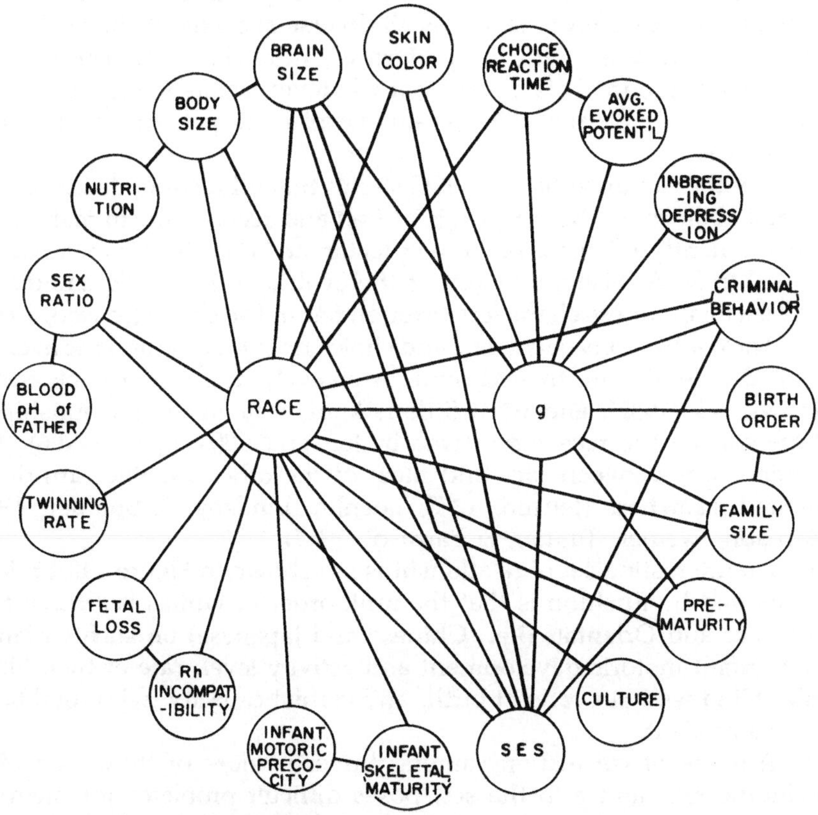

Figure 1. Reliable correlations (indicated by straight lines) between a number of biological and behavioral variables which have been reported in the scientific literature.

Enough research evidence is already at hand to support conclusions with some reasonable assurance that certain factors are most probably not the causes of the average white–black IQ difference, such as test bias or factors in the testing situation, educational inequality, teacher expectancy, socioeconomic status, and nutrition (Jensen, 1973, 1980a, 1981). Profitable search is more apt to be directed toward other variables. A few words about some of the variables shown in Figure 1 are in order.

A factor analytic study of the Wechsler Intelligence Scale for Children-Revised (WISC-R) by Jensen and Reynolds (1982) indicates that the mean white–black IQ difference is primarily a difference in *g* rather than in other ability factors or subtest-specific sources of variance, a hypothesis originally put forth by Spearman (1927, p. 379). This study also showed conclusively that when *g* is statistically held constant the *pattern* of white–black differences across 13 diverse subtests of the WISC-R is *negatively* correlated with the pattern of social class differences *within* either racial group. This means that whatever factors make for SES differences in the pattern of abilities cannot explain the pattern of white–black differences.

Another study with the WISC shows that inbreeding depression of mental abilities (in the offsprings of first and second cousin matings) is predominantly a depression of the *g* factor, and this effect is theoretically related to R. A. Fisher's theory of the evolution of genetic dominance in Darwinian fitness characters through natural selection (Jensen, 1983).

The connections between *g* and choice reaction time and the average evoked potential are treated quite extensively in a recent book edited by Eysenck (1982), and what little is known about the relationship of these variables to race is reviewed by Jensen (1980a, pp. 704–706). The relationships between race and most of the other variables are documented elsewhere (Jensen, 1973; Loehlin, Lindzey, & Spuhler, 1975; Harrison, Weiner, Tanner, & Barnicot, 1964).

An interesting set of relationships not shown in Figure 1 but briefly mentioned by Rushton is that the rank order of whites is *between* that of blacks and Orientals (i.e., Chinese and Japanese) on such variables as *g*, infant motoric development and activity level, rate of twin birth, color blindness, sex ratio at birth, and cranial capacity (with total body size controlled).

Any set of correlations among three or more of these variables, including race and *g* in the set, poses difficult problems for theoretic interpretation if it is to advance beyond superficial plausibility for support of a biological explanation of the observed racial differences in *g* or IQ.

2. Race, Brain Size, and IQ

That there is a significant correlation between brain size (with total body size statistically controlled) and intelligence across various species of animals and between brain size and IQ in humans is now quite well substantiated in recent reviews and studies by Van Valen (1974) and Passingham (1975, 1979). The relationship holds within sexes and when body height and weight are controlled for, although in a recent study (Passingham, 1979) the correlation between cranial capacity and IQ was nonsignificant after height was partialled out (but not when weight was partialled out). But the propriety of partialling total body size or height out of the correlation betwen IQ and brain size is debatable, as I shall attempt to show. Until this theoretical and methodological problem is adequately resolved, it is hard to see how any reasonable interpretation can be made of the race–IQ–brain size intercorrelations.

That there is a correlation between race (i.e., white–black) and IQ is not in doubt (Loehlin, Lindzey, & Spuhler, 1975; Osborne & McGurk, 1982).

The evidence is also now quite solid that adult whites and blacks (of both sexes) differ in cranial capacity and brain weight, the difference being about 100 grams. A recent study (Ho, Roessmann, Straumfjord, & Monroe, 1980b) shows that a significant racial difference remains even after various body parameters (height, weight, and total body surface area) are controlled. Ho *et al.* (1980b) note that neither race nor sex differences in brain size are present at birth, for full-term neonates, but the differences become apparent at a statistically significant level by age six years. Premature neonates, however, show a race difference in brain weight (whites heavier) (Ho, Roessmann, Hause, & Monroe, 1981), but there seems to be no evidence that by adulthood the racial difference in adult brain weight is greater for persons who were born premature than for those born at term. Therefore, the conjecture by Ho *et al.* (1981) that the racial difference in brain weight for older children and adults in general is possibly the result of the higher incidence of prematurity in the black population than in the white population does not appear to be well supported.

In any case, the theoretical significance (as contrasted with the statistical significance) of the race difference in brain size must remain in limbo until some quite fundamental questions have been answered about the IQ–brain size correlation itself, granting its reliability. The fact that the questions I will now explicate have not yet even been raised, much less answered, in any of the research literature on brain size and intelligence will come as rather a surprise to many.

Correlation between traits can represent either a true causal-functional relationship or a merely adventitious relationship due to the common assortment of the genes for both traits because they are both subject to positive assortative mating in a given culture. The correlation between height and IQ appears to be an example of such adventitious correlation. In American and European Caucasian populations there is a significant, low within-sex correlation (ranging in various studies from about .1 to .3, with an average of about .25) betwen IQ and physical stature. This correlation almost certainly involves no causal or intrinsic functional relation due to common processes between stature and intelligence but is a result of the common assortment of the genetic factors for both

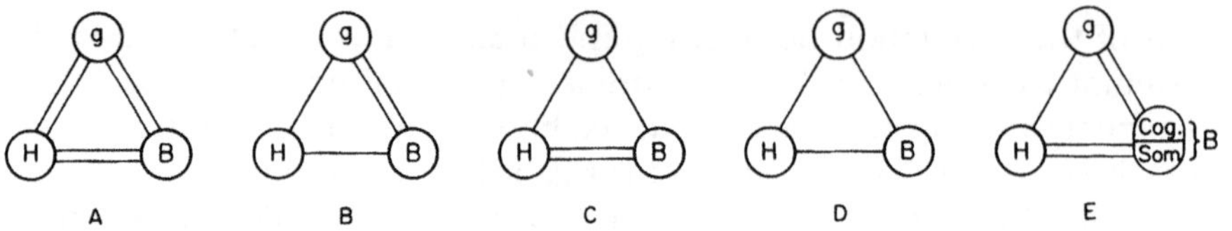

Figure 2. Five of the possible models of the nature of the intercorrelations among intelligence (*g*), height (H), and brain size (B). Single line indicates only an adventitious between-families correlation; double lines indicate both a between- and a within-families correlation, a necessary (but not always sufficient) condition for intrinsic causal-functional relationship.

height and intelligence. These are both perceived in our society as desirable characteristics, and there is a fairly high degree of positive assortative mating for both characteristics. This results in a *between*-families genetic correlation between the traits, and the best evidence we have indicates that there is no *within*-family correlation between height and IQ (Jensen, 1980b; Laycock & Caylor, 1964). The statistical logic and methodology for determining whether a given correlation represents an adventitious, nonfunctional relationship (i.e., a correlation between two traits which exists between-families but not within-families) or an intrinsic causal-functional relationship (i.e., a correlation which exists both between-families and within-families), and the theoretical significance of this important distinction have been fully explicated elsewhere (Jensen, 1980b). (Within-families correlation always implies between-families correlation, but the reverse is not true.)

Now, the apparent lack of a within-families correlation between height and IQ raises an important question about the correlation between brain size and IQ, or *g*. The several most likely possiblities are depicted in Figure 2. Double lines between the variables intelligence (*g*), height (H), and brain size (B) indicate the existence of both within-family and between-families correlation; single lines indicate only a between-families correlation. Figure 2A would present no real problem, theoretically, but it is most likely untrue, as we have no evidence of a within-family correlation between height and *g*. Figure 2B is more likely true, although at present we have no evidence at all as to the nature of the correlations H-B and *g*-B. If Figure 2B were true, the logical propriety of partialling height out of the *g*-B correlation would be questionable, since height is only adventitiously related to both *g* and B. Figure 2C is another likely possibility—*g* is adventitiously related to both H and B. If this crucial hypothesis cannot be rejected with a reasonable degree of statistical confidence, the IQ–brain size correlation would have no theoretical significance for either the understanding of intelligence or of race differences in intelligence. The same thing can be said for Figure 2D, although it is most improbable that the H-B correlation would be adventitious if there were no intrinsic correlation between *g* and B (i.e., Figure 2B). Figure 2E shows a more complicated likelihood, that is, *g* is intrinsically correlated only with parts of the brain that subserve cognitive functions, whereas H is intrinsically correlated only with those parts of the brain involved in somatic and motoric functions, while the *g*-H correlation is purely adventitious (i.e., only between-families).

At present, we can reject only the model in Figure 2A with some degree of confidence. All of the others are real possibilities, and the theoretical meaning of the observed IQ–brain–race intercorrelations, which I consider quite well established just as (between-families) correlations *per se*, hinges on which one of the other models is correct. Obviously, the crucial information we now lack, and which future studies must provide, is whether the correlation between intelligence and brain size exists within-families. This information should be obtained for both racial populations. Until such knowledge is forthcoming, the IQ–brain–race intercorrelations will remain isolated as an intriguing but theoretically uninterpretable triad, which mere partial correlations (holding height constant) cannot enlighten.

Readers will not have overlooked the fact that the very same questions arise in connection with practically all other correlations shown in Figure 1. It is an almost daunting realization in terms of the future task for research in this field. Although it may seem disappointing to discover at this late date that we actually know so little about the nature of the brain–IQ correlation, scientific progress also depends, in part, on achieving a clear recognition of the extent and precise nature of our ignorance.

3. References

Eysenck, H. J. (Ed.), *A model for intelligence.* New York: Springer-Verlag, 1982.

Harrison, G. A., Weiner, J. S., Tanner, J. M., & Barnicot, N. A. *Human biology: An introduction to human evolution, variation, and growth.* London: Oxford University Press, 1964.

Ho, K-c., Roessmann, U., Straumfjord, J. V., & Monroe, G. Analysis of brain weight: I. Adult brain weight in relation to sex, race, and age. *Archives of Pathology and Laboratory Medicine,* 1980, *104,* 635–639. (a)

Ho, K-c., Roessmann, U., Straumfjord, J. V., & Monroe, G. Analysis of brain weight: II. Adult brain weight in relation to body height, weight, and surface area. *Archives of Pathology and Laboratory Medicine,* 1980, *104,* 640–645. (b)

Ho, K-c., Roessmann, U., Hause, L., & Monroe, G. Newborn brain weight in relation to maturity, sex, and race. *Annals of Neurology,* 1981, *10,* 243–246.

Jensen, A. R. *Educability and group differences.* New York: Harper & Row, 1973.

Jensen, A. R. *Bias in mental testing.* New York: Free Press, 1980. (a)

Jensen, A. R. Uses of sibling data in educational and psychological research. *American Educational Research Journal,* 1980, *17,* 153–170. (b)

Jensen, A. R. *Straight talk about mental tests.* New York: Free Press, 1981.

Jensen, A. R. The debunking of scientific fossils and straw persons. *Contemporary Education Review,* 1982, *1,* 121–135.

Jensen, A. R. The effects of inbreeding on mental ability factors. *Personality and Individual Differences,* 1983, *4,* 71–87.

Jensen, A. R., & Reynolds, C. R. Race, social class and ability patterns on the WISC-R. *Personality and Individual Differences,* 1982, *3,* 423–438.

Laycock, F., & Caylor, J. S. Physiques of gifted children and their less gifted siblings. *Child Development,* 1964, *35,* 63–74.

Loehlin, J. C., Lindzey, G., & Spuhler, J. N. *Race differences in intelligence.* San Francisco: Freeman, 1975.

Osborne, R. T., & McGurk, F. C. J. (Eds.). *The testing of Negro intelligence* (Vol. 2). Athens, Georgia: Foundation for Human Understanding, 1982.

Passingham, R. E. The brain and intelligence. *Brain, Behavior, and Evolution,* 1975, *11,* 1–15.

Passingham, R. E. Brain size and intelligence in man. *Brain, Behavior and Evolution,* 1979, *16,* 253–270.

Van Valen, L. Brain size and intelligence in man. *American Journal of Physical Anthropology,* 1974, *40,* 417–424.

1

Sociobiology, Personality, and Genetic Similarity Detection

Robin J. H. Russell, J. Philippe Rushton, and Pamela A. Wells

We offer here two extensions to Rushton's paper. First, we shall attempt to develop some ideas which suggest why personality traits are to be expected. In addition, we shall extend the line of speculation with which Rushton ends his paper by arguing for the notion, drawn from sociobiological theorizing, that genetic similarity is a variable which may predict and explain a large part of human behavior.

1. Supergenes and Personality Traits

Suppose that there are two mutations at different chromosomal locations. If either of these new genes on its own increases the chances of its possessor's successfully reproducing, then it will increase in frequency until every member of the species possesses it. If either on its own is disadvantageous, it will disappear. In either case, after an interval of time, there will be no individual differences due to differential possession of that gene. If, however, the possessor of one of these genes

The editors have agreed to permit Professor Rushton to participate in a discussion of his contribution. The intent of commentary is to enhance the original arguments, and the Russell, Rushton, and Wells critique serves that end.

Robin J. H. Russell and **Pamela A. Wells** • Department of Psychology, University of London Goldsmiths' College, Lewisham Way, London SE14 6NW, England. **J. Philippe Rushton** • Department of Psychology, University of Western Ontario, London, Ontario, Canada N6A 5C2.

is at a disadvantage compared to the person who possesses neither of them or both of them, different consequences will ensue. If the genes are on the same chromosome, then their inheritance is linked: in time, some people may inherit both and some neither, but few people will possess only one. If they are on separate chromosomes, translocation may put them on the same chromosome, and, as before, the possessors of neither or both will be at an advantage compared to the people with only one. The argument can be extended to more than two genes. Thus a collection of linked genes, a "supergene" (Ford, 1976), will have been formed. Individuals in the population will tend to differ from each other in terms of their possession of supergenes, each of which will produce a tendency to exhibit more than one behavioral quality, the cluster constituting a useful collection of attributes.

We do not wish to imply that this is the only way in which traits can be formed. When a large number of genes all have an additive effect on some useful quality, the increase in frequency of each gene will be very slow. Until all of them have replaced their less useful alleles, the population will have a variable amount of the quality influenced by those genes. An example of a variable human quality probably resulting from this process is cranial capacity, which has been increasing for a long time (Passingham, 1982).

In short, where a mutation has an additive effect on reproductive success, it will eventually prosper or die out, and a state will be reached wherein individuals do not differ. When two or more mutations have an interactive effect on reproductive success, stable individual differences will result, with each extreme consisting of an adaptive pattern of correlated behavioral tendencies. Thus general psychological traits or types may be expected to arise. It is of interest to psychology to discover not only whether such traits exist but also the ways in which they are functional.

2. Genetic Similarity Detection and Evolutionary Theory

Recent evolutionary views can generate additional hypotheses which are both interesting and counterintuitive. We shall focus on the development of one. It is reasonable to assume that genes prevalent in a population are ones which have had the effect of ensuring their own survival. The idea of kin selection is not new (Haldane, 1932; Hamilton, 1964), but it has only recently become more widely known (Dawkins, 1976) and forms a key element in sociobiological theory. Kin selection essentially means that genes may ensure their own survival, not only by causing the organism of which they form a very small part to reproduce but also by causing it to act in such a way that its relatives reproduce more than they would have done without its action. It is strange that attention has been focused on this particular example (kin selection) of a more general principle which may be stated as follows: *A gene or supergene may ensure its own survival by acting so as to bring about the reproduction of any organism in which copies of itself are to be found*. In order to pursue this general strategy, it must, in effect, be able to detect copies in other organisms. This could most readily happen in an organism capable of sophisticated judgments about other individual members of its own species. Humans preeminently fall into this category.

This, then, brings us to the point made by Rushton that individuals may favor others who are genetically similar to them. One way of accomplishing this is for individuals to be able to perform genetic similarity detection (GSD) and then exhibit favoritism toward individuals whom the GSD mechanism judges to have a higher than average proportion of genes or supergenes in common with them. The remainder of this commentary will examine the validity of this hypothesis.

3. Assortative Mating

A well-known phenomenon which is readily explained by the GSD hypothesis is that of assortative mating. It could be argued that assortative mating has nothing to do with genetic similarity but occurs as a result of common environmental influences. This argument has difficulty accounting for the incidence of assortative mating in natural and laboratory settings in species ranging from insects through birds to primates (see Thiessen & Gregg, 1980, for a review; they reach a theoretical position very similar to ours). In the case of humans, it is widely accepted that assortative mating occurs on the basis of such characteristics as socioeconomic status, ethnic background, social attitudes, level of education, IQ, and personality variables. Not only the occurrence, but also the quality of marriage can be predicted by close matching on several personality characteristics (Cattell & Nesselroade, 1967; Meyer & Pepper, 1977). It is perhaps more surprising that assortative mating coefficients are quite high for several physical features: for example, forearm length (.43), middle finger length (.61), maximum lip circumference (.22), minimum wrist circumference (.55), interpupillary breadth (.20), and ear length (.41) (Thiessen & Gregg, 1980, p. 120, Table 3). These coefficients are sufficiently high to lead to rejection of the explanation that they result from assortative mating for size (stature, .29; weight at present marriage, .23). In contrast to the studies cited above, it is interesting to note that humans do not appear to choose spouses on the basis of similarity of ordinal position within the family (Kemper, 1966), an important but nongenetic variable.

Human marriages, however, are not the best testing ground for the GSD hypothesis, for several reasons. First, it is possible to befriend and exhibit altruism toward many individuals of either sex but not to marry them. Second, whereas no harm is done by exhibiting favoritism toward those who are genetically very similar, it may be harmful to mate with them because of the possibility of inbreeding depression; even apes seem to exhibit "incest taboos" (Mellen, 1981). Third, it must be harder to make accurate judgments of genetic similarity concerning someone of the opposite sex than of one's own. But can accurate estimates of genetic similarity be made at all?

4. Genetic Similarity Detection in Animals

Blaustein and O'Hara (1981) found that tadpoles, separated before hatching and reared apart, preferred to associate with siblings rather than nonsiblings. Grau (1982), using a complicated design, found that deermice could discriminate nonrelatives from relatives they had never previously encountered. Wu, Holmes, Medina, and Sackett (1980) found that unacquainted half-sibling macaques showed more interest in each other than in nonrelatives. Bateson (1982) found that quail reared with siblings and tested with individuals of the opposite sex preferred first cousins to third cousins and both of these to unrelated conspecifics. Siblings appeared comparable to unrelated individuals, a result which Bateson attributes to inbreeding avoidance. It is perfectly reasonable to suppose that a GSD mechanism operates in humans as well. If so, the cues, which include olfaction (Porter & Moore, 1981), are probably complex and multifarious. Taken together, the results of these studies indicate that animals, including humans, detect and respond preferentially to others who share many of their genes.

5. Family Relationships

What happens when the level of genetic similarity is low? When this occurs within families, the question can be answered. When a male lion takes over a pride he is likely to kill the existing cubs (Bertram,

1975). The Bruce effect (Bruce, 1959) provides a more subtle example of an incoming male disposing of offspring containing few of his genes: If a female mouse who has just conceived is exposed to an unfamiliar male mouse, the pregnancy often fails. In the case of humans, children dissimilar to a parent are at risk. A disproportionate number of battered babies are stepchildren (Lightcap, Kurland, & Burgess, 1982). Adoptions are more likely to be successful where the parents perceive the child as similar to them (Jaffee & Fanshel, 1970). Finally, anthropological data show that when paternity is uncertain (that is, when there is a considerable risk of low genetic similarity between a father and his wife's children), extreme measures may be taken: more resources may be invested in a sister's than a wife's children; in 15 out of 60 societies studied, adultery constitutes grounds for infanticide (Daly & Wilson, 1981).

6. Friendship

It is plausible to argue that relationships outside the family are also affected by GSD. Friendships appear to be formed on the basis of similarity. This holds for similarity as perceived by the friends (La Gaipa, 1977). It also holds for similarity on a variety of measured characteristics. For example, Berkowitz (1969) found that friends tend to be of similar height. It has been more usual to assess similarity by questionnaire. Using such methods, friendship or liking has been linked to similarity of activities (Karylowski, 1976), needs (Seyfried & Hendrick, 1973), personal constructs (Neimeyer & Neimeyer, 1981), and attitudes (Newcomb, 1961). Having reviewed available data, Richardson (1939) concluded that friends were of generally similar personality. Recent data tend to support this view (e.g., Gibson, 1971). Experimental studies in which perceived similarity has been manipulated have shown it to be a powerful predictor of liking (Byrne, 1971). Apparent similarity of personality, or of any of a wide range of beliefs, has been found to be positively related to liking in subjects of varying ages and from many different cultures (Byrne, 1971; Berscheid & Walster, 1978). On the strength of the above studies, and on the assumption that friends benefit each other, it seems reasonable to hypothesize that friendship is one of the mechanisms which lead people to sacrifice willingly their own reproductive potential for the sake of other individuals who share many of their genes. Certainly in young children it has been demonstrated that friendship sociograms correspond closely to sociograms based on altruism patterns (Strayer, Wareing, & Rushton, 1979).

7. Altruism

We have implied that the function of friendship is to promote altruism. The most direct test of the validity of the GSD hypothesis is to see if genetic similarity produces altruism. We know of no appropriate direct test. However, it is possible to ask whether or not altruism is generally increased by actual or perceived similarity. Stotland (1969) reported studies in which subjects observed another person apparently receiving electric shocks. By manipulating the subjects' beliefs about similarity to the confederate, Stotland demonstrated covariations in physiological reactions and in reported empathy. Subsequently, Krebs (1975) found that apparent similarity increased not only physiological measures indicating empathy but also willingness to reward the victim. Other studies cited by Rushton (1980, and this chapter) also lead to the conclusion that similarity promotes altruism.

8. Conclusions

We have referred to a number of empirical studies, many of which could be explained in a variety of different ways. It is much harder to explain the whole range of findings. The idea of GSD can do so. It does not necessarily exclude detailed accounts of results from particular paradigms, because it is a theory about the distal rather than the proximal level of causation. Given its explanatory power, it can readily be extended from the study of individual behavior to the study of families, tribes, and any groups of intermediate size, as well as to the relations between such groups.

9. References

Bateson, P. Preference for cousins in Japanese quail. *Nature,* 1982, *295,* 236–237.

Berkowitz, W. R. Perceived height, personality, and friendship choice. *Psychological Reports,* 1969, *24,* 373–374.

Berscheid, E., & Walster, E. *Interpersonal attraction* (2nd ed.). Reading, Mass.: Addison-Wesley, 1978.

Bertram, B. C. R. Social factors influencing reproduction in wild lions. *Journal of Zoology,* 1975, *177,* 463–482.

Blaustein, A. R., & O'Hara, R. K. Genetic control for sibling recognition? *Nature,* 1981, *290,* 246–248.

Bruce, H. M. An exteroceptive block to pregnancy in the mouse. *Nature,* 1959, *184,* 105.

Byrne, D. *The attraction paradigm.* New York: Academic Press, 1971.

Cattell, R. B., & Nesselroade, J. R. Likeness and completeness theories examined by Sixteen Personality Factor measures on stably and unstably married couples. *Journal of Personality and Social Psychology*. 1967, *7*, 351–361.
Daly, M., & Wilson, M. I. Child maltreatment from a sociobiological perspective. *New Directions for Child Development*, 1981, *11*, 93–112.
Dawkins, R. *The selfish gene*. Oxford: Oxford Universiy Press, 1976.
Ford, E. B. *Genetics and adaptation*. London: Edward Arnold, 1976.
Gibson, H. B. The validity of the Eysenck Personality Inventory studied by a technique of peer-rating item by item, and by sociometric comparisons. *British Journal of Social and Clinical Psychology*, 1971, *10*, 213–220.
Grau, H. J. Kin recognition in white-footed deermice (*Peromyscus leucopus*). *Animal Behaviour*, 1982, *30*, 497–505.
Haldane, J. B. S. *The causes of evolution*. London: Longmans, Green, 1932.
Hamilton, W. D. The genetical evolution of social behaviour: I and II. *Journal of Theoretical Biology*, 1964, *7*, 1–52.
Jaffee, B., & Fanshel, D. *How they fared in adoption: A follow-up study*. New York: Columbia, 1970.
Karylowski, J. Self-esteem, similarity, liking and helping. *Personality and Social Psychology Bulletin*, 1976, *2*, 71–74.
Kemper, T. D. Mate selection and marital satisfaction according to sibling type of husband and wife. *Journal of Marriage and the Family*, 1966, *28*, 346–349.
Krebs, D. L. Empathy and altruism. *Journal of Personality and Social Psychology*, 1975, *32*, 1134–1146.
La Gaipa, J. J. Testing a multidimensional approach to friendship. In S. Duck (Ed.), *Theory and practice in interpersonal attraction*. London: Academic Press, 1977.
Lightcap, J. L., Kurland, J. A., & Burgess, R. L. Child abuse: A test of some predictions from evolutionary theory. *Ethology and Sociobiology*, 1982, *3*, 61–67.
Mellen, S. L. W. *The evolution of love*. Oxford and San Francisco: Freeman, 1981.
Meyer, J. P., & Pepper, S. Need compatibility and marital adjustment in young married couples. *Journal of Personality and Social Psychology*, 1977, *35*, 331–342.
Neimeyer, G. J., & Neimeyer, R. A. Functional similarity and interpersonal attraction. *Journal of Research in Personality*, 1981, *15*, 427–435.
Newcomb, T. M. *The acquaintance process*. New York: Holt, Rinehart & Winston, 1961.
Passingham, R. E. *The human primate*. Oxford and San Francisco: Freeman, 1982.
Porter, R. H., & Moore, J. D. Human kin recognition by olfactory cues. *Physiology and Behavior*, 1981, *27*, 493–495.
Richardson, H. M. Studies of mental resemblance between husbands and wives and between friends. *Psychological Bulletin*, 1939, *36*, 104–142.
Rushton, J. P. *Altruism, socialization and society*. Englewood Cliffs, N. J.: Prentice-Hall, 1980.
Seyfried, B. A., & Hendrick, C. Need similarity and complementarity in interpersonal attraction. *Sociometry*, 1973, *36*, 207–220.
Stotland, E. Exploratory investigation of empathy. In L. Berkowitz (Ed.), *Advances in experimental social psychology* (Vol. 4). New York: Academic Press, 1969.
Strayer, F. F., Wareing, S., & Rushton, J. P. Social constraints on naturally occurring preschool altruism. *Ethology and Sociobiology*, 1979, *1*, 3–11.
Thiessen, D., & Gregg, B. Human assortative mating and genetic equilibrium: An evolutionary perspective. *Ethology and Sociobiology*, 1980, *1*, 111–140.
Wu, H. M. H., Holmes, W. G., Medina, S. R., & Sackett, G. P. Kin preference in infant *Macaca nemestrina*. *Nature*, 1980, *285*, 225–227.

1

Interaction between Biological and Cultural Factors in Human Social Behavior

Philip E. Vernon

Professor Rushton has produced a scholarly, well-argued, and up-to-date account of the present state of sociobiology—that is, the view that not only the abilities but also the main traits underlying human social behavior have a substantial genetic component and that they have evolved among animals and humans according to Darwinian theory of natural selection. One of the reasons why other expositions by Wilson (1975) and Campbell (1975) have been criticized is that they are hghly speculative, lacking in experimental and other forms of scientific evidence. Rushton has culled the literature very widely, all the way from animal and cross-cultural psychology and historical anthropology to statistical studies of individual differences. Thus most of his claims are accompanied by relevant confirmatory evidence. The paper, therefore, should be of considerable interest and value to psychologists and students.

Sociobiology has had a checkered history in psychology. One of the major early exponents was William McDougall, whose first book: *An Introduction to Social Psychology* (1908) had phenomenal sales in the United Kingdom. McDougall believed that one could not make a start in social psychology until the nature and role of genetic factors, namely instincts, had been determined. He failed to make any impact on American psychology, partly because of the behaviorist influence of Thorndike, but also because the word *instinct* implied to most psychologists "wired-in" reflex behaviors (as among insects), whereas he used it to name about

Philip E. Vernon • Department of Educational Psychology, University of Calgary, Calgary, Alberta, Canada T2N 1N4.

15 general drives, many of which happen to overlap with Rushton's choice of partly heritable traits. Much later, Darlington, in *The Evolution of Man and Society* (1969), showed the importance of genetic group differences (mainly in abilities) which underlay the rise to prosperity, and the decline, of ethnic groups (for example, the effects of conquest, absorption of foreign cultures, cross-breeding, migration, and the like on the achievements of nations). He relied almost entirely on historical evidence combined with genetic theory.

During the 1960s and early 1970s, any suggestion of biological differences between human groups was anathema to the majority of social scientists. However, in 1975 Wilson's book appeared; and Campbell's presidential address to the American Psychological Association, together with Wispé and Thompson's (1976) critique, made the topic respectable again. But Campbell differed from most writers in regarding social behavior as due in part to biological evolution but probably still more to cultural (nongenetic) evolution. The latter has been mainly responsible for the buildup of the traditions, values, norms, and morals of any group. Biologically man is basically sinful, egotistic, and aggressive. But in addition altruistic characteristics such as parent–offspring defence and willingness to sacrifice oneself for the good of others have become inbred by Darwinian natural selection over the past 300,000 years or so. It is the combination of biological with cultural evolution that has turned man into a reasonably balanced and social being, even if this balance is often unstable.

Now Rushton does not follow this line. He admits that man shows considerable plasticity and that differences between individuals and groups are to a large extent a matter of social learning, which modifies the expression of the basic traits. But he also maintains that conditioning and social learning do not provide an adequate understanding of human socialization because they ignore the role of evolutionary biology. Also he does not point out that man's adaptation takes place far more rapidly through cultural buildup and transmission to succeeding generations by language than through genetic change. It is effective because it is Lamarckian, whereas biological evolution is Darwinian. I would suggest that his paper might have been strengthened if he had admitted that social behavior patterns are phenotypes, and like all phenotypes they derive from the interaction between genes and the physical and cultural environment. No particular genes can be regarded as solely responsible for any single type of behavior. I would have to admit, though, that there is even less satisfactory evidence available on how different features of the environment operate; so that a simultaneous discussion of both aspects of social development might become too complex and speculative to be worthwhile. Because Rushton has limited himself to genetic influences, he is able to cover a much wider range of behavior than altruism, including activity level, dominance, sexuality, emotionality, aggression, intelligence, and locus of control.

I also find it difficult to discuss sociobiology without mentioning Freud's contributions, which provided a reasonable explanation of the control of biological impulses by the superego, that is, the introjected system of cultural norms. Although his own writings on group psychology appear rather bizarre, he has inspired others such as Dollard and Miller (1950). Thus Durbin and Bowlby (1938) provided an insightful and plausible explanation of the origins of aggression between human groups. True, psychodynamics is not a science, but neither is the historical approach to human evolution.

It would be useful at this point to consider a particular example of genetic-cultural interaction, namely, the achievements over the past 100 years of Chinese and Japanese immigrants in North America (cf. Vernon, 1982). The original immigrants were of peasant stock and were hated and persecuted by white Americans and Canadians. In view of their language difficulties and their persistent attachment to their own cultural values, it was not until the end of World War II that they became sufficiently acculturated to be accepted as full citizens. Yet even in the 1920s their children were scoring more highly than whites on some nonverbal intelligence and spatial tests, and by now they have caught up on verbal tests. Their educational and occupational achievements, on average, are actually superior to those of whites. Similarly in Japan itself technological growth and harmonious labor relations are admitted to have outstripped those of western industrial nations.

I attempted to sift the reasons for this success as impartially as possible and concluded that oriental intelligence was unlikely to be genetically better than that of Caucasians, though the persistent spatial versus verbal bent might be partly genetic. But there is very strong evidence for oriental newborn children to be more placid and docile, less excitable than white babies. The manner of bringing up oriental children tends to reinforce this basic temperamental difference; they grow up to be intensely motivated for education and possessing social cooperativeness and drive which, in my view, are mainly responsible for their achievements.

Returning now to Rushton's exposition: The bulk of his evidence for genetic variance in personality derives from comparisons of monozygotic and dizygotic twins, or other kin correlations. Almost all of this is positive, but the great variations in heritability reported by different authors are noticeable. Typically they run from about .20 to .70, though part of this irregularity might well be due to the small population samples used in most of the investigations. Loehlin and Nichols (1976) and Nichols (1976), who tested 850 twin pairs in high schools, did obtain fairly stable figures around .40 for a wide range of personality variables. Despite Rushton's and other writers' rejection of the argument, it still appears to me likely that the environments of monozygotics tend to be more similar than those of dizygotics or siblings. At least it is clear that the heritability of personality traits is lower than that of IQs, implying that such traits are more affected by family and cultural environment.

Another type of evidence is cited which is unacceptable: namely, the parent–offspring regression to the mean, both in height, intelligence, or other attributes. Eysenck (1973) states that this phenomenon is proof of heredity, since no environmental theory could explain why the offspring of bright parents have lower IQs than their parents whereas the children of dull parents have on average higher IQs than their parents. But in fact regression to the mean is merely a necessary consequence whenever two sets of scores, such as parent and offspring IQs, are imperfectly correlated. Although the moderate level of correlation in this instance could be explained genetically, it could also be partly due to differences between parents and children in their upbringing and environments.

One other important point on which I differ from Rushton is in his use of the term *trait* to describe variations in people's patterns of social behavior. I discussed the weaknesses of this concept in 1964, particularly the lack of consistency between different measures of the same trait; and this was strongly criticized by Mischel in 1968. Rushton is quite justified in pointing out that one of the main reasons for such low intercorrelations is the unreliability of personality tests. He and his colleagues (1981) have shown that by aggregating several test scores the combined measure achieves higher reliability and better validity for predicting social behavior. But he does not mention the tendency described by Campbell and Fiske (1959) for correlations between different traits assessed by the same method to be higher than the correlation between different methods of assessing the same trait; nor the studies that have been published which indicate higher situational variance than personal variance. However I agree that we have not yet arrived at a more satisfactory taxonomy for talking about people's social behavior.

Sometimes, though rarely, Rushton falls into the trap of accepting evidence that supports his position and neglecting evidence that contradicts it. For example, writing on genetic factors that underly social status, he claims correlations of .50 to .90 with IQ. The more typical value from a large number of studies is .35, though this would of course be boosted if SES levels are correlated with group mean IQs (as Jensen sometimes does, 1973).

But to me the striking thing is the infrequency of such lapses. Given that Dr. Rushton is concentrating on the genetic aspects of sociobiology, he seems to me overall to be a highly reliable guide.

1. References

Campbell, D. T. On the conflicts between biological and social evolution and between psychology and moral tradition. *American Psychologist*, 1975, *30*, 1103–1112.

Campbell, D. T., & Fiske, D. W. Convergent and discriminant validation by the multitrait-multimethod matrix. *Psychological Bulletin*, 1959, *56*, 81–105.

Darlington, C. D. *The evolution of man and society*. London: Allen & Unwin, 1969.

Dollard, J., & Miller, N. E. *Personality and psychotherapy*. New York: McGraw-Hill, 1950.

Durbin, E. F. M., & Bowlby, J. *Personal aggressiveness and war*. London: Kegan Paul, 1939.

Eysenck, H. J. *The inequality of man*. London: Temple Smith, 1973.

Jensen, A. R. *Educability and group differences*. New York: Harper & Row, 1973.

Loehlin, J. C., & Nichols, R. C. *Heredity, environment, and personality*. Austin: University of Texas Press, 1976.

McDougall, W. *An introduction to social psychology*. London: Methuen, 1908.

Mischel, W. *Personality and assessment*. New York: Wiley, 1968.

Nichols, R. C. Heredity and environment: Major findings from twin studies of ability, personality, and interests. *Conference of the American Psychological Association*, invited address, 1976.

Rushton, J. P., Jackson, D. N., & Paunonen, S. V. Personality: Nomothetic or idiographic? A response to Kenwrick and Stringfield. *Psychological Review*, 1981, *88*, 582–589.

Vernon, P. E. *Personality assessment: A critical survey*. London, Methuen, 1964.

Vernon, P. E. *The abilities and achievements of orientals in North America*. New York: Academic press, 1982.

Wilson, E. O. *Sociobiology: The new synthesis*. Cambridge, Mass.: Harvard University Press, 1975.

Wispé, L. G., & Thompson, J. N. The war between words: Biological vs. social evolution. *American Psychologist*, 1976, 31, 341–386.

1

Group Differences, Genetic Similarity Theory, and the Importance of Personality Traits

Reply to Commentators

J. Philippe Rushton

Jensen and Russell, Rushton, and Wells provided extensions to my paper on the sociobiology of individual and group differences, while Vernon placed the earlier paper in historical context, and qualified some of the points made. In this paper I shall concentrate on three issues: Group differences, genetic similarity theory, and the importance of personality traits.

1. Group Differences

Jensen proposed that behavioral differences between races are "the bugbear of both differential psychology and sociobiology" (p. 50). He suggested that we "confront head-on what is probably the chief focus of anxiety about sociobiology" so that, if we can overcome this resistance, other difficulties would fall. In his paper Jensen outlined ways in which causal explanations could be advanced for the many correlational data sets involved in ethnic group differences. One possibility involves examining, for each ethnic group separately, the *within*-family correlations on the variables of interest, in addition to the more usual *between*-family correlations. If significant correlations are found to exist both between-

J. Philippe Rushton • Department of Psychology, University of Western Ontario, London, Ontario, Canada N6A 5C2.

families and within-families then the relationship is more likely to be causal (Jensen, 1980). Jensen accepted that the task of uncovering the functional relations among all the variables in his Figure 1 was a daunting one.

One reason why the finer-grained analyses advocated by Jensen have so infrequently been conducted is because so few behavioral scientists are currently researching group differences. Many psychologists believe that group differences are unimportant. Explaining group differences, however, may provide a useful catalyst for understanding individual differences, for the former constitutes an aggregate of the latter. A good example of this is provided by Symon's (1979) analysis of male/female differences in sexual behavior. Rather than focus only on individuals he also focused on the aggregated male and female "cultures" generated by homosexuals. When the necessities to compromise required by the presence of the opposite sex are removed, males and females are freer to construct the norms of behavior most compatible with their genotypes. Thus homosexual male culture is typically promiscuous (i.e., involving a large number of sexual partners in a detached manner) and emphasizes youthful attractiveness. Female homosexual culture, on the other hand, typically emphasizes stable, long-term monogamous relationships with a more supportive set of social norms.

One useful way of exploring group differences in behavior is to seek historical and cross-cultural evidence. To the degree group differences are generalizable across time and situation, the genetic hypothesis becomes more plausible. Consider the question of ethnic group differences in intelligence. The within-United States comparisons suggested an ordering of Asian > European > African (see main paper). When the comparisons are made internationally, and across time, by examining the cultural attainments of these groups on their home continents (e.g., by dating such inventions as written language, numbering systems, calendars, astronomical systems, codified rules of law, domestication of plants and animals, and metal technology), the rank ordering remains the same (Baker, 1974). A similar historical consistency between inventiveness and intelligence is demonstrated by Jewish people who have made substantial contributions to science and art across many different time periods and countries, and are found to score higher than other Euro-Americans on standardized IQ tests, and particularly so on tests of verbal ability (Loehlin, Lindzey, & Spuhler, 1975).

Other group differences may also be illuminated by historical and cross-cultural analyses. One striking figure from Vernon's (1982) compilation of the achievements of the Chinese and Japanese in North America concerns their remarkably low incidence of violent crime. Afro-Americans, however, are currently overrepresented in such crimes. Although they constitute only 12% of the U.S. population, they make up 48% of the prison population. They are also the main victims of crime. For example, over 60% of all homicide victims are black (*Newsweek*, 1981). Similar figures are found for African-descended people in Britain. For example, while comprising 13% of the population of London, they account for 50% of the crime (*Daily Telegraph*, March 24, 1983). Asian immigrants to Britain, on the other hand, are under represented in crime figures. These findings hold regardless of whether they are based on official police records or on victimization surveys. It should be possible to collect additional cross-cultural data to examine whether these ethnic-group differences are specific to current Anglo-American culture or whether they represent more general trends. Clinard and Abbot (1973) provided an illustrative study. They compared the crime rate in selected developing countries. India, for example, was found to have a low incidence of crime (i.e., 165 instances per 100,000 people). Uganda had one of the highest; 874 per 100,000 (homicides, assault, and forcible rape being 276 per 100,000).

Putting the results on law-abiding behavior together with those reported in the earlier article on intelligence, activity-level, and physical coordination, we find that Europeans consistently fall midway between Asian and African groups. This rank ordering of the races raises interesting theoretical questions. Are these traits correlated *within* racial groupings, as well as across them, and are they related to still other characteristics that distinguish racial groups? Does a single dimension underlie these behavior patterns? Can the comparison of existing racial groups be used to further our understanding of the evolution of human behavior? And, what are the evolutionary origins of the races? (Coon, 1962; Loehlin, Lindzey, & Spuhler, 1975; Rushton, 1984).

One possibility is that individual differences in "K" underlie many of the racial group differences, a proposal introduced here as *Differential "K" Theory*. K refers to one end of the hypothetical r/K continuum evolutionary biologists use to differentiate the reproductive strategies organisms engage in (Wilson, 1975). At the r end, organisms produce many offspring but invest little energy in any one. Oysters, producing 500 million eggs a year, exemplify this extreme. At the K end, organisms produce few offspring but invest a great deal of energy in each. The great apes, producing only one infant every 5 or 6 years, exemplify this extreme. The r/K continuum organizes data on several correlated characteristics pertaining to between-species differences in life-history traits, social behaviors, and physiological functioning. The more K the species is, the smaller the litter size, the greater the spacing of births, the fewer the total number of offspring, the better developed the parental care, the lower the rate of infant mortality, the slower the rate of physical maturation, the older the age of reproduction, the longer the life span, and the greater the degree of intelligence, social organization, and altruism.

As a species, humans are at the K end of the continuum. Some people, however, are postulated to be more genetically K than others. The more K a person is, the more likely he or she is to be intelligent, altruistic, law abiding, behaviorally restrained, maturationally delayed, lower in sex drive, and longer lived (Rushton, 1984). In terms of the racial differences discussed above, therefore, it is hypothesized that, due to different selection pressures, and on average, Orientals are more K than Europeans, who, in turn, are more K than Africans. This ordering accords well with data on dizygotic twinning rates, which could be taken as an index of "litter size." The rate per 1,000 among Orientals is 4; among Europeans, 8; and among Africans, 16 (Bulmer, 1970). If the differential K theory of race differences is correct, numerous other indices of K will be found to correlate both between and within races and, following Jensen, between and within families.

2. Genetic Similarity Theory

Russell, Rushton, and Wells ordered several disparate sets of data within the hypothesis that genetically similar individuals detect each other and thereafter have a tendency to congregate together and provide mutual support. What can be referred to as Genetic Similarity Theory (GST) states that there is an alternative means by which genes can propagate themselves to those usually discussed in sociobiological theorizing: Rather than behaving altruistically only toward kin, organisms could have a behavioral tendency to detect other genetically similar organisms and to exhibit favoritism and protective behavior toward these "strangers," as well as toward their own relatives. GST may provide a significant extension to sociobiological theorizing.

Aspects of genetic similarity theory have been outlined by others (e.g., Dawkins, 1976, 1982; Hamilton, 1964). Dawkins (1976), for example, proposed a "thought experiment" in which a gene had two effects: It causes individuals possessing it to have a green beard, and to behave altruistically toward green-bearded individuals. The green beard serves as a recognition cue (not necessarily conscious recognition, of course) for the altruistic gene. Altruism, therefore, could occur with no necessity for the individuals to be related. Thus with humans, dimensions of similarity (personality, attitudes, physical appearance) may constitute a "green beard" effect (Sorrentino & Rushton, 1981).

If GST is correct it follows that similarity based on genetic traits would predict altruism more than similarity based on non-genetic causes. This deduction could be tested in the context of friendship for, as Russell *et al.* suggested, friendship is a mechanism that leads to altruism. Freedman (1979) cited studies in which respondents reported their intention would be to help close kin over distant kin and distant kin over strangers. GST would predict that friends would be responded to at least as altruistically as distant kin and that the more genetic similarity there was between friends, the more altruism would occur. To test this prediction, estimates of genetic similarity are needed. Biological assays (e.g., chromosome analysis) would be ideal, while blood antigen analysis may provide a reasonable approximation. Cruder estimates are also possible. For example, similarity on traits known to have high heritabilities should be more predictive of friends' altruism than similarity on traits of low heritability. Unfortunately, the differential heritability of personality traits is not established (Loehlin, 1978). Alternatively, it may be possible to construct two alternative tests of the same personality trait, one composed of items of high heritability and another of items with low heritability (Buss, 1983).

Given a genetic basis to friendship and that friends reciprocate in altruistically benefiting each other, GST also implies an interaction with Trivers's (1971) explanation of the natural selection of reciprocal altruism (see p. 6–7, this volume). GST predicts that the more genes shared by strangers, the easier reciprocal altruism will be to develop. There would be no necessity for strict reciprocity.

A different test of GST could be made by examining preferences within families. Although each parent will have a minimum of 50% of his or her genes in common with each offspring, upward variations on this percentage will be expected. Some children will be genetically more similar to one parent than the other. This can be readily demonstrated in the hypothetical case in which two parents shared 10% of their genes in common. Suppose the father gives the child 50% of his genes, 2% of which are shared with the mother, and the mother gives the child 50% of her genes, 8% of which are shared with the father. If this occurred, the child would share 52% of his genes with the mother (50% from mother, 2% from father) and 58% of his genes with the father (50% from father, 8% from mother). By analogous reasoning, it is expected that while siblings, on average, will be 50% genetically similar to each other, fluctuations around this figure will occur. Parents and siblings can be expected to favor the child who is more similar to them. Favoritism within families is an unexplored topic. GST may render it an important one.

A related expectation involves parental care of their offspring as a function of the degree of genetic similarity between spouses. Russell *et al.* documented data on assortative mating and the effects of genetic dissimilarity that bear on this issue (e.g., stepchildren are more at risk for abuse). A more general proposition can be stated: The more genetically similar the parents are to each other, the more genetically similar they and their children will all be to each other, and the more within-family altruism will occur. Conversely, the less genetically similar the parents are to each other, the less genetically similar they and their children will all be to each other, and the less within-family altruism will occur. This proposition could be tested in at least two ways: Parents who are first or second cousins should be more protective to their children than less related parents. Moreover, in multi-ethnic countries the prediction would be that the greater the disparity in ethnicity between parents (and, presumably, the lower the genetic similarity), the less protectiveness and care for the children there will be.

Indeed the very notion of ethnicity is based on the idea of extended kinship. Any two individuals *within* an ethnic group will be more genetically similar than any two *between* ethnic groups. The implications of this for relations between ethnic groups may be far reaching. There will be, for example, a biological basis for what van den Berghe (1981) has characterized as "ethnic nepotism." Ethnic nepotism is manifest in many ways. It explains why different ethnic groups often prefer to congregate in the same geographical areas. Ethnic nepotism also predicts clear patterns of altruism—charitable donations, for example, are predicted to be made in greater quantities within ethnic groups than across them. Many studies have found that people are more likely to help members of their own race or country than members of other races, or foreigners (Brigham & Richardson, 1979; Feldman, 1968).

3. The Importance of Personality Traits

Vernon raised several qualifications to my paper, two of which I will respond to. In addition, I will elaborate on the usefulness of the concept of personality traits for understanding human behavior.

First, Vernon suggested that: "At least it is clear that the heritability of personality traits is lower than that of IQs, implying that such traits are more affected by family and cultural environment" (p. 70). This however is not known. Some estimates of the heritability of intelligence are as low as .50 (Plomin & DeFries, 1980), while estimates of the heritability of some personality traits prove as high as .80 (Eysenck & Eysenck, 1976). The differential heritability of traits is far from established (Loehlin, 1978). Second, Vernon (p. 70) cited as "unacceptable" the evidence on regression to the mean as support for the heritability of IQ. He suggested regression effects are due to error of measurement. What his argument fails to consider, however, is that regression effects vary with the genetic similarity between the two groups being compared. The more genetically similar the groups, the less regression will be expected. Thus parent–child comparisons demonstrate less regression than those between grandparent–grandchild. Regression effects also explain upward and downward social mobility within families. Such findings pose a challenge to purely environmental explanations of the distribution of intelligence (Eysenck & Kamin, 1981).

Let us now turn to a discussion of the usefulness of the trait concept. As Vernon noted, he was one of those who critically evaluated the state of the field of personality in the 1960s and concluded that: "The real trouble [with the trait approach] is that it has not worked well enough, and despite the huge volumes of research it has stimulated, it seems to lead to a dead end" (1964, p. 239). Vernon (1964) rested much of his argument on the low correlations of .2 and .3 apparently found across different measures of the same trait. This argument was reiterated by Mischel (1968) who also provided a theoretical alternative to trait theory. Mischel's social-learning reconceptualization of personality emphasized the modifiability of behavior within highly discriminable situations. According to this viewpoint, correlations of .2 and .3 would be all that one could expect, on average, between different behaviors. Partly as a result of Vernon's (1964) and Mischel's (1968) reviews, the trait concept fell out of favor among many researchers.

Both the critiques of trait theory as it existed, and the social-learning reinterpretation of personality can be said to have provided a service. It is correct, for example, to emphasize that people alter their behavior to suit different situations, and that this intra-individual variance is a function of both an individual's social-learning history and of his or her encoding strategies and information-processing capacities. Unfortunately this has sometimes been interpreted as meaning that cross-situational consistency does not exist, or at least that it does not exist in sufficient quantity to make the concept of traits very useful. If .2 or .3 is indeed representative of the degree of cross-situational consistency, thus accounting for only 4% to 9% of the variance, it can be doubted whether the trait concept is substantial enough to provide the base for a scientific understanding of personality.

It is now known, however, that the belief that traits only account for 4% to 9% of the variance is erroneous. When the principle of aggregation is adhered to and traits are more representatively assessed (e.g., over multiple-response forms) and are then validated against the more appropriate multiple-act criteria, validity coefficients rise to .5 to .7, thus accounting for between 25% and 49% of the criterion variance (Epstein, 1979, 1980; Rushton, Brainerd, & Pressley, 1983). Vernon graciously conceded this point although he still feels that a more satisfactory taxonomy than traits would be preferable for describing and explaining personality. Mischel, too, now acknowledges that when traits are measured using aggregated assessments that "stable mean levels of behavior" (as he prefers to refer to traits) ensue (Mischel & Peake, 1982, pp. 747–748).

Hopefully, therefore, the disillusionment with trait theory is on the wane. If so, perhaps renewed energy will go into theorizing, as well as producing more reliable and valid assessment techniques. From a sociobiological perspective, social behavior traits are as important for understanding human nature as is intelligence. Ultimately, knowledge about consistent patterns of individual differences may lead to profound predictions concerning the very structure of society. Different genotypes may generate different social norms and create different types of societies (as well as have differential reproductive success).

Psychological studies often constrain individual differences. To investigate the full impact of personality, a more useful strategy might be to provide people with an array of choices to examine preferences as a function of personality. Achievement oriented individuals, for example, are more likely to select environments in which hard work and industriousness is rewarded than are less highly achievement oriented people. In addition it would be illuminating to bring together different groups of people with similar constellations of traits to see what kinds of social norms they generate. If different groups reliably generate different "cultures" that are in keeping with their genotypes, a microcosm of society would be created. In this way the genetic basis of culture could be studied.

4. References

Baker, J. R. *Race.* London: Oxford University Press, 1974.

Brigham, J. C., & Richardson, C. B. Race, sex and helping in the marketplace. *Journal of Applied Social Psychology,* 1979, *9,* 314–322.

Bulmer, M. G. *The biology of twinning in man.* Oxford: Clarendon Press, 1970.

Buss, D. M. Evolutionary biology and personality psychology: Implications of genetic variability. *Personality and Individual Differences,* 1983, *4,* 51–63.

Clinard, M. B., & Abbott, D. J. *Crime in developing countries: A comparative perspective.* New York: Wiley, 1973.

Coon, C. S. *The origin of races.* New York: Knopf, 1962.

Dawkins, R. *The selfish gene.* Oxford: Oxford University Press, 1976.

Dawkins, R. *The extended phenotype.* San Francisco, Ca: W. H. Freeman, 1982.

Epstein, S. The stability of behavior: I. On predicting most of the people much of the time. *Journal of Personality and Social Psychology*, 1979, *37*, 1097–1126.
Epstein, S. The stability of behavior: II. Implications for psychological research. *American Psychologist*, 1980, *35*, 790–806.
Eysenck, H. J., & Eysenck, S. B. G. *Psychoticism as a dimension of personality*. London: Hodder and Stoughton, 1976.
Eysenck, H. J., & Kamin, L. *The intelligence controversy*. New York: Wiley, 1981.
Feldman, R. E. Response to compatriots and foreigners who seek assistance. *Journal of Personality and Social Psychology*, 1968, *10*, 202–214.
Freedman, D. G. *Human sociobiology: A holistic approach*. New York: The Free Press, 1979.
Hamilton, W. D. The genetical evolution of social behavior: I. and II. *Journal of Theoretical Biology*, 1964, *7*, 1–51.
Jensen, A. R. Uses of sibling data in educational and psychological research. *American Educational Research Journal*, 1980, *17*, 153–170.
Loehlin, J. C. Are CPI scales differentially heritable? How good is the evidence? *Behavior Genetics*, 1978, *8*, 381–382.
Loehlin, J. C., Lindzey, G., & Spuhler, J. N. *Race differences in intelligence*. San Francisco: W. H. Freeman, 1975.
Mischel, W. *Personality and assessment*. New York: Wiley, 1968.
Mischel, W., & Peake, P. K. Beyond deja vu in the search for cross-situational consistency. *Psychological Review*, 1982, *89*, 730–755.
Newsweek Magazine, March 15, 1981.
Plomin, R., & De Fries, J. C. Genetics and intelligence: Recent data. *Intelligence*, 1980, *4*, 15–24.
Rushton, J. P. *Do "r" and "K" apply to individual differences in humans?* Paper presented at the 14th Annual Meeting of the Behavior Genetics Association, Bloomington, Indiana, May 23–26, 1984.
Rushton, J. P., Brainerd, C. J., & Pressley, M. Behavioral development and construct validity: The principle of aggregation. *Psychological Bulletin*, 1983, *94*, 18–38.
Sorrentino, R. M., & Rushton, J. P. Altruism and helping behavior: Current perspectives and future possibilities. In J. P. Rushton & R. M. Sorrentino (Eds.), *Altruism and helping behavior: Social, personality, and developmental perspectives*. Hillsdale, N. J.: Lawrence Erlbaum Associates, 1981.
Symons, D. *The evolution of human sexuality*. New York: Oxford University Press, 1979.
Trivers, R. L. The evolution of reciprocal altruism. *Quarterly Review of Biology*, 1971, *46*, 35–57.
van den Berghe, P. L. *The ethnic phenomenon*. New York: Elsevier, 1981.
Vernon, P. E. *Personality assessment: A critical survey*. New York: Wiley, 1964.
Vernon, P. E. *The abilities and achievements of Orientals in North America*. New York: Academic Press, 1982.
Wilson E. O. *Sociobiology: The new synthesis*. Cambridge, Mass.: Harvard University Press, 1975.

2

Psychoanalysis as a Scientific Theory

Benjamin B. Wolman

Abstract. This article presents an analysis of Freud's psychological theory. Freud's system is comprised of six classes of propositions. Four of them are empirical, dealing with (1) observable behavior, (2) introspectively observable behavior, (3) inferrable behavior, and (4) empirical generalizations. The other two classes are (5) theoretical or hypothetical constructs and (6) praxiological or applied propositions.

Hypothetical constructs form the backbone of all theoretical systems. Freud introduced seven of them, namely, (1) epistemological realism, (2) monism, (3) energetism, (4) determinism, (5) economy, (6) pleasure–unpleasure continuum, and (7) the constancy principle.

There are good reasons for developing fresh concepts in keeping up with the progress of scientific research, but Freud's original contribution must be preserved and respected as a pioneering work that has inspired fruitful psychological research by followers, dissidents, and opponents.

Scientific systems are usually comprised of two parts, namely, (1) description of empirical data and (2) interpretation of those data. The first part of the system operates with a set of synthetic propositions ascertaining fact. Factual data are descriptions of bodies and whatever is happening with them. Such descriptions can deal with individual cases or with categories or classes; in the latter case they are called *empirical generalizations*. The following proposition describes an individual case: "Mr. A., a catatonic schizophrenic, was committed to a mental hospital." The following proposition describes a class of events and is therefore an empirical generalization: "Catatonic schizophrenics are usually committed to mental hospitals."

Benjamin B. Wolman • Suite 6D, 10 West 66th Street, New York, New York 10023.

The second part of a scientific system, called *theory*, does not describe individual cases nor classes of cases. A theory is a set of analytic and synthetic propositions. The analytic propositions, that is, definitions, are exceedingly useful logical tools. A theory cannot be of much use unless its basic concepts are carefully and clearly defined.

The entire psychoanalytic system, as developed by Sigmund Freud, and all of its propositions can be divided into those describing (1) observable behavior, (2) introspectively observable behavior, (3) inferrable behavior, and (4) empirical generalizations. In addition to these four empirical (synthetic) propositions, psychoanalysis uses two nonempirical propositions, namely, (5) theoretical or hypothetical constructs (unfortunately called metapsychology) and (6) praxiological propositions, dealing with psychoanalysis in its therapeutic mode (Wolman, 1964).

The main body of a theory is formed by a set of hypothetical propositions. These propositions are not statements of facts; thus, a theory cannot be true or false in the empirical sense. These hypothetical or theoretical constructs cannot be proven or disproven; however, the philosophy of science has developed certain formal, logical rules applicable to the formation of theories.

Freud introduced four theoretical principles, namely, (1) epistemological realism, rejecting Kantian and *Wiener Kreis* philosophies; (2) monism, originally physicalistic (topographic theory), and in the structural theory, the id serves as a bridge between mental and physical processes; (3) energetism; and (4) determinism. Freud's three ancillary principles are (5) economy, (6) pleasure–unpleasure, and (7) constancy principles.

1. Epistemological Realism

For centuries philosophers exercised considerable influence if not control over scientific inquiry. A special branch of philosophy called epistemology was devoted to the analysis of the relationship of the scientist as an observer and the objects he observed. Epistemologists have scrutinized the conditions under which cognition takes place and maintained that they can determine the meaning of the term *truth*.

Immanuel Kant and John Locke suggested diametrically opposed answers. According to Kant (1929), sensory evidence is no evidence at all, for whatever we perceive is colored by our *a priori* set mental framework called by Kant *absoluter Geist*. According to Locke (1894), there is no other source of knowledge but our sensory apparatus. However,

Kant's categories of cognition led to Schopenhauer's epistemological solipsism, and Locke's sensualism was amply criticized by Hume, and Mach reduced the universe to a bundle of perceptions.

One may assume (and only assume) that what we perceive is or is not related to things as they are, and proceed from that point onward or accept that the world is as perceived by us, or offer any other solution or combination of solutions.

An assumption that the universe is a cluster of perceptions leads to contradictions. Mach said:

> Nature is composed of sensations as its elements. . . . Sensations are not signs of things. . . . The world is not composed of things as its elements, but of colors, tones, pressures, spaces, times, in short what we ordinarily call sensations. (Mach, 1960, p. 482)

Mach's world consists of Mach's perceptions; there is no way to prove (using Mach's perceptions) that the perceptions of other people are the same as his. One may follow Schopenhauer at this point and assume that the world is the way "I" see it. In such a case, any further pursuit of truth is futile.

Small wonder that Auguste Comte's (1864) system of positivism did not include any specific epistemological theory and that William James's pragmatism related the craft of pursuing truth to realistic achievement. "My thinking," James wrote, "is first and last and always for the sake of my doing" (James, 1890, vol. 2, p. 333).

It seems that the only possible solution was that of *epistemological realism,* which assumes (and nothing more than assumes) that whatever exists, exists irrespective of whether someone perceived it or not. In other words, the earth was round before Copernicus, and America was where it is before Columbus discovered it.

Were the world a cluster of sensations, all sciences would have shrunk to psychology. However, such a panpsychologism, as developed by Berkeley or Mach, would render scientific psychology impossible. A psychologist would have to observe the way in which he observes, and so on *ad infinitum.* He and his own perceptions would have become the sole source of information. There is no other way but to assume that other individuals exist independently of those who perceive them. This is merely an assumption, for there is no empirical evidence for empirical evidence. Thus a *radical realism* is the only assumption free of inner contradictions.

A clinician observes the behavior of disturbed individuals. More psychologists may join in the process of observing and all observers may observe the same phenomenon. Their observations can be checked against one another, and the knowledge thus acquired becomes objective, provided certain canons of scientific procedure are observed.

The first principle of epistemologic realism is that of transcendent truth. A proposition conveys transcendent truth whenever, and only when, its content corresponds to reality. The principle of transcendent truth requires that all propositions of empirical science be checked against reality.

Radical realism is not naive and does not assume an infallibility of human perceptions. Human perception can be improved by the use of scientific apparatus, by precision, by control, and by several other devices. The aim of all these devices is to prove the correspondence between scientific propositions and the objects of their inquiry. The objective of all these inquiries is to establish transcendent truth.

The logical principles, namely, identity, contradiction, and excluded middle (Cohen & Nagel, 1934, pp. 181ff.), offer together the necessary overall principle. This principle of immanent truth means that scientific propositions must be free of inner contradictions and, within a given system, must not contradict one another. Thus, a scientific system is formally true whenever it is free of inner contradictions; then and only then does it meet the requirements of immanent truth. A scientific system is empirically true when it does not contradict the body of well-established empirical data and meets the requirements of *transcendent truth.*

In contradistinction to logical positivism, epistemological realism requires both immanent and transcendent truth. Here is what Freud wrote about the Viennese circle philosophers:

> No doubt there have been intellectual nihilists of this kind before, but at the present time, the theory of relativity of modern physics seems to have gone to their heads. It is time that they start out from science, but they succeed in forcing it to cut the ground from under its own feet, to commit suicide, as it were. . . . According to this anarchistic doctrine, there is no such thing as truth, no assured knowledge of the external world. . . . Ultimately we find only what we need to find and see only what we desire to see. . . . And since the criterion of truth, correspondence with an external world, disappears, it is absolutely immaterial what views we accept. All of them are equally true and false. (Freud, 1932/1933, p. 240)

Freud developed his theoretical framework on the grounds of epistemological realism. While neopositivists and operationists insisted that theory be merely a superstructure of sensory data, Darwin, Huxley, Sechenov, Pavlov, Brüke, Freud, Einstein, and other empirical scientists neither accepted the subjective sensory data as the sole basis for science nor insisted on presenting a theory in terms of sensory data.

2. Monism

Freud was always aware of the organic foundations of mental life. At best, Freud believed, one can assume—and never do more than assume—that mental processes utilize a form of energy that is at the disposal of the living organism. This energy is analogous to any other energy, and that is all we know.

> We assume, as the other natural sciences have taught us to expect, that in mental life some kind of energy is at work; but we have no data which enable us to come nearer to a knowledge of it by analogy with other forms of energy. (Freud, 1932/1933, p. 44)

Being a monist, Freud never gave up hope for a monistic interpretation that would combine both physical and mental processes in one continuum. But at the present stage of scientific inquiry, a radical reductionism must be rejected. Psychology must continue to do what Freud actually did: develop new hypothetical constructs independent of the physical sciences. Freud knew that his later theoretical constructs were nonreductionistic and irreducible to any of the constructs of physics or chemistry. Although he believed that the future might prove that chemical substances influence the amount of energy and its distribution in the human mind, work on such an assumption would not be too productive at the present time.

Quite late in his life Freud arrived at the conclusion that psychology must develop its own conceptual system, since the processes with which psychology is concerned

> are in themselves just as unknowable as those dealt with by the other sciences, by chemistry or physics, for example; but it is possible to establish the laws which those processes obey and follow over long and unbroken stretches, their mutual relations and interdependences. . . . This cannot be effected without framing fresh hypotheses and creating fresh concepts. . . . (Freud, 1938/1949, p. 36)

3. Energetism

Freud believed that there is one kind of energy in nature and that all observable actions are either produced by this energy or exist as its variations or transformations. If this holds true in physics, it holds true also in other sciences, such as chemistry, biology, and psychology. This must not be construed in a radical reductionistic vein, for human thoughts are not electrical processes and cannot be reduced to terms of amperes, or watts, or volts. Mental processes cannot be reduced to anything that is not mental, but they develop from the same physical source as everything else in the world. Mental energy is energy in the physical meaning of the word, that is, something that can be transformed into another kind of energy in a manner analogous to the transformation of mechanical into electric in generators. Energy can be accumulated, preserved, discharged, dissipated, blocked; but it cannot cease to exist. The law of preservation of mental energy, its transformability, and its analogousness to physical energy is one of the guiding principles of psychoanalysis.

> Among the psychic functions there is something which should be differentiated (an amount of affect, a sum of excitation), something having all the attributes of a quantity—although we possess no means of measuring it—a something which is capable of increase, decrease, displacement, and discharge, and which extends itself over the memory-traces of an idea like an electric charge over the surface of the body. We can apply this hypothesis . . . in the same sense as the physicist employs the conception of a fluid electric current. (Freud, 1894/1962, p. 61)

Freud postulated that psychic energy is not an entirely new or a completely different type of energy. Mental energy is a derivative of physical energy, though no one can really tell how the "mysterious leap" takes place either from body to mind or vice versa.

4. Determinism

Quantum theory has rendered the causal principle useless in theoretical physics. The elaborate structure that nineteenth-century science built seems to be dismantled. Moreover, as Einstein pointed out:

> All attempts to represent the particle and wave features displayed in the phenomena of light and matter, by direct course to a space–time model, have so far ended in failure. . . . At the present, we are quite without any deterministic theory directly describing the events. . . . For the time being, we have to admit that we do not possess any general theoretical basis for physics which can be regarded as its logical foundation. (Einstein, 1940, p. 488)

Contemporary physics cannot present its data in a deterministic continuum, but some physicists, among them, Einstein himself,

> cannot believe that we must abandon, actually and forever, the idea of direct representation of physical reality in space and time; or that we must accept the view that events in nature are analogous to a game of chance. (Einstein, 1940, p. 491)

Spinoza defined causation as follows:

> *Ex data cause determinata necessario sequitur effectus, et contra si null detur causa, impossibile est, ut effectus sequetur.* (The effect follows necessarily from a given determined cause; and to the contrary it is impossible for an effect to follow if there is no cause.) (Spinoza, 1919, Axiom 3)

John Stuart Mill (1879, Book IV, Chapter 3) maintained that the causal principle is "the ultimate major premise of all inductions." He defined *cause* as "the unconditional, invariable antecedent," and *effect* as the "invariable, certain and unconditional sequence."

The element of necessary, inevitable temporal sequence is included in practically all definitions of causality. This necessary evolvement of effects out of causes inspired Spinoza to say that "*Res aliqua nulla alia de causa contingens dicitur, nisi respectu defectus nostrae cognitionis*" (We are calling a thing coincidental only because of deficiency of our cognition).

A similar idea was expressed by Laplace:

> We ought to regard the present state of the universe as the effect of its antecedent state and as the cause of the state that is to follow. An intelligence knowing all the forces acting in nature at a given instant, as well as the momentary positions of all things in the universe, would be able to comprehend in one single formula the motions of the largest bodies as well as of the lightest atoms in the world, provided that its intellect were sufficiently powerful to subject all data to analysis; to it nothing would be present to its eyes. The perfection that the human mind has been able to give to astronomy affords a feeble outline of such an intelligence. Discoveries in mechanics and geometry, coupled with those in universal gravitation, have brought the mind within reach of comprehending in the same analytical formula the past and the future. (Laplace, 1820, p. 120)

Einstein took a definite stand on causation, assuming that its validity had not been finally abolished by the recent developments in theoretical physics.

It seems that the causal principle can be defined as *temporal sequence, ontologically necessary,* and *genetic.* Certainly there are other noncausal temporal sequences, such as winter and spring. Furthermore, there are logically necessary, nontemporal sequences, such as if a = b, and b = c, then a = c. There are certain ontologically necessary (yet not causal) sequences, such as day and night. There are genetic (productive) sequences, such as a bud and a flower, a seed and a sapling, that are noncausal. Causality is a certain type of relationship that includes sequence in time, *ontological* necessity, and *genetic* elements (Wolman, 1973a).

The causal principle cannot be empirically proven. It may be postulated if it proves useful and helps the organization of factual data in a coherent system.

Obviously, theoretical physics has substantial difficulties with causation, especially in quantum theory and the concept of time, but there is no reason for importing concepts from one area of science into another. Since Einstein, theoretical physics has had difficulty with the concept of time. However, in biology and psychology, the concept of time is self-evident, for all living organisms have clearly established beginnings and ends of life, and whatever goes between these two points is neither A. Einstein's *fourth dimension* nor R. P. Feynman's *negative time*.

Moreover, while the concept of causality, defined as temporal, ontological necessity, and genetic, that is, produced, may not apply to physics and may sound anthropomorphic, it certainly applies to human behavior.

According to Freud, objective and verified observation is the sole source of knowledge. The results of these observations can be "intellectually manipulated" and put together into a system of generalizations and laws that form a system of propositions that explain empirical data.

One of these general principles is the principle of causation. Natural sciences, especially microcosmic physics, struggle with the difficulties arising from a strict application of the causal principle. No such difficulties have been encountered in any of the areas of scientific psychology. All students of psychology apply a more or less strict deterministic point of view. Freud preferred a rigorous determinism that accepts no causes without effects, no effects without causes.

Once determinism is postulated, it forces the research worker to continuous efforts in seeking for causes and predicting outcomes. Every successful case serves as evidence that one is on the right track, encouraging further efforts that promise to bring additional evidence. Lack of success indicates that one has to check and double check his methods and look for additional data. Strict determinism helped Freud in the study of the most irrational areas of dreaming and symptom-formation in neuroses. The principle of "whatever is, has its causes" forced Freud to give up the early theory of instincts that juxtaposed sex to self-preservation and to assume the existence of destructive instincts. Causal considerations also put him constantly on guard in searching for minute details that might have been partial determinants of mental health and of mental disorder (Wolman, 1973b).

5. The Economy Principle

Freud's theory follows the principle of preservation of energy, and this principle is applied to mental energy. Mental energy can be transformed, released, or accumulated; but it can never disappear entirely. When a degree of energy is invested into something, this object becomes loaded or charged with a certain amount of mental energy in a manner analogous to that in which bodies become charged with electricity. This process of charging ideas of objects with mental energy was called by Freud *cathexis*, and objects in which mental energy was invested were cathected. Cathexis can be applied to external objects as well as to one's own organism.

Energy is transformable and displaceable. Mental processes are processes of mental energy economics, that is, quantitative processes of transformation, accumulation, investment, and discharge of mental energy. Some processes consume more energy, some less. When powerful instinctual drives press for an immediate discharge of energy, a great amount of energy is needed for anticathexis. Individuals torn by inner conflicts cannot be very efficient because considerable amounts of their energy are being tied in inner struggle.

Mental economy depends on the comparative strength of the external stimuli, instinctual drives, and the inhibitory forces. Human behavior can be presented as a series of reflex-arcs. A stimulus acts on the organism causing a disequilibrium (perceived as tension), and the tension leads to an action, that is, to a discharge of some amount of energy. The discharge of energy restores the equilibrium and is experienced subjectively as relief and pleasure.

Between the tension and discharge of energy two contradictory types of forces step in, one facilitating the discharge of energy that brings relief, the other preventing or postponing this discharge. The forces that urge and facilitate discharge are called by Freud *drives* or *instincts*. The instincts, or instinctual drives, press for discharge of energy, for lowering the level of excitation, and reduction of the tension in the organism. Thus these forces help to restore the equilibrium. Since homeostasis, or the tendency to keep equilibrium, seems to be a general property of living matter, the instinctual drives must be basic, innate, and primary biological forces.

6. The Pleasure–Unpleasure Principle

Freud followed Fechner with the idea of pleasure and unpleasure as related to the mental economy of excitation. Freud quoted Fechner as follows:

> Insofar as conscious impulses always have some relation to pleasure or unpleasure, pleasure and unpleasure too can be regarded as having a psycho-

> physical relation to conditions of stability and instability. This provides a basis for a hypothesis [that] every psycho-physical movement crossing the threshold of consciousness is attended by pleasure in proportion as, beyond a certain limit, it approximates to complete stability, and is attended by unpleasure in proportion as, beyond a certain limit, it deviates from complete stability; while between the two limits, which may be described as qualitative thresholds of pleasure and unpleasure, there is a certain margin of aesthetic indifferences. (Freud, 1920/1962, p. 8)

The ideas of constancy and economy were derived from clinical observations of pleasure and unpleasure, though from a logical point of view the pleasure–unpleasure continuum would follow the principle of constancy. The mental apparatus endeavors to keep the quantity of excitation low, and any stimulus that increases the stimulation is felt as unpleasant.

> We have decided to relate pleasure and unpleasure to the quantity of excitation that is present in the mind but is not in any way "bound"' and to relate them in such a manner that unpleasure corresponds to an increase in the quantity of excitation and pleasure to a diminution. (Freud, 1920/1962, p. 8)

7. The Principle of Constancy

The principle of constancy serves as the general framework of Freud's theory of motivation. It represents a tension–relief continuum and explains the compulsion to repeat first experience. This "repetition compulsion" is responsible, and manifests itself in several aspects of human life.

> The attributes of life were at some time evoked in inanimate matter by the action of a force [of] whose nature we can form no conception. . . . The tension which then arose in what had hitherto been an inanimate substance endeavored to equalize its potential. In this way the first instinct came into being: the instinct to return to the inanimate nature. It was still an easy matter at that time for a living substance to die. For a long time, perhaps, living substance was thus being constantly created afresh and easily dying. (Freud, 1920/1962, p. 40)

Freud invoked the constancy principle also in regard to the sexual instinct. Since "science has little to tell us about the origin of sexuality," Freud reported a myth that "traces the origin of an instinct to a need to restore an earlier state of things" (Freud, 1920/1962, pp. 57–58). In Plato's Symposium, Aristophanes relates: "Everything about these primeval men was double; they had four hands and four feet, two faces, two privy parts, and so on. Eventually Zeus decided to cut these men in two." After the division had been made, "the two parts of man, each desiring his other half, came together, and threw their arms about one another eager to grow into one" (Freud, 1920/1962, p. 57–58).

Freud hypothesized that living matter was broken down into small particles "which have ever since endeavored to reunite through sexual instincts." Several biological processes may be interpreted in the light of this tendency to restore an earlier state of things.

8. Theoretical Continuities and Discontinuities

Hardly any other idea has influenced and encouraged as many original thinkers and scientists as psychoanalysis. Sigmund Freud's theories have inspired and stimulated the development of new ideas and new approaches to the enigma of the human mind. Instead of repeating and rehashing the words of the master, many of his disciples paved new roads and opened new vistas in psychology and psychopathology. Heinz Hartmann, Anna Freud, Paul Federn, Franz Alexander, Melanie Klein, and many others have added new wings to the house Freud built. Margaret Mahler, Erik Erikson, Sandor Rado, Heinz Kohut, Benjamin B. Wolman, and others, leaning on Freud's work, have developed new and original ideas. These theories, while rooted in Freud's theoretical framework, represent fresh approaches to psychology and psychopathology, quite often in disagreement with this or another aspect of the master's blueprint.

Freud was an iconoclast, and the vitality shown by followers and dissidents (such as Karen Horney, Harry Stack Sullivan, Eric Fromm, and others) bears witness to the tremendous impact of Freud's work. Some of Freud's *words* have been abandoned by disciples and dissidents, but his *spirit* is very much alive (Wolman, in press).

The fact that some of Freud's empirical data and theoretical concepts have been modified or rejected does not justify the tampering with the master's theory which, indeed, offered the very source from which all of us, disciples and dissidents, borrowed. It is perfectly justified to develop new ideas based on Freud's theory, but there is no reason to "reform" or "modernize" Freud to make his work fit into current and *ever changing* empirical data in neurology, psychology, and psychiatry. Freud's classical work need not be rewritten nor edited. It gave impetus to certain new developments and new theories, but it must be preserved and respected the way Freud formulated it.

9. References

Cohen, M. R., & Nagel, E. *An introduction to logic and scientific method.* New York: Harcourt Brace, 1934.

Comte, A. *Cours de philosophie positive.* Paris: Baillère, 1864.

Einstein, A. The fundaments of theoretical physics. *Science,* 1940, *91,* 487–492.

Freud, S. The neuropsychoses of defense. In J. Strachey (Ed. and Trans.), *The standard edition of the complete psychological works of Sigmund Freud* (Vol. 3, pp. 45–61). London: Hogarth Press, 1962. (Original work published 1894)

Freud, S. Beyond the pleasure principle. In J. Strachey (Ed. and Trans.), *The standard edition* (Vol. 8, pp. 7–64). London: Hogarth Press, 1962. (Original work published 1920)

Freud, S. *New introductory lectures on psychoanalysis.* (J. Strachey, Trans.) New York: Norton, 1933. (Original work published 1932)

James, W. *Principles of psychology.* New York: Holt, 1890.

Kant, I. *Critique of pure reason.* London: Macmillan, 1929.

Laplace, P. S. *Théorie analytique des probabilités.* Paris: Alcan, 1820.

Locke, J. *An essay concerning human understanding.* Oxford; Oxford University Press, 1894.

Mach, E. *The science of mechanics.* Chicago: Open Court, 1960.

Mill, J. S. *A system of logic* (8th ed.). London: Longmans, Green, 1879.

Spinoza, B. *Ethics.* London: Bell, 1919.

Wolman, B. B. Evidence in psychoanalytic research. *Journal of the American Psychoanalytic Association,* 1964, *12,* 717–733.

Wolman, B. B. Concerning psychology and the philosophy of science. In B. B. Wolman (Ed.), *Handbook of general psychology.* Englewood Cliffs, N. J.: Prentice-Hall, 1973. (a)

Wolman, B. B. *Call no man normal.* New York: International Universities Press, 1973. (b)

Wolman, B. B. *The logic of science in psychoanalysis.* New York: Columbia University Press (in press).

2

The Biological Origins of Psychological Phenomena

David A. Freedman

The perception, not only of psychoanalysis but of all psychology as occupying a place both within and without the natural sciences is a perpetual source of frustration and fascination for those of us who come to the field as physicians. We share with Freud and other spiritual ancestors of similar background a sense of continuing challenge. How can we reconcile our neurologic and physiologic frame of reference with the clinically inescapable fact that in our day-to-day efforts we seem to be working on an entirely different plane? Our plight—and I use the word advisedly—as psychoanalysts involves dealing with extremely substantial—even quantifiable—physiological processes, while at the same time working in a medium which commits us to the use of explanatory devices which have to do only with relationships. We may detect evidence of what we consider rage, aggression, sensuality, and so forth. What we interpret, however, is the complexities of feelings as they are manifested in the context of human relations. No one has ever profited from being told that he suffers from an excessive aggressive drive. Our patients, however, regularly benefit from becoming aware of covert *object*-directed impulses and fantasies. Properly timed interpretations can be powerful in affecting not only perception of self and others but physiological processes as well.

The difficulties inherent in the situation I have outlined has led many members of our profession simply to throw up their hands. Highly respected analysts have come to the conclusion that our field is best regarded as a hermeneutic science. Freud himself came to the conviction

David A. Freedman • Department of Psychiatry, Baylor College of Medicine, Texas Medical Center, Houston, Texas 77030.

that he would have to work exclusively within the psychological domain. Yet he always held out the hope that at some point in time all psychology would be reducible to chemistry. Even more significantly, it has been pointed out by Jones (1953) and Strachey (1966), among others, that the model he attempted to develop in the uncompleted 1895 essay now called *A Project for a Scientific Psychology* influenced all his later theoretical work. How to reconcile mental and neurologic phenomena remained a challenging leitmotif throughout his working life. It continues to be one for those of us who consider ourselves to be his intellectual and professional descendants.

Nowhere are the complexities which confront us more apparent than in our formulation of the ultimate epistemological question. What is the relation between that which we experience and perceive and the universe around us? It is a platitude to say that this is not a question for psychoanalysts alone. Yet it has particular relevance for us, as it does for workers in any other discipline which is concerned with the workings not only of the structural givens of the perceiving organism (the limited sensitivity of transducers, noise within systems, and the like are after all problems with which even builders of optical and electronic devices must deal), but also with those idiosyncratic peculiarities which reflect both the point in biological development our patients have achieved (cf. Piaget) and the highly idiosyncratic conditions under which any given individual has come to any given point in his developmental progress. As Wolman proposes, Freud took for granted the data of perception as *reflecting* a world which exists independently of our attention to it. But—and this is a critical caveat—they only reflect. The correspondence is never complete and is always limited by organismic factors (cf. above) as well as by both ongoing social and idiosyncratic individual considerations. Closely cropped hair, that epitome of conformity and conservatism in the 1950s and 1960s, was introduced around the turn of the century by the International Workers of the World. The Wobblies did not wish to be confused with the bearded, long-haired hated capitalist exploiters of the working man.

The distortions we uncover in the analysis of the transference underscore this issue from yet another perspective. We tend to lose sight of the ultimate definition of energy. It is, after all, nothing more nor less than the ability to do work or produce change. So defined, there should be only one problem with the concept of mental energy. It must be conceded that mental energy does not follow the law of conservation. However potent it may become in the workings and influence of a given individual, it ceases to exist with his death. More specifically, there would be only this one problem if we could isolate ourselves from the nineteenth-century model which deals, after all, with mass effects. Control and regulatory systems, including the cerebrum, can be viewed as the repositories of virtually unlimited energy. This proposition does not violate the essentially monistic position which Wolman ascribes to Freud. Neither is it inconsistent with McCarley and Hobson's (1977) characterization of Freud's epistemological position as one of mind–body isomorphism. At the same time it also does nothing to fill in the gaps with which the monistic view of the relation of mind to its substrate leaves us. We continue unable to specify the idiosyncratic physiological characteristics which specify what each one of us *as an individual* experiences as a particular mental content. We can, of course, go a long way toward specifying the circuitry whose operation is experienced as a particular affect. The content, however, with which the affect is associated remains beyond our capacity to specify in an anatomic sense.

The situation would appear to be analogous to that in physics. Heisenberg's principle does not reject the assumption that a particle passes a given point at a given time and with a given velocity. It simply states that all of these parameters cannot be specified simultaneously. It also introduces the observer as a factor. For those of us who practice analysis his role has long been recognized and labeled as countertransference. The parallels to the situation in physics are striking in another respect. So long as one works with extremely ill (psychotic) individuals one can view his patients as very different from himself. An illusion of "objectivity" inheres in the ability to describe and alter behavior by a multitude of methods, including the use of drugs. The psychiatrist can maintain a distance not unlike that of the classical physicist or engineer who works with static objects and concepts such as the principles of mechanics. Like the physicist who deals with ultimate particles, the psychoanalyst is in a very different position. The challenge of differentiating between himself, his feelings, his values, and those of his patient is constant and probably never entirely to be resolved. It would certainly appear to be the case in psychology, as it is in other branches of science, that ultimate causes are not to be determined. As Wolman insists, however, to say this is not to eschew the convictions (a) that they exist and (b) that continuing diligence can be expected to bring us closer to them in our approximations.

That all human behavior involves compromise between simultaneously existing, albeit mutually contradictory and countervailing, forces is as probable as the analogous assumption in the physical world. Phenomena as we see them are not reducible to simple, unidirectional cause-

and-effect relations. To the extent that this apparent truism has validity we must invoke a quantitative economic principle, an assumption concerning the balancing of forces in the determination of any resultant phenomenon. Whether as Wolman proposes, human behavior can be adequately represented as a series of reflex arcs is, however, another matter. I suspect the problems are much too complex. The same argument which attempts to define mental energy in terms which make it analogous to control systems requires that one go beyond the relatively simple concept of the reflex arc. Complex patterns of neuronal activity imply a whole new order of phenomena, one for which the simple model of the reflex arc appears to me to be inadequate.

I agree with Wolman that mental economics can be characterized *at the clinical level* as involving the comparative strengths of the external stimuli, the instinctual drives, and the inhibitory forces. I would, however, prefer to substitute "endogenous sources of motivation" for "instinctual drives." My preference is based on the importance, as I see it, to distinguish between that which is gene-determined and therefore inborn and that which reflects, in addition, the influences of idiosyncratic experiences. It is the latter, especially during critical developmental periods, which lend uniqueness to each individual's developed personality.

Finally, a word about the pleasure principle. It seems to me that it is with regard to this concept that modern psychoanalysts must differ most drastically with Freud. The concept of homeostasis did not play a role in his thinking. For him there was no dynamic equilibrium in the sense Cannon explicated when he introduced the term. Pleasure in its ultimate sense meant complete quiescence, and the task of the nervous system, according to Freud, was to rid itself of stimuli. That such a notion should have lead ultimately to the egregious concept of the death instinct, while logically understandable, is one of the less happy facts of his theorizing. Like his adherence to Lamarckian evolutionary theory, it seems to me to be the consequence of his decision, once the "Project" foundered, to limit himself to working in psychological terms alone. To pursue the study of the mind, I speculate, he found it necessary to limit himself in other areas. As a consequence, he remained wedded to biological data and theories which became progressively more outmoded during the nearly half century he spent developing psychoanalysis. At the same time it is impressive how little the consequences of these obvious deficiencies have affected the general usefulness of Freudian theory. The basic principles continue both to have enormous clinical utility and to serve as the basis for the continuing development of psychological theory and research.

1. References

Freud, S. *The origins of psychoanalysis*. New York: Basic Books, 1950.

Jones, E. *The life and works of Sigmund Freud* (Vol. 1). New York: Basic Books, 1953.

McCarley, R. W., & Hobson, J. A. The neurobiological origins of psychoanalytic dream theory. *American Journal of Psychiatry* 1977, *134*, 1211–1221.

Strachey, J. Editor's introduction. Project for a scientific psychology. *The standard edition of the complete psychological works of Sigmund Freud* (Vol. 1). London: Hogarth Press, 1966.

2

Structure, Function, and Meaning

Rudolf Ekstein

> It is one of the most interesting paradoxes that Freud, who started with the determined attempt to find out the mechanics of psychic life, has discovered that the psychic life is full of meaning.
>
> Paul Schilder, (1935)

I have followed Wolman's work for many years, and I am proud that he invited me to contribute to his encyclopedic work several times. He is not only an encyclopedic but also a very poetic man. Once, after our work conference was completed, he took me to one of New York's museums. Both of us, educated in different parts of central Europe but having originally lived in the same Austro-Hungarian monarchy, found how similar our education had been. As we walked through the museum rooms we tried to recall German poetry which we had to learn by heart during our high school days. Each of us was often able to carry on when memory failed the other. I do not know who won, but I remember the mutual delight in our recalling common educational background and the erudition of our teachers with whom we identified.

I can only agree with Wolman's purpose that demands of us not to relinquish the continuity between the ideas of a Freud, his theories, and their connection with the intellectual and cultural atmosphere of that time and new ideas often springing up like short-lived new fashions. Such intellectual dilemmas almost force us to repeat the eternal struggle between fathers and sons, teachers and students, old schools of thought and new ways of thinking. After all, it was Freud (1925) who spoke about "the patchwork of my labors," never suggesting that his theories and techniques were meant to be complete and untouchable (p. 70). It is for this reason that he often stated that he did not want to think of

Rudolf Ekstein • 536 South Westgate Avenue, Los Angeles, California 90049.

psychoanalysis as a *Weltanschauung*. He never committed himself to write a complete technique and he would have agreed with the philosophical attitude expressed by Bronowski (1951): "By its nature it [science] must prize the search above the discovery and the thinking (and with it the thinker) above the thought. . . . The act of judging (is) more critical than the judgement." We would then speak not about *the* psychoanalysis but rather about psychoanalyzing; not about a closed system but about a process.

Wolman wishes to reemphasize the basic constructs of Freud's theoretical system which he feels must be preserved in order to continue to inspire fruitful psychological research regardless of whether the people engaged in this research consider themselves as followers, dissidents, or opponents. Must one be the one or the other? Or, should one rather follow a way of endless clarification beyond the need of refutation? Wolman's paper seems to be written in the spirit of defense, and I wish to add a few thoughts which are meant to strengthen the task that I think he gives, namely, to see scientific process as continuity, a widening of our insights, and an increased capacity to learn by identification, a process which grows out of the struggle against identification. Perhaps I can best do this, having had my original philosophical training with the *Wiener Kreis*, if I were to elaborate in which way this group of philosophers saw each task. When Wolman quoted Freud, I felt a little puzzled. I knew that 1932 quotation also, but it never occurred to me to relate it to the philosophy of Wittgenstein and his followers. It is true that many popular opinions used some of the thinking of Einstein, or rather misread it as a kind of eclectic position according to which "everything is relative" but that had actually little to do with the philosophical work that I refer to and wish to describe (Ekstein, 1959, 1978).

The basic philosophical view of this Viennese group of philosophers (I refer to Wittgenstein, Schlick, Carnap, Feigl, Neurath, Popper, and many others) suggests that it was the task of philosophy not to find the truth but rather to examine the propositions of science and/or philosophy in order *to clarify* them. The task of philosophy was to clarify the meaning of a proposition, but it was the task of science to search for and to establish truth, to establish criteria of evidence, of verification, and/or falsification. In older days, so this school contends, philosophy was seen as search for truth rather than simply as love of wisdom. The philosopher's activity could be compared more with the activity of a Socrates who examined the meaning of words and propositions and was not to be taken as a substitute for science. In an earlier phase, approximately at the time when Freud wrote the words Wolman quotes, this school dismissed as meaningless all those statements that did not allow verification or falsification. In a later phase, the *meaning of meaning* was enlarged so that we would not need to dismiss from the system of communication all nonscientific statements; rather, the task was now to think about their deeper meaning.

Wittgenstein (1922) suggested that it was the task of philosophy to clarify all these problems which are hidden behind and within philosophical systems, behind metaphysics, and he suggested a process of clarification would make the problems disappear:

> *The riddle* does not exist.
>
> If a question can be put at all, then it *can* also be answered. . . .
>
> We feel that even if *all possible* scientific questions be answered, the problems of life have still not been touched at all. Of course there is then no question left, and just this is the answer.
>
> The solution of the problem of life is seen in the vanishing of this problem. (p. 187)

He then only saw scientific problems which could be solved with scientific means, and he thought that the nature of philosophical problems often did not require an answer, as was true for scientific problems, but rather that they needed clarification.

In a more recent communication (Ekstein, 1978) I tried to show an interesting parallel between the thinking of Wittgenstein and the thinking of Freud. Freud (1909) said that the patient who brings problems to us will discover during the process of psychoanalysis that his problem, his symptom, was related to repressed, forgotten conflicts, and that the discovery of these problems through the means of interpretation, *Deutung*, will bring the conflict to light, and thus it will vanish. Freud often used metaphors from archaeology. For example, Pompeii, buried by the volcanic ash of Vesuvius, preserved what was now slowly being destroyed when brought to the light of day. He suggested that after the meaning of a symptom is made clear through the interpretive process it will disappear and thus permit adaptive forces to prevail rather than the symptoms of neurotic conflicts.

The philosopher suggests that the philosophical puzzlement, the riddle, disappears upon (logical) clarification, whereas the psychoanalyst suggests that the interpretive process, another form of (emotional) clarification, will make the problem, the symptom, the conflict, disappear.

Both then speak about a process of change, and thus they are less concerned with the system of philosophy, or the metapsychology of psychoanalysis, but more with the process, the activity toward change.

Different psychoanalytic schools, stressing different aspects of theory or metapsychology, are really committed to a way of working, a way of speaking, a way of interpreting, and thus it is important to read psychoanalytic authors in such a way that one always has in mind whether they wish to explain or whether they refer to these propositions which make for the technical armamentarium, the language of psychotherapy. I suggested once (Ekstein, 1959) that every theoretical system really hides a decision about the intervention, the technique to be employed. The different school systems in psychoanalysis or psychotherapy do not pay enough attention to their *technical* meaning and are too much involved with the idea that they have found the final explanation, *the* theoretical foundation.

I think that Wolman's intent was to remind us of the basis which allows us to develop new concepts, improved ways of working with our patients, and to reattach ourselves to the history of our science, the history of thought, and not to yield to the temptation to give in to fragmentation or eclecticism which may serve the fashion of the day. He is identified, I believe, as I am, with the word of Goethe (1808) which Freud quoted as well:

> What thou hast inherited from thy fathers, acquire it to make it thine and thee.

1. References

Bronowski, J. *The common sense of science.* Boston: Little Brown, 1951.

Ekstein, R. Thoughts concerning the nature of the interpretive process. In M. Levitt (Ed.), *Readings in psychoanalytic psychology.* New York: Appleton-Century-Crofts, 1959.

Ekstein, R. Further thoughts concerning the nature of the interpretive process. In S. Smith (Ed.), *The human mind revisited.* New York: International Universities Press, 1978.

Freud, S. Notes upon a case of obsessional neurosis. *The standard edition of the complete psychological works of Sigmund Freud.* (Vol. X, pp. 155–318) London: Hogarth Press, 1955. (Original work published 1909)

Freud, S. An autobiographical study. *The standard edition* Vol. X, pp. 3–74). London: Hogarth Press, 1959. (Original work published 1925.)

Goethe, Johann Wolfgang von. *Faust, Part One.* Translated by Walter Kaufmann. Garden City, New York: Doubleday 1961. (Original work published 1808)

Schilder, P. Psychoanalysis and philosophy. *Psychoanalytic Review*, 1935, *22*, 274–487.

Wittgenstein, L. *Tractatus logico philosophicus.* London: Kegan Paul, Trench, Trubner, 1922.

2

The Heuristic Value of Freud

Gordon F. Derner

Wolman approaches the examination of psychoanalysis as a scientific theory by a brief course in philosophy of science. He demonstrates his noted skill at attacking problems in a careful, scholarly, step-by-step process. His enormous breadth of scholarship, which extends far beyond psychoanalysis and psychology, is excellently used in the present paper. By defining the philosophical ground rule of evaluating Freudian theory on the basis of epistemological realism, he negates much of the objection of the demand of logical positivism that the description must entail directly observable phenomena.

Wolman encourages the reader to accept that Freud knew that psychological theory could not be reduced to a physiological substrate, but he fails to note Freud's considerable effort to the contrary. Freud's most serious attempt to explain mental phenomena as physical–chemical actions was in his so-called *Project for a Scientific Psychology,* although he apparently had thought to call it *Psychology for Neurologists.* He submitted a draft to Fliess but later withdrew it with the comment that it was "pure balderdash." Yet this obtuse, condensed presentation may have had more useful significance if it had been further developed, knowing what we now know with the breakthrough of research on the neurotransmitters.

The section on energetism is weakened not by Wolman's scholarship but by the changing tides of Freud's theoretical convictions. The chosen paper, *The Neuropsychoses of Defense,* as a statement of Freud's position on energy needs to be matched with his later thoughts about a physical-chemical explanation as "balderdash." Unquestionably it is possible to view all energy and energy transformation with the same theoretical

Gordon F. Derner • Late of the Institute of Advanced Psychological Studies, Adelphi University, Garden City, New York 11530.

model, but as Wolman notes, "Mental processes cannot be reduced to anything that is not mental." Still the physical and psychological effect of the diminution of pain in the so-called placebo effect, when a person is told that a drug which is in fact benign does relieve the pain, raises questions about the interlocking interface of mind and body. In this regard, Freud may have had more to offer to science than his major concepts of the unconscious, transference, and resistance.

Wolman notes that, as with much of psychology, psychoanalysis is deterministic. Effects have causes. As with Freud, many psychoanalysts continue to search for deeper and deeper motivations as the cause of an action. If a person says he or she loves his or her mother, the analyst may very well interpret the reason for the statements as hatred of the mother and vice versa. A person stating that he or she hates his or her mother may really mean that love is the real feeling. This glibness in causal explanation is part of the problem of unverifiable statements in psychoanalysis. Although Wolman says that "every successful case serves as evidence that one is on the right track," the statement loses in its generalization power. A successful case is a complexity which often even in gross detail is not applicable in another instance. The persistent effort to search for minute details on an observational level falters when interpretations are made from some forced fund of possible interpretations. It is noteworthy how much differently several analysts will interpret the same set of data and the same observations. It is the factor of the individual in the participant–observer role that complicates developing a quantifiable, measurable, testable theory in psychoanalysis. Each equation must have many subvariables so that parentheses within parentheses become necessary. It is difficult, therefore, to establish the details of a theory with such fluidity.

Wolman elaborates the basic concepts of energy transformations with facilitation or inhibition of discharge incorporating the signficance of the concepts in Freudian theory. The principle of preservation of energy, however, opens the question as to the original source within the individual and secondly, suggests a quantum notion of an amount of energy which needs to be discharged. The later idea leads to the clinical concept of symptom substitution if a behavior is reduced because it is viewed as a symptom of something else and its reduction will result in a new symptom. The failure of research to support the concept has not deterred many clinicians. They apparently find it difficult to accept Freud's statement that sometimes a cigar is a cigar.

As Wolman notes, Freud was much influenced by Fechner. Freud once stated that he had followed him on many important points. The pleasure–unpleasure principle was one such concept, and one to which Fechner felt a mathematical psychology could apply. Freud was either not aware of or else disregarded that criticism of Fechner by William James that his work led to nothing and was not even worthy of a footnote. Still the Herbartian concept through Fechner that an idea once in the consciousness and repressed could return is a key to much of psychoanalytic theory and clincial practice.

It is curious that Wolman accepts Freud's use of Aristophanes' myth of the split of the double-bodied person to explain sexuality. It should carry no more weight in the theory than the explanation for dogs smelling each other's behinds because they had hung them up at a party which was raided, grabbed the first one handy and now smell to see if they can find their own. The science of psychoanalysis has a stronger base than Artistophanes' legend for sexuality. Modern analytic theorists have emphasized the overpowering cultural learning which supersedes biology in sexuality. Further, the theory of sexuality is diminished by concepts such as that of small particles in living matter endeavoring to reunite through sexual instincts.

Wolman correctly gives full recognition to Freud's genius and the influence of psychoanalytic thinking on all phases of human activity. Psychoanalytic theory serves as a monument to the careful working and reworking of ideas from a variety of sciences, intuitive thinking, careful observations and clinical work combined in one of the greatest thinkers of modern history. Wolman is quite right that Freud's spirit and even a sizeable portion of his words are very much alive. The basic concepts of the unconscious and its influence on behavior, the development of the person through interpersonal interaction and its influence on subsequent relationships, the tenacity of the return of formed behavior patterns even when the patterns result in emotional disorder were conceived, amplified, interrelated, and subjected to constant review by Freud. In fact, it is difficult to describe a "Freudian theory," since he made innumerable adjustments and changes throughout his life. From his monumental *The Interpretation of Dreams* to his last papers, he kept the theory in flux. It is hard to accept Wolman's admonition not to "tamper with the master's theory" when he notes changes which have occurred with new observations, new perspectives, and new research. Even Wolman has participated in many other papers in emendations and alterations. It would be preferable to view Wolman's warning as preserving Freud's papers intact and to read them for inspiration and as a basis for further developments rather than abandon Freud. I could not agree more.

Although Wolman's careful statement of philosophical principles and demonstration of their application in psychological superior scholarship are valuable, it would have been helpful if Wolman had included some of the excellent research on Freudian theory. Some of it supports

Freud's theoretical concepts and some does not. An emphasis on a return to parsimony in theory, rather than wallowing in untestable and arguable abstractions, could clarify some of the morass which gives rise to the challenges to the scientific stature of psychoanalysis. Wolman, with his exceptional scholarship, could have moved in that direction more fully than in the present paper. The scholarship in philosophy he shows affords a mini-course in the field, but at least one reader would have hoped for more emphasis on the contributions of the newer theoretical attempts and research findings to bolster an exposition of psychoanalysis as a scientific theory.

It is true that much of the research on psychoanalysis may be challenged as in Engel's (1968) statement "Currently it is customary to discuss such work with a contemptuous, 'But it's not analysis.'" The same view apparently was held by Freud in relation to experimental investigation of psychoanalytic theory. In a letter to Rosenzweig in 1934 (see MacKinnon & Dukes, 1962, p. 702), Freud writes,

> I have examined your experimental studies for the verification of psychoanalytic assertions with interest. I cannot put much value on these confirmations because the wealth of reliable observations on which these assertions rest make them independent of experimental verification. Still it can do no harm.

A number of surveys have been done on the experimental investigation of psychoanalytic concepts. Sears (1942) presented an extensive review but generally found the evidence wanting, even though he had a good grasp of the theory. Generally the studies reported were minor and segmented. However, a more recent review by Fisher and Greenberg (1977) is more positive concerning the validity of psychoanalytic theory. Many studies, however, suffer from what Masling and Schwartz (1979) cogently identify as "confusing the map and the landscape." And one might conclude with Wolman, as Fisher and Greenberg conclude, that the theory is so robust and insightful in its description of human behavior that it can continue to stand even with constructs which are difficult to operationalize and research investigations that evidence many faults. So as Wolman concludes we should still look to Freud, the authentic genius of the mind, but for this writer we should see the theory more as heuristic than as final.

1. References

Engel, G. L. Some obstacles to the development of research in psychoanalysis. *Journal of American Psychoanalytic Association*, 1968, *16*, 195–204.

Fisher, S., & Greenberg, R. P. *The scientific creditibility of Freud's theories and therapy*. New York: Basic Books, 1977.

MacKinnon, D. W., & Dukes, W. F. Repression. In L. Postman (Ed.), *Psychology in the making*. New York: Knopf, 1962.

Masling, J., & Schwartz, M. A critique of research in psychoanalytic theory. *Genetic Psychology Monographs*, 1979, *100*, 275–307.

Sears, R. R. *Survey of objective studies of psychoanalytic concepts*. New York: Social Science Research Council, 1942.

2

Psychoanalysis as a Scientific Theory

Reply to Commentators

Benjamin B. Wolman

I feel honored by having my article reviewed by three distinguished scholars, Drs. Derner, Ekstein, and Freedman. I would like to express my profound gratitude for their thought-provoking commentaries and thank the editors of the *Annals,* for sending my contribution to these leading authorities.

Let me start with Dr. Derner's scholarly, precise, and succinct commentary.

I could not agree more with Dr. Derner that many tenets of Freud's theory are no longer valid. Certainly in Freud's times no one knew much about genetics, neuroscience, and the social psychology of family life. Psychoanalysis, as any other thought system, must keep up with the progress of scientific research and need not become a fossil. There is plenty of room for innovations and modifications, and the works of Hartmann, Erikson, Mahler, Kohut, and many others bear witness to the vitality of the psychoanalytic school of thought. My objections were directed not against those who modified or totally rejected some or many of Freud's ideas, but against the efforts of editing or rewriting Freud. Present and future research may invalidate many of Freud's theories, but the house that Freud built should remain intact, and Freud's work must be respected as a major breakthrough in psychology. In other words, let us discover facts unknown to Freud and develop fresh hypotheses, but I see no reason for trying to modernize Freud's writing.

Benjamin B. Wolman • Suite 6D, 10 West 66th Street, New York, New York 10023.

The purpose of my essay was not to prove Freud right, but to pay tribute to his genius that opened new horizons in psychology. Dr. Derner is right to criticize my phrase "Mental processes cannot be reduced to anything that is not mental." Of course they can, and I have elsewhere introduced the hypothesis of *monistic transitionism*. However, it was Freud and not I who maintained that "in mental life some kind of energy is at work; but we have no data which enable us to come nearer to a knowledge of it by analogy with other forms of energy" (Freud, 1932, p. 44).

Dr. Derner mentions Freud's *Project for a Scientific Psychology*. This *Project* attracted the attention of several scholars, among them Karl Pribram. However, the fact remains that while in 1894 Freud believed in reductionism, the state of science at his time forced him to abandon reductionism without giving up hope that the future might prove that chemical substances influence the amount and distribution of mental energy. It was, therefore, a sort of "hoped-for-reductionism."

Dr. Derner quotes several experimental studies that only partially or not at all confirmed Freud's hypotheses. There is no doubt that some of Freud's hypotheses were refuted by experimental research, and some were never proven. However, the experimental research on anxiety failed to prove or disprove anything, for the experimentalists' definition of anxiety had very little to do with any of Freud's definitions (by the way, Freud changed his mind more than once). Freud never rejected the idea of experimental verification of his hypotheses; it seems to me that the best support of his theories of the unconscious came from Russian neuroscientists. Anna Freud (1951) was quite critical of the Barker, Dembo, and Lewin experiment which was supposed to prove that frustration leads to regression; she could not agree that a mild removal of toys could compare to severe frustration in the psychoanalytic frame of reference.

By the way, I mentioned the Upanishad myth in relationship to Freud's constancy principle and *not* as an explanation of sexuality.

In conclusion, I am grateful to Dr. Derner's erudite comments and I wholeheartedly agree with him that we should see Freud's theory "more as heuristic than as final."

Dr. Rudolf Ekstein's comments carry the flavor and enticement of the Old Vienna. He has chosen to "speak not about *the* psychoanalysis but rather about psychoanalyzing; not about a closed system but a process." I am delighted to join him in this journey. Dr. Ekstein is not only a leading psychoanalyst but also an authority on philosophy of science. He received his philosophical baptism with the world renowned group of neopositivists, the Viennese Circle. For reasons beyond my control,

I received my philosophical training from the Warsaw School, that was opposed to Carnap and Wittgenstein, but met halfway with Moritz Schlick, the independent and somewhat dissident member of the Viennese Circle. However, it would be a futile task to prove who was right and who was wrong.

Dr. Ekstein endeavors to find certain proximity between Freud and Wittgenstein. He certainly has a point, especially in regard to the concept of *Deutung* (interpretation). His further remarks about the process of change make perfect sense. However, I firmly believe that Freud's epistemological realism stands clearly in opposition to the neopositivism and physicalism of the Viennese Circle.

This controversy reminds me of Wilhelm Reich's early efforts to reconcile psychoanalysis and Marxism. Certainly one can find some areas of communality, but Freud himself was opposed to Marx's ideas.

Dr. Ekstein suggests that we avoid fragmentation and eclecticism and reattach ourselves to the history of our science. I followed his advice in my book *The Logic of Science in Psychoanalysis* (Wolman, in press), where I tried to trace the roots of psychoanalysis and follow its contemporary branches and briars.

Dr. David A. Freedman's penetrating comments point to the ever growing gulf between Freud's psychogenic theories and the growing body of physiological data and their relevance in therapeutic endeavors. There is no question that in Freud's times there was need for "framing fresh hypotheses and creating fresh concepts," for the neurosciences of his time could hardly offer much help. This situation has changed radically, and at the present time it is rapidly changing. In 1977 the *International Encyclopedia of Psychiatry, Psychology, Psychoanalysis and Neurology* was published. A few years later, as the editor-in-chief of the encyclopedia, I wrote to all the authors and editors of it requesting their assistance in preparing a progress volume. In 1983 the *First Progress Volume* was published, reporting changes and modifications related to several areas of human mind, its ills, and their remedies. It is my impression that the most spectacular progress took place in the neurosciences.

Dr. Freedman reminds us that Freud himself came to the conviction that "he would have to work exclusively within the psychological domain," whereas "to reunite mental and neurological phenomena" remained a challenging leitmotif throughout his working life. Freud's epistemological realism is certainly no more than an assumption, for, as I wrote, there is "no empirical evidence for empirical evidence."

Dr. Freedman introduces the concept of observer versus observed in an analogy to Einstein's relativity theory. This concept is pretty close to the psychoanalytic concept of countertransference. It may, however, transcend countertransference in a situation of objective research, as pointed out in field-theoretical studies of Kurt Lewin.

I did not suggest a simple, reflex-arc theory. I only pointed to the differences between the application (or failure of application) of the causal principle in psychology as opposed to molecular physics. "The causal principle," I wrote, "cannot be empirically proven. It may be postulated if it proves useful and helps the organization of factual data in a coherent system." Freud's strict determinism has a definite heuristic value and has greatly contributed to Freud's relentless search for truth.

I would also beg to disagree with Dr. Freedman's statement that "the concept of homeostasis did not play a role in his [Freud's, B.B.W.] thinking." I believe that Freud's *constancy* concept conveys practically the same meaning as Cannon's homeostasis and Kurt Goldstein's equipotentiality.

In conclusion, I am in full agreement with Dr. Freedman's concluding remark that Freud's basic principles continue both to have enormous clinical utility and to serve as the basis for the continuing development of psychological theory and research.

3

The Nature and Challenge of Teleological Psychological Theory

Joseph F. Rychlak

Abstract. Teleological theory is shown to rely upon final causation, which in turn also makes use of formal-cause patternings as the 'that' for the sake of which events are being intended. In the rise of science over the sixteenth and seventeenth centuries, the belief crystallized that it was possible to explain events by reducing them to underlying material and efficient causation. Cartesian mathematics made it appear that motion caused patterns to come about and hence was basic to patterns. Modern physics has changed all this, placing the formal cause at the center of explanation. The unseating of material and especially efficient causation in science makes it possible for psychology to formulate telic theory. Formal causation is germane to meaning, and human beings can be seen to behave for the sake of such meaningful patterns. Mechanism is shown to be an instrumentality rather than a basic cause of behavior. Logical learning theory is presented as an example of telic theorizing. It is argued that unless psychology meets the challenge of teleological description it will never emerge as a distinctive area of study with a unique contribution to the family of the sciences.

1. Teleology as Final Causation

The Greek word *telos* translates as "end" or "purpose," and hence we should expect that a teleological explanation is going to have something to do with the ends or purposes involved in behavior. To what purpose? and What end is being sought? are the sorts of questions which spring to mind when we begin to think in a telic or teleological manner. But now, if there are such things as ends or purposes in behavior, are they observable? Authorities differ on this question, some holding that purpose as goal-oriented or end-oriented behavior is observable (Rosenblueth, Wiener, & Bigelow, 1943, p. 18) and others contending that it

Joseph F. Rychlak • Department of Psychology, Loyola University of Chicago, Chicago, Illinois 60626.

is not (Taylor, 1950, p. 328). In psychology, Tolman (1932/1967) was to take the former view in his concept of 'docility' relative to an observable end (p. 14). We repeatedly observe the food-deprived animal run a maze and see that it is able to improve its running time to the goal-box containing food. It is learning to attain its end. But surely in human behavior a person's purpose is not always observable, if for no other reason than that, unlike the palpable goal box, it may not yet exist. Thus, one enters the stock market with the end in hopeful view of making money and instead *loses* money—even repeatedly! In Tolman's view, this would amount to purposeless behavior because the person would *fail* to (a) attain the goal and (b) reflect docility in profiting from (trial-and-error) experience.

There is an important issue involved in such accounts, having to do with the slant or perspective from which our theory 'accounting for' behavior is to be written. If we take our descriptive stance exclusively 'over here' as observers of events taking place 'over there', we invariably frame theoretical explanations in an *extraspective* manner (Rychlak, 1981a, p. 27).Extraspective explanations are written in third-person phrasings, describing how that, it, he or she, they, or those organisms or events move along under our controlled observation. This was Tolman's perspective, and hence he looked for an observable course of improvement in behavior (*docility*) to name as the best objective evidence supporting purpose. But it is also possible to write theoretical explanations from the perspective of the organism under observation, as reflected in terms of first-person terminology, *I, me, my, us, we,* and so forth. This *introspective* theoretical slant (p. 27) does not require that we have some overtly observable manifestation of purpose since, as with our hapless stock investor, not all ends are actualized and a facility in achieving them is not always attained. Some people fail, learning to err with an alarming facility. Traditional personality theory in the dynamic tradition of psychoanalysis provides us with ample examples of this kind of learning (see Rychlak, 1981b).

I contend that telic theory demands an introspective perspective in order to present accurately a theory of intentionality. Thus, the theologian who professes belief in a divine plan as supposedly unfolding (extraspectively) 'in nature' may not presume to speak for the deity, but in effect this is precisely what he or she does in claiming to have a sense of the (introspectively framed!) divine intention creating the natural order. At quite another level of explanation, when the philosopher of science Thomas Kuhn (1970) tells us that scientists rely on *paradigms* to order their thinking, he is advancing an introspective conception. The paradigm acts as a framing assumption through which each scientist at a given time in history or within a school at some point in history, orders and understands that which is under investigation. When we shift perspective from looking 'at' people (extraspection) to looking 'with' people (introspection), as their life looms up before them, we realize that an end, goal, or purpose does not function "down the road of time." Considered introspectively, such *reasons* for behavior in the present are intellectually active in the present. Whether we think of the scientist's paradigm, the theologian's divine plan, or the stock investor's dreams of financial security, all such premising frameworks function in precisely the same *teleological* manner. Behavior described in this fashion is usually termed *intentional*. Intentions bring forward into observed reality (successfully or unsuccessfully) those ends, assumptions, goals, dreams, wishes, and so forth which the agent is seeking to further (i.e., the reason for or purpose of his or her behavior). It is common to use the terms *purpose* and *intention* interchangeably.

If we now glance back to Aristotle, we can frame the issues we have been considering in terms of *causation*, a much abused and misunderstood construct today. The word Aristotle used for what we translate as "cause" was *aitiā*, which has the meaning of *responsibility;* in following his usage we would be trying to assign responsibility for why anything exists or takes place in nature. Aristotle based his constructs of causation (which are Kuhnian-like paradigms in actual practice) on previous philosophical-scientific writings and asserted: "We think we have scientific knowledge when we know the cause, and there are four causes [in nature]" (Aristotle, 1952, p. 128).[1] As is widely known, these four "sources of responsibility" for events and things include: the *material cause* or substance that "makes things up"; the *efficient cause,* or impetus that assembles things or brings events about instrumentally; the *formal cause,* or pattern in events as well as the various shapes that things assume and the ordered sequence that logicomathematical proofs take on; and the *final cause* or "that [reason, purpose] for the sake of which" events are intended to happen or exist. It is evident that teleological description always entails final causation.

It would be a mistake to reify the four types of causal descriptions. These terms are best conceptualized as paradigms or models encompassing certain assumptions that we simply have to make if we are going to 'account for' anything in our experience. Even if we do not want to

[1] We are foregoing Aristotle's specific theory of primary and secondary substance, as well as various uses he made of final causation. It is not our intention to present Aristotelian philosophy, but merely to acknowledge the roots of causal description in the history of thought.

use the word *cause,* the meanings encompassed by these four terms still apply. Hence, it is simply true that in assigning purpose or intentionality to behavior the psychologist is—whether he or she realizes it or not—employing final causation in the descriptive account. At least, this is the case if the psychologist believes that the person is a true agent and not simply a manipulated mediator serving as a conduit of past efficient causes operating now in the present to 'determine' behavior. As we have noted in other contexts (Rychlak, 1977, p. 245; 1981b, p. 265), there are determinisms to be subsumed by *each* of the causal meanings. Determinism is not limited to efficient or material causation. Thus, a person who can *in essence* determine his or her own behavior as an agent will always be seen to do so thanks to a combination of formal and final causation *in addition to* material and efficient causation. Free will is a popular way of referring to the fact that human beings are telic organisms, determining their own course of life as agents. Before we can delve more deeply into this matter of agency and free will, we must gain an appreciation of the varying fortunes that our causes have had in the history of science.

2. The Search for 'Basic Causation' in the History of Science

It is surely no accident that natural science—which we usually trace to the fifteenth and sixteenth centuries in the work of Copernicus and Galileo—did not begin to advance until Aristotle had been rediscovered in the twelfth century. Indeed, Aristotelian theory of science remained dominant till the advent of Descartes, Locke, and Newton. Randall (1940) has shown how, in the universities of northern Italy, a series of debates raged in the fifteenth century over how causal conceptions were to be formulated whether qualitatively or mathematically. Gradually, the concept of 'cause' was identified as a *force:* "that is, in a definite way of behaving or in something that acted in a definite way" (Randall, 1940, p. 182). Galileo was to equate 'cause' with 'force,' even though his celebrated mathematical proofs of the heliocentric theory of the universe (*à la* Copernicus) relied on the formal causation of mathematics. The idea of tracking a force mathematically was to become central in the rise of natural science. In the closing decades of the sixteenth century, a debate took place concerning whether final causes had any place in natural philosophy (i.e., science). Gradually, a partitioning of causation usage occurred so that the efficient cause, as most descriptive of forces, impetus, motions, and so forth, was to emerge as the most basic of all causes in nature. As we shall see, this assumption is changing in our time based upon solid empirical evidence.

In the seventeenth century the efficient-cause bias was greatly furthered in British philosophy, led by Sir Francis Bacon's attack on Aristotelian scientific description (which was heavily teleological). Aristotle had suggested that nature operated for a purpose, as when we note that trees provide leaves to shade the fruit on their limbs. Bacon found such speculations to be frequent among what William James would later call armchair philosophers. One cannot address the purposes of nature as a scientist, argued Bacon; one can only study nature's workings through manipulative intervention and the seeking of lawful regularities (Farrington, 1949, p. 109). Bacon (1952) flatly states that a scientific explanation is *not* being rendered if one claims that "the bones are for [have the 'end' of acting as] the columns or beams, whereupon the frames of the bodies of living creatures are built" (p. 45). Bones do not have it as their 'aim' to hold up the muscles of our bodies any more than leaves have it as an 'aim' to shade fruit. Such 'that for the sake of which' explanations must be eschewed by the scientist in favor of explanations based strictly on material and efficient causation.

Of course, Bacon was not contending that natural science (natural philosophy) could answer all questions about life that confronted the human being. He framed not only a natural but a human and a divine philosophy and included formal and final causes in his metaphysics. Indeed, Bacon properly recognized that mathematics as a discipline was more related to formal and final causation than to material and efficient causation. Even so, as British philosophy was to be propounded through Hobbes, Locke, the Mills, Berkeley, and Hume, an increasing emphasis was placed upon the interpretation of cause as strictly and only efficient causation, which in turn referred to antecedents impelling, cuing, or otherwise moving events along. This style of explanation was to become the heritage of American psychology.

But probably an even more important development in the rise of natural science took place in the seventeenth century when Descartes conceptually united the formal-cause discipline of mathematics with the by then completely efficient-cause conception of motion (as force, thrust, impetus, etc.) (Simon, 1970). In Aristotelian physics we attempt to account for things and events based on their 'nature.' Natural objects like rocks or people or donkeys were viewed by Aristotle as 'in place,' set within the natural order. The challenge here was to explain why movement occurred, and Aristotle postulated qualitative, quantitative, and what he called 'local' motions. His accounts were heavily teleologized, for he viewed motion as the "act of a thing in potency" (Simon, 1970, p. 59). Aristotle always related the 'nature' of anything to its ends. Thus, a rock fell to earth in order to draw closer to that with which it shared a nature. Thanks to his reliance on mathematics, Descartes dismissed such intentionality from his approach and dealt exclusively with what he considered was 'natural' causality (Simon, 1970, p. 15). Only 'local' motion (shortened to 'locomotion') was considered by Descartes in his conceptualization of nature. And, to this very day, locomotion is viewed strictly in efficient-cause terms.[2]

Descartes' significant contribution to this winnowing of the causal meanings down to one (i.e., efficient-causation) was the fact that he underwrote his metaphysics with a new conception of geometry. Cartesian geometry begins on the assumption that objects are in motion from the outset. The scientist had to deal with spatio-temporal characteristics that were fluid. In essence, as Cartesians we account for why an object may be said to be at rest rather than having to account for why an object is in motion since everything is moving by definition. Euclidian geometry does not have its figures defined 'in motion' as Cartesian geometry does, where, for example, rather than the shortest distance between two points a straight line becomes a moving point defining a straight-line 'function.' Newtonian science capitalized on this Cartesian view, which in turn made a totally mechanical or nontelic description of the universe possible. As Simon (1970) notes

> Mechanism is nonteleological and antiteleological precisely insofar as it remains faithful to the great Cartesian ideal of understanding by motion nothing else than what geometricians do when they say that a line is generated by the motion of a point. (p. 75)

The assumption made by the Newtonians was that their empirical measurements and the mathematics at the core of such description track an ongoing course of efficient causation as motion, which was interpreted as the observation of a body traversing a certain distance over a fixed unit of time (Wightman, 1951, pp. 273–274). 'Time and motion' were thus inextricably united in traditional science. Although initially concepts of 'absolute' time and motion were entertained, it eventually became clear that the standard against which the mathematician applies his measure of 'the' rate of movement is arbitrary. To speak of absolute time as 'passing' in some ethereal realm is to voice an unwarranted assumption. There is no absolute time *or* motion. About as close as we

[2] Local motion is also called *local causation* in science. The main idea being advanced here is that the motion of an object is under a continuing line of efficient causation—essentially being pushed along by a factor "in the vicinity."

can get to a 'fixed' frame of reference against which to assess motion in time's passage is the fixed (albeit expanding) *pattern* of our galaxy (Frank, 1957, p. 143). But if this is so, then time and motion have a formal-cause component (patterned galaxy) without which they make no sense.

We can see a related patterning issue in Galileo's unfortunate clash with the clergy of the Inquisition. The churchmen realized that a heliocentric theory of the universe was mathematically possible to conceptualize. They viewed such "mathematical suppositions" as intellectual exercises, suitable for discussion among scholars but not representative of the 'reality' of the heavens (we are back to Aristotle's 'nature' again). The issue here was primarily on the side of formal causation, and Galileo was asked, then forced to relent on what was reality and what was a mathematical demonstration of man's imagination. At quite another level, however, we can view this incident as a dramatic confrontation of teleological theory (i.e., religious dogma) with the then emerging, nontelic, mechanistic theory of 'natural' science. This and other such incidents during the Inquisition made a lasting impression on the scientists of the seventeenth and eighteenth centuries. There was even greater incentive now to avoid those theorists who would profess to believe in a presence (identity) or spiritual force (entelechy) that moved within events according to some prearranged design or plan. Teleology (final causation) abruptly dropped from view in scientific explanation.

Newtonian physics, which took root from Galilean and Cartesian precedents, looked to the underlying material and especially efficient causes which moved events along. This is what *reductionism* actually comes down to—winnowing a scientific account down to one or two causes rather than using all four. But note: In framing his laws Newton relied on the spatially generated arguments of mathematics. Mathematics does *not* rely upon efficient causation. It is, as we have noted above, a strictly formal-cause discipline, and, if we take into consideration the shifting grounds on which the mathematician reasons, there is a flavor of final causation in mathematical activity as well. In his construct of gravity, which he introduced in the context of bodies moving in fixed relations to other bodies, Newton dealt mathematically with a force of attraction which he could not literally see, much less picture in imagination. Though Newtonian science presumed to study the underlying efficient-causes (locomotion, local causes) of events, the truth of the matter was quite otherwise. Newton once frankly admitted that he did not know "the cause of gravity" (Wightman, 1951, p. 48). He meant, of course, a cause in the efficient-cause sense. In time, the theory of ether was proposed as a medium through which quasi-efficient causes might be said to take place.

Another historical development accounting for why the efficient cause was taken as the 'basic' (if not the 'only') cause in scientific description stemmed from the use of experimental manipulations in conducting research. William Gilbert is often credited with fathering the scientific method of "control and prediction" in laboratory experimentation (Zilsel, 1957). Rather than relying on the philosophical proofs of reason or common sense (procedural evidence), Gilbert's scientific method asked that scientists submit their ideas to empirical test through observed manipulations of antecedents which bore a measurable relation to consequents that might be repeated, hence 'predicted' (validating evidence; Rychlak, 1981a, pp. 74–79). In combination with the Baconian philosophy of 'acting' on nature to secure empirical knowledge, this experimental attitude promoted an efficient-cause bias. Newtonian scientists believed they were tapping into the underlying motions of objective reality, as so-called 'laws' of nature, and that the only thing which prevented them from achieving perfect prediction in the mechanical (mathematized) progression of observed events was their inability to get all of the objective facts aligned at one time. As Laplace was to frame it at the outset of the nineteenth century, "a superhuman intelligence acquainted with the position and motions of the atoms at any moment could predict the whole course of future events" (Burtt, 1955, p. 96).

By the close of the last century, Ernst Mach was denying that such things as atoms existed (Bradley, 1971, p. 13). The smallest units of (uncuttable) reality (material cause) that moved about and formed into 'things' through motion (efficient cause) were being brought into question at precisely that time in history when psychology as a science was "being born." As historical luck would have it, our founding fathers placed their bets on Newtonian mechanism rather than on the relativistic explanations to grow out of the line of thought from Mach to Einstein, Bohr, Heisenberg, and many others, leading to the ironic situation today of a physics highly psychologized by a psychology inadequate to subsume the psychology of physical explanation—to explain the behavior of physicists!

That is, what we witness in the rise of modern (or so-called new) physics is the gradual replacement of material- and especially efficient-cause reductionism by an acceptance of the *formal* cause as basic to all scientific description. It is not motion *per se*, but the pattern of motion which matters. Indeed, as noted above, what we know as motion is only a judgment of the changing pattern of discernible items in a field of relationship. The dream harbored by the Newtonians of someday tracing events in uniform steps of efficient causation has long since been abandoned. Units of measurement vary, depending upon location of the observer in a field. It is even possible to think of such traditional efficient-cause movements of passing through past, present, and future as existing 'all at once,' as a mosaic pattern with distinctive features known by these names (de Broglie, 1949, p. 114). What one experiences as past, present, and future is dependent upon the slice of space–time within which an observer frames his or her assumption about that which is being 'observed.' Modern scientists no longer even think of themselves as observers of a passing series of events. They recognize that their role is always that of a participator in what is being 'seen' empirically. They are an agent in what is being formulated and known. They are creators of their observations.

Bohr's (1934) complementarity principle justified the fact that separate and distinct experiments can lead to mutually exclusive findings yet retain validity. Light can be shown to have wave or particle properties, given only that we perform one or another type of experiment. Einstein's celebrated "thought experiments" challenged the Gilbert formulation for, although his speculations were indeed put to validation over the years, he was not dealing in efficient causation (manipulation) when he presented his arguments (Kondo, 1969, p. 38). Once again, mathematical argumentation of a formal-cause variety was being substituted for the reductive aspirations of those scientists who hoped someday to put their hand on the 'basic' (efficiently caused) laws which ran the universe according to Laplace. The special and general theories of relativity were also important influences on the *theories* of physics (considered independently from methodological validation). Though Einstein retained motion as a fundamental construct in his theorizing, it is clear that he did not think of this as an efficient-cause conception. He rejected Newtonian gravity altogether.

It was in the subatomic realm that some of the clearest indications of the primacy of formal-cause description were to be seen in physics. First of all, it was becoming apparent to the physicists early in this century that they could not hope to 'picture' the world through mechanisms of the sort which Newtonianism had proposed (Cassirer, 1950, p. 110). Subatomic particles are not 'things' in the palpable sense of a material cause. Nor do their motions follow the unilinear course implied by the tracking of an efficiently caused sequence of events. Subatomic particles make jerky movements, jumping across units of uniform, linear measurement. Indeed, efficient causality breaks down completely in the realm of the subatomic, where physicists are prone to speak of particles as "complex structures" (Bohm, 1957, p. 33) that change in patterned organization without discernible efficient-cause sequences intervening.

Thus, when an electron changes orbits (or shells) around the nucleus of an atom, there is no way *in principle* to observe the electron 'moving over' from one orbit (shell) to the next in efficient-cause fashion. All we can observe is the changing *pattern* of what Bohr (1934) called a "given stationary state" (p. 108) of the atom from one organization (state, pattern) to the next (state, pattern). Each of these states (patterns) was to be taken as an "individual process," with its unique probabilities, and the change from one stationary state to another is not and never will be amenable to "detailed description" (p. 109) in the sense of tracking the electron's path from one orbit to the next as the succeeding patterns make their appearance. Having now taken a formal-cause theoretical stance, Bohr startles us by going a step further in using the final-cause meaning as well. He addresses the fact that efficient causation will never be applicable to subatomic description and then adds: "We are here so far removed from a[n efficient-] causal description that an atom in a stationary state may in general even be said to possess a free choice between various possible transitions to other stationary states" (p. 109).

The Heisenberg principle, which states that we cannot determine the position and momentum of subatomic particles in the same experimental procedure, underscored two aspects of the altered view of science which was underway early in this century (Feuer, 1974, p. 176). First, experimentation is an aid to knowledge, but it does not answer questions by tracing an independent Newtonian reality lying substrate to all that can be seen in the macroscopic world of everyday events. Second, knowledge is never entirely free of presumptions or standards of comparison, so that what we can know is intimately related to what we can or do 'begin with' in the relational sense of given A we can establish B. The latter (B) cannot stand independent of the former (A), for knowledge is relational, framed within a network of interlacing or interacting *patterns*.

Subatomic particles move about 'in a relationship with' or 'in relation to' the instruments we use to observe and measure them, the mathematical assumptions we frame in laying out our experiments, and the actions of events in other regions of energy fields which somehow enter into the "decisions" (Bohr) made by the atom to rearrange its particulate organization. There are proven examples of subatomic particles in one area of the field somehow influencing particles in another area which are far beyond the influence of local (i.e., efficient) causes (Schrödinger, 1935). It is as if one particle 'knows' what the other particle is manipulated to do (e.g., moving in a certain way) and counters this action by doing the opposite (i.e., moving in an opposite direction) without manipulation. Once again, the only way in which we can understand such phenomena—mathematically or however clumsily in a conceptual (i.e., picturable) sense—is to think of them in terms of a pattern, field, or similar organization of changing or interacting relationships. There can be no doubt: In the history of science, formal causation has emerged as the basic cause after all! Einstein's famous formula, $E = Mc^2$, demonstrates most clearly how energy (efficient causation) and matter (material causation) are two ways of expressing the *same patterned totality* (formal causation) in relation to a grounding standard, that is, the speed of light[3] (see Kuhn, 1977, pp. 26–27).

Probably the most important recent development in physics is the theorem advanced in 1964 by the physicist J. C. Bell (Zukav, 1979, p. 282). We often hear it said that subatomic theories have no relevance for the everyday, macroscopic world that we live in. Mechanistic theorists like to point out that "In the realm of observed behavior, Newtonian mechanics are accurate, so why do we need concern ourselves with subatomic conceptualizations?" Though it is true that Newtonian principles 'work' in the macroscopic realm, it is no longer true that there are no implications for this realm in the theories of subatomic physics. Bell's theorem focuses on the patterned interrelationships to be found in subatomic events, extending this—what we have now argued is a formal-cause organization of mutually interdependent, interacting events—to the world at large. As Zukav (1979) summed it up, Bell's mathematical theorem

> projects the "irrational" aspects of subatomic phenomena squarely into the macroscopic domain. It says that not only do events in the realm of the very small behave in ways which are utterly different from our commonsense view of the world, but also that events in the world at large, the world of freeways and sports cars, behave in ways which are utterly different from our commonsense view of them. (p. 290)

3. Psychology and the New Physics

When John Watson (1924) asked that we think of the person as "an assembled organic machine ready to run" (p. 216), he voiced a New-

[3] The speed of light as a standard brings a final-cause meaning to the Einsteinian formula, for it is this fixed measurement rate as a 'that' (formal cause) which enables the mathematician to tautologize E and M. In traditional terms, 'for the sake of' (final cause) a grounding point of reference, we can see how mass is energy and energy is mass. Note that in order to speak this way we must take an introspective perspective. Rather than thinking of $E = Mc^2$ extraspectively, as being written on a blackboard, we must think of the mathematician's reasoning processes as he or she literally grasps that two concepts can be proven interchangeable given only that we have a fixed grounding within or against which to frame the relativity.

tonian aspiration. At about the same time (circa 1920) Edward Lee Thorndike wrote a three-act play (*The Miracle*) in which he essentially contrasted the "miracle" of human knowledge and skill with the other-world pretensions of the religionists (Jonçich, 1968, p. 63). The fundamental clash in this play between Thorndike's mechanistic conception of science and his disdain for such "magical potencies" (p. 590) as will power and reason in human behavior dramatizes the prevailing attitude we have seen reflected in psychological science over this century. To propose teleological conceptions of behavior is to violate scientific description in favor of spiritualism. This attitude trails back well into the nineteenth century, as in the 1845 pact formed by Ludwig, du Bois-Reymond, Brücke, and Helmholtz to "fight vitalism" in their scientific careers (Boring, 1950, p. 708). Vitalistic theory as a spiritualistic formulation of Galenic medicine is indeed worthy of opposition within empirical science. But the tradition that Helmholtz fostered in the views of the young Wundt, however confusingly (Blumenthal, 1979), and Brücke in the young Freud, also confusingly (Rychlak, 1981c), was that a true scientist opposes telic description in *any* form.

And so it has happened in psychology that a mixture of Newtonian science with antireligious sentiments has resulted in an active suppression of telic description, even when such description is aimed at the human being and not at a presumed deity. Though telic theory may indeed lend encouragement to deistic world views, we must never forget that an antideist like Nietzsche was also a committed teleologist in his view of human behavior. Teleology and theology are separate endeavors, and, in any case, it is not for the psychological scientist to decide what will or will not be advanced as a description of human behavior based upon theological considerations. Our legitimate province as scientists is the realm of validating evidence (Rychlak, 1981a, p. 77). Deities may be postulated by people correctly or incorrectly depending on what we accept as evidence for such views, but it must be of fundamental interest to all psychologists that it takes a certain kind of 'animal' to make such otherworldly postulations. Can we really understand this evaluating, moralizing animal on a totally mechanistic theory of behavior? For those of us who think we cannot do so, it is clearly *not* properly objective to reject our efforts to use a final-cause explanation as somehow in violation of basic scientific rules. Science advances on *rules of evidence* and not on rules of theoretical usage. Surely Galileo's classic encounter with the churchmen of the Inquisition has taught us this lesson.

The religious issue may, one hopes, now be set aside as an irrelevancy in the debate over telic or nontelic psychological theorizing. Returning to the question of what makes good science, however, we can now appreciate how in the rise of the 'new' physics scientists were free to develop a line of (mathematical) theorizing completely devoid of material- and efficient-cause considerations (reductions). Through his daring thought experiments, Einstein not only departed from the Newtonian ideal of tracking an independently 'functioning' series of natural laws moving in efficient-cause regularity over time, but he actually proposed a conception of reality which was to that point in time not only unseen but unthinkable. This is an aspect of the revolution in physics for which physicists need have no concern but which psychologists can hardly ignore—the fact that sheer abstract thinking could somehow frame a reality not yet experienced (see Einstein's own thoughts on the matter: 1934, p. 18). In psychology we have traditionally assumed that what is 'in' mind has been put there through stimulus inputs of one sort or another. Yet again and again in the rise of modern physics we see the human being construing new realities. Whereas in psychology the use of theory has been limited by Newtonian decree (avoid the hypothetical!) to the tracking of observables, in quantum physics theory has been used to generate unobservables which bring about what can then be observed.

Quantum theory totally invalidates the view of psychologists who believe that there are S–R and R–R laws in operation (Spence, 1956, pp. 16–17). This distinction rested squarely on the Newtonian assumption that formal-cause patterns (R–R laws) were reducible to underlying efficient-cause motions (S–R laws) in the tradition of Cartesian geometry (see above). This is why Spence could confidently suggest that: "These R–R laws represent only one small segment of the total framework of a science of behavior, and unfortunately not a very *basic* [italics added] one at that" (p. 9). As we have noted above, when we get to the fundamentals of subatomic reality, correlational laws of a relativistic cast are—fortunately or unfortunately—the *most basic conceptions* we have to work with. Indeed, the concept of fundamental lawfulness in which antecedents determine consequents across time in uniform steps—as demanded by S–R theory—is no longer theoretically defensible (see Slife, 1981).

Subatomic events are not motored by continuous gradations of energy (force) expenditure, "running down" as does a clock or a related "organic machine" (Watson, 1924) performing as a system. Energy is emitted in discontinuous spurts, and when subatomic particles interact the 'cause' is less a push or shove in impetus fashion than a *mutual* pattern of withdrawal or repulsion (Popper & Eccles, 1977, p. 37). When these particles collide they also create or 'alter into' new forms of mass/energy. And, most amazing of all, subatomic particles behave as if they were agents, making decisions to behave independently of experimental intervention. Of course, when Bohr spoke of the free choice of an atom to select its steady-state condition, he was framing, however seriously, an extraspective teleology. Recall that we have said (above) that experts in teleology differ as to whether intentionality can really be observed in behavior. We have suggested that a telic account must at some point be framed introspectively, because final causes 'exist' in the premises of the agent and hence to understand intentional behavior we must have some way of assessing such premising assumptions, hypotheses, points of view, paradigms, and so on. Even so, it is implied in Bohr's very willingness to use telic language that he believed human behavior *is* telic (reflected in comments on p. 100 of Bohr, 1934). He was analogizing to a human capacity which he took seriously. Yet at that point in history (the 1920s) a psychologist who would have advanced a theory of free choice in human behavior would have been accused of putting forth anthropomorphic claptrap.

There are those who believe that the current trend to so-called cognitive theory has moved psychology along significantly beyond what Watson and Thorndike had contended. This belief is hard to accept when we look more closely at what is involved in such cognitive theorizing (Neisser, 1966; Weimer & Palermo, 1974). Basically, all cognitive theories are mediational formulations; they focus on the presumed factors which 'come between' what has always been thought of as the stimulus-input and the eventual response-output of a behaving organism. Mediation theory is therefore an effort to frame processes 'within' the flow of efficient causation. Because cognitive theories use concepts of *encoding* and *feedback*, the impression is given that there is something more active going on in the organism than earlier mediational models—such as those of Tolman (1967) or Hull (1943)—were able to capture. When we look at these modern cognitive theories in terms of causal conceptions, it is clear that there is no fundamental difference in the style of explanation compared to the earlier mediational conceptions.

Mediation models do place emphasis on formal causation in the sense of a 'cognitive map' or 'program' which supposedly intervenes between the inputs and outputs. Modern mediation models, which draw from cybernetics or information-processing conceptions (Wiener, 1954), add a concept of feedback which beguiles psychologists into believing that they have accounted for such higher-level reasoning capacities as choice, self-direction, or goal attainment in using such a construct in their explanations of behavior. Feedback occurs when some of the (efficiently caused) output returns as input (positive feedback), or when after a definite 'target' has been set and the mechanism (e.g., a missile) is drifting off course in moving toward that target is signaled to realign its course (negative feedback).

The feedback conception has been stretched far beyond its technical meaning in psychology, to where it presently subsumes literally any form of new information received by one person, in interpersonal contact with another person. But even more misleading is the fact that psychologists seem to believe that mediating feedback mechanisms can account for transcendent or self-reflexive reasoning processes thanks to the reciprocity they make possible between the mechanism and its environment (e.g., Bandura, 1978, p. 345). Such reflective "thoughts about thoughts" (Bandura, 1979, p. 439) are presumed to run off as mechanisms, albeit dynamically interacting mechanisms which perform as "mediated transactions with the environment" (Bandura, 1978, p. 348).

It should be clear from our review of science that such theories of mediational information 'stored' in a person's cognitive processes (or biological substances), and reciprocally interacting with ever-arising environmental stimuli, must fall short of the kind of reasoning we witness the physicists of this century employing. This is the sort of reasoning which the computer scientist Weizenbaum (1976) has called "instrumental" rather than being an example of "authentic human rationality" (p. 253). It is predicated on the unilinear conception of time (p. 204), with graded levels of interactions rather than abrupt and irregular occurrences; and it completely underestimates the creative role of human reason in framing knowledge (p. 222). Einstein did not simply shift "thoughts about thoughts" around in a newly generated interaction. He repremised *all* thoughts *at once,* shifting their meanings in a breathless, creative reconceptualization. All mediational conceptions come down to explanations of behavior in terms of yesterday's push acting as today's shove. The person is construed as acting within a succession of efficient-cause units of time. The description is totally extraspective. The dynamic interaction (reciprocal determinism) referred to draws primarily from these efficient-cause assumptions.

An ironic outcome of such mediational theorizing in psychology today is attribution theory. When Heider (1958) first used the term *attribution,* it seemed a neo-Kantian formulation, as a kind of ongoing tendency for the human being to structure the phenomenal field. He relied on Lewinian theory, and Marrow (1969) among others has characterized Lewin as a teleologist. However, when Heider (1958) spoke of "the linking of an event with its underlying conditions" (p. 89) his followers apparently took this to mean "reduce things to the basic efficient causation in all behavioral events." Thus, even though he put intentionality at the heart of what he called "personal causation" (p. 100), Heider's followers were to translate such theorizing into information-processing, cybernetic terminology. For example, we find Kelley (1973), a leading attribution theorist, saying "It must be emphasized that attribution theory deals *only* [italics added] with the processes by which attributions are derived from informational input" (p. 126).

Why should this be the case? Why should not input *per se* be a form of attribution, a framing assumption (formal cause) for the sake of which (final cause) the individual can understand reality (see Kant, 1952, p. 14)? Surely the concept of an attribution implies some such ascriptive, propositional, meaning-lending quality in the vein of formal/final causation. Nothing in the empirical research findings would contravene such a theoretical interpretation of 'the facts' of causal attribution. Kelley (1973) has cited Cicero and Virgil, noting that they both eulogized man's delight in knowing 'the causes' in things (p. 127). Cicero was instrumental in bringing Greek philosophy to Rome, and he personally favored Aristotle to Plato. Not only was Virgil educated in the Epicurean school—a heavily teleological philosophy—but his references to 'the causes' throughtout the *Aeneid* (Virgil, 1952) are predominantly of the final-cause, telic variety (see, e.g., p. 103, p. 111, p. 176, and p. 230). Thus, what Kelley failed to point out was that when the Roman poets referred to causation in the plural they did not mean a myriad of efficient causes. They were referring to *all four* of the Aristotelian causes in events. But, ironically, in his cybernetic formulations the modern attribution theorist does indeed employ exclusively efficient cause accounts to explicate the "between input and output" determinisms of behavior.

To think of determinism in behavior as *also* based upon formal-cause or final-cause factors is to violate the Newtonian style of explanation which such accounts continue to rely upon. Yet, as we have seen in our review of science above, the proofs which mathematicians offer are totally devoid of efficient causation. There may be physical mechanisms taking place in the cerebral cortex of the mathematician, who may also employ efficient causes in manipulating a chalk to draw his or her line of argument on a blackboard. But in the mathematics *per se* we *cannot* understand what is being proven in anything approximating an efficient-cause meaning. If the mathematician were scribbling random marks on the blackboard we would have efficient causation taking place. But we would not have a mathematical proof under demonstration. In order to understand why one use of the chalk is a proof whereas another is scribbled nonsense, we must move to an *exclusively* formal-cause analysis, with ancillary final-cause considerations in the fact that assumptions are being made in the line of mathematical development.

It is this need to go beyond material- and efficient-cause meanings that has led noted brain scientists to postulate "mind [as] . . . a distinct and *different essence* [from the brain]" (Penfield, 1975, p. 62) and/or a "self-conscious mind [which] . . . need not itself have the property of spatial extension" (Eccles, in Popper & Eccles, 1977, p. 376). Though psychologists continue to seek the mechanism of cognition and denigrate the search for a "psychic agent" (Bandura, 1979, p. 440), our fellow scientists unashamedly acknowledge that though the brain may be a cybernetic processor or computer, the mind is not reducible to such mechanisms. The emphasis on choice, purpose, and intention in the dualistic theories of Penfield (1975, p. 61) and Eccles (Popper & Eccles, 1977, p. 496) testifies to the need they see for a teleological explanation of behavior. They are not cowed by the thought of transcending the observables of palpable reality for they have learned the lessons and follow the lead of modern science. Human reasoning capacities seem to be something totally different from such 'picturable' mechanisms as a cybernetic machine. Mentation is more a matter of logical assumption, decision-making, and intentionality than anything else. And all such mentally unobservable and nonpicturable conceptions point to a teleological account of human behavior. But how are we to think of this teleology as coming about? Fortunately, the fact that formal causation has emerged as the *central cause* in modern science enables psychology to begin postulating theories of a suitably telic nature.

4. Formal Causation, Meaning, and Teleological Theory

Subatomic physics and Bell's theorem (see above) have pulled the rug from beneath those traditionalists in psychology who insist that the ideal of science is to reduce everything that takes place in human behavior—including thoughts—to a substrate reality of antecedents impelling consequents in efficient-cause fashion. This Newtonian remnant has been the bane of the teleologist's existence, even though over 25 years ago the eminent physicist Robert Oppenheimer (1956, p. 134) cautioned psychologists against taking such an outmoded view of science. If a theoretical construct is proposed and researched in which it is held that a human being *intends* to act in a certain way, the traditionalist insists on knowing the antecedents that 'caused' these consequent intentions to come about. If the teleologist then names antecedents such as reasoned plans, a style of life, preferred assumptions framed earlier, and so forth, the traditionalist is not satisfied because according to the nineteenth-century scientific assumption on which he or she proceeds, a plan (formal cause) or preference (final-cause choice) cannot be the cause of anything. Hence the teleologist is put in the unreasonable position of having to conform to a norm in which neither he or she nor scientists in other disciplines believe (Rychlak, 1980).

A psychologist who is paying attention to developments in modern science must surely know that there *is no* efficient-cause substrate to reality. It follows as a theoretical necessity that there *is no* efficient-cause substrate to behavior! What, then, are we left with to investigate as psychological scientists? I believe that psychology's future is immensely important in the family of the sciences. Though others see the intimate relationship between, for example, physics and psychology (Zukav, 1979, p. 31), we psychologists seem incapable of framing a view of the person which will cement and enrich this necessary union. For what the non-psychologist wants from psychology is a teleological presentation of human behavior, a way of being more in consonance with the spectacular intellectual machinations of the theoretical physicist. Put in other terms, it is not 'science' which is our point of focus as psychologists, but 'the scientist'.

Doubtless it is no accident that the formal cause has emerged as the most significant causal meaning in science. As our discussion of mathematics (refer above) suggests, the formal cause is readily cast in either an extraspective or introspective theoretical perspective. The material and efficient causes are essentially extraspective formulations. We cannot address their influence from the point of view of the palpable substance or 'blind' motion which is their essence. And, as we also have noted above, though there are extraspective theories relying on final causation, the clearest telic usage occurs when we "look through the conceptual eyes" of the intending identity (who frames the 'that, for the sake of which'). But the formal cause has this marvelous facility of, for example, describing logicomathematical steps which can be sketched upon a blackboard—standing alone, in extraspective, third-person fashion—or conceived of introspectively as the actual steps in reasoning employed by the mathematician. The formal cause is also a necessary (but not sufficient) aspect of final causation, for it represents the 'that' (reason, purpose, etc.) in the final-cause definition. The football coach sketches a play or the 'that' (formal cause) on the strategy board, and his players, behaving for its sake (final cause) enact the strategy in the athletic contest to follow. This great versatility of formal causation undoubtedly makes it the central concept that it is.

The mistake of the Newtonians was to assume that the motion postulated in Cartesian mathematics *created* the geometric form being traced. Since the moving point defined the straight line or circle, it seemed self-evident that it (efficiently) 'caused' the form to come about. Euclidian conceptions, defined in terms of *unmoving* points and lines, did not convey this suggestion. Motion is readily conceptualized in efficient-cause terms, if we forget about the problems of having to delineate 'which' motion is taking place from among the many possible, depending upon where we place our coordinates of measurement, and the lack of an absolute time frame within which to 'clock' this motion. When we do take such factors into consideration, motion 'reduces' to relative patternings of displacement tracked according to an arbitrary scale framed by an assumption-taking human intelligence. What Mach and then Einstein recognized was that Descartes first knew the geometric *form* he was striving to define, that he had fixed it mentally in mathematical space, and then brought it into overt motion as an aspect of his definition. Rather than motion creating pattern, pattern was basic to and 'shaped' or created motion! Efficient causation thus becomes an *instrumental action,* reducible to formal/final causation. Motion is a way of tracing the patterns that nature presents to us, or, phrased another way, that we as telic organisms first presume (fix) and then bring forward into reality when we select standards against which to trace the patterns we know presumptively beforehand.

In human affairs, patterns always take on meaning. The word *meaning* derives from the Anglo-Saxon roots of "to wish" and "to intend." Meanings point to that to which they make reference, or form a *pattern of relationship* with ('relate to,' etc.). Thus, when we say that something like a word or a sensory impression of seeing or smelling 'has meaning' it would only have this quality because of what it pointed to relationally. In a final-cause sequence, we would view the pattern of meaning as being the 'that' for the sake of which behavior is intended. When the football players grasp their coach's strategized play they understand its meaning, and having this understanding, they enact the end being sought with resultant success or frustration in the contest to follow.

Even though meaning relations are probably never this singular, it is possible to think of just two poles or referents tying together in the relationship. For example, *bread* can be thought of as one pole of a meaning relationship, tying into *butter* as a second pole. Bread can also relate to other words like *milk* and *oven,* so the relational patterning may be very complex. A recipe for bread printed in a cookbook is framed extraspectively, much as the mathematician's derivation on the blackboard. We can also see introspective relational patterns in human reason, as when a young man is thinking about how fresh bread (pole) reminds him of his mother (pole), who used to bake bread on special family occasions. The question which arose quite early in psychology is: Why do certain words (ideas, thoughts, etc.) become patterned into meaningful relations with other words (ideas, thoughts, etc.)?

Psychology's answer to this question followed British associationism, which in turn analogized to the Newtonian mathematical conceptions of gravity. It was therefore assumed that the closer two idea-inputs (poles) occur together 'in mind' and the more associatively 'large' these congeries of ideas might be, the *stronger* is the resultant associative bond of meaning between them. Largeness of an associative mass was held to occur through repetitive experiences with the idea in question. Hence, the two principles of explanation on which associationistic psychology has always rested are *contiguity* and *frequency*. In fact, all theories in psychology today come down to the frequency of past contact (contiguity) with certain behaviors which have supposedly been stored in memory, combined with related behaviors, and then retrieved in the present (with reconstructions included, etc.). The underlying fluidity (motion) of this conception can be traced directly to the Newtonian/ Laplacian view of scientific description. The concept of 'shaping' behavior follows from this erroneous notion that motions (efficient causes) pattern (formal causes) behavioral habits.

But now, assume that we let the lessons of our review of science influence us at this point. Can we see patterns in the behavioral and/or mental associations of people? Surely S–R habits form into discernible stylistic patterns of behavior. Subjects in paired-associates learning tasks are notorious for the so-called mediators they concoct in order to facilitate recall of the 'associated' CVC trigrams. Recent research in the stimulation of the cerebral cortex suggests that even the most rudimentary "raw feel" type of experience requires 0.5 seconds of patterning among the neurons of the cortex before it is experienced consciously as a 'sensation' (Libet's work, as cited in Popper & Eccles, 1977, p. 259). Mountcastle (1975) has found that cortical neuronal action in the motor areas brings about movements in a holistic (patterned) manner. Without doubt, the predominant view today is that neuronal activity in the cortex does not create thought but rather reflects the patternings of thoughts (Popper & Eccles, 1977, pp. 361–362). If we now draw a parallel to Cartesian geometry, we might suggest that associations in the human experience are like the 'motions' of the points defining a figure. That is, associations cannot arise unless they are already presaged in an underlying pattern which is 'there' presumptively to begin with.

The intellectual stance of a 'learner' is thus to 'take a position' in relation to that which can be known, presumptively affirming one from among the many possibilities in a learning situation. The infant who associates "da da" to the image of a male parent, again and again, may not be framing an *attachment* (Dollard & Miller, 1950, p. 249) between these poles over time. The child may associate "da da" to the image of a male parent only following the establishment of an interpersonal tie to (pattern with) the parent, and in addition, after there is an emerging 'intuition' (vaguely delimited pattern) of what the learning task entails (i.e., forming a tie between father as one pole and "da da" as the other pole of the meaning-relation). We know that infants can be encouraged ('conditioned') into emitting *only* those sounds which they have already verbalized spontaneously (Mussen, 1979, p. 32). If this is true, then infants may not be 'receiving' efficient-cause manipulations in their verbal conditioning but rather may be initiating the process by bringing already *known* patterns of sound ("da da") into further patterned relations with experience in a more formal/final-cause sense. Infants may extend what they already know to relate to that which they are being asked to know—in this case, that "da da" and a certain person's visage 'go together.' The traditionalist assumes that the frequency/contiguity factors in the situation create the patterned habit of speech, but this Newtonian precedent is not the only position to be taken. It may be that human mentation involves placing oneself as an identity in relation to the passing scenes of life, the multiple possibilities of experience, lending it order based on such fixed groundings—groundings that 'work' by lending personal meaning to lived experience.

But how is it possible for a person essentially to 'take a position' on various (patterns of) meanings in this fashion? How are we to conceptualize this reasoning process? First of all, we must, as did the physicists, relinquish all hope of characterizing this activity in quasi-mechanistic fashion, as a picturable, extraspectively framed 'mechanism' which moves events along. We must instead frame an introspective theory relying on a more abstract formulation—more in the sense of a logic, a patterning of meaning according to styles of reasoning—akin to the mathematical abstractions of the physicists. Second, we must recognize that relational ties between poles in meaning are not all the same. Some poles are unipolar designations which can stand alone but are brought into meaningful relationship with other unipolarities as when we tied the noun *bread* to the nouns *butter, milk, oven,* and so forth. But other relational ties cannot be described in terms of uniting such independent designations one into another. Some relational ties are *oppositional* in that there are bipolarities involved. In this case, one pole literally defines or shares in the implicit meaning of the other pole so that, for instance, to know 'good' the reasoning intelligence must necessarily already have an understanding of 'bad.' Such words, as poles of meaning-relations, cannot stand alone independent of one another.

Aristotle's (1952, p. 143) distinction between demonstrative and dialectical reasoning encompasses the alternative forms of meaning relations. A demonstratively framed meaning would encompass logical patternings of unipolar designations. Demonstrative reasoning presumes a one-to-one relationship between a word and the referent for which it stands. Such unipolarities can be multiplied as more and more meaningful relations are brought (denotatively and connotatively) into the complex totality of meaning. If words have an oppositional relation, such as *high–low, hot–cold,* and so forth, according to an exclusively demonstrative explanation, this occurs through the frequent bonding of unipolarities so that we come to think of them as intrinsically related even though they are not.

Dialectical reasoning presumes that some patterns of meaning are intrinsically bipolar, so that rather than conjoined they are as if pulled apart from a common core into opposite poles in which one side is essential in the definition of the other side. On this view, the meaning of *left* is not a unipolar designation which has through frequent repetition been joined to *right*. Left is only left due to its relation to right, so that in a true sense left must *also* participate in the meaning of right, and vice versa. There are many such word relations in all languages—indeed, the more ancient the language, the more dialectical it is likely to be (see Rychlak, 1976)—and, a point not to be overlooked, such oppositional relations usually take on meanings which might be considered evaluative, judgmental, and comparative. Put another way, dialectical relations in meaning are often concerned with qualitative issues in contrast to the quantitative implications of demonstrative relations.

We have been speaking of the poles of meaning as framed by words, but of course it is also the case that concepts, ideas, visual impressions, and all manner of human experiences take on the characteristics of either a demonstrative or dialectical pattern of meaning. When the prehistoric individual living in a valley looked up to view a mountain top, it was not the words he or she used for "down–up" or "here–there" but the conceptual possibility of being (living) in one region or another that was brought into question once this dialectical meaning was framed. It is implied in such a conceptualization that, having settled down in the valley, it is also possible to seek a residence up on the mountain. Not everyone living in the valley need frame such an implication. The point a dialectician would make, however, is that the primitive individual who thought of such an alternative place of residence and relocated up the mountain did *not* have to receive this suggestion through a fortuitous 'input' association, acquired in demonstrative (unipolar) fashion from the environment. It was possible for this individual to reason from within the bipolarities of dialectical meaning and affirm an alternative that was *not* associatively input!

And right here is where we find our theoretical justification for teleological theory in human description. We build our case on the two assumptions concerning human mentation already suggested: (a) Meanings as formal-cause patterns are not 'made up' by underlying motions that shape them through frequent, contiguous repetitions of antecedents impelling consequents in efficient-cause fashion; and (b) Some of these irreducible ('basic') patternings are bipolar so that in order to enact the meaning implied by one pole or the other the individual must serve as an agent, taking a position on or affirming either of the two poles—or an intermediate position between these poles—which present themselves as alternatives in the 'one' pattern. Indeed, the 'one' pattern often serves as a pole of meaning against which an oppositional pattern of complex alternatives can be framed. We are in essence speaking always of multipolarity in the machinations of dialectical reasoning. And the observed motions made by people in their overt behavior are the resultant *instrumentalities* fashioned by the ends (reasons, purposes) of the patterned meanings which are being actively brought forward by the agency or executive-reasoning capacities of the people involved.

This sense of agency and self-determination is experienced by everyone in daily living. Indeed, the dialectical capacity for self-reflexive thought and transcendence—turning back on ourselves as 'opposite identities' to know that we are a factor (agent) in what will or will not be enacted in our life—is the source of the popular phrase 'free will' (Rychlak, 1979). This dialectical capacity to question, challenge, and thereby alter the grounding premise (plan, reason, i.e., formal-cause pattern) for the sake of which we are behaving (final-cause enactment) is what the average person means by *free will*. Psychologists who investigate free will are prone to frame it in positive terms, as, for example, combining intelligence and efficacy or 'imagination' which leads to successful attainment of constructive goals (see Easterbrook, 1978, pp. 25, 69, 73).[4] But there is no need for such a one-sided interpretation of free will. Even less intelligent and ineffectual individuals can be seen to behave with agency, intending and bringing forward purposes of dubious quality. And highly intelligent, successful people can also intend to do themselves or others harm.

[4] It would appear that the centuries of theological usages of the free-will conception have biased psychology's attitude here. It is assumed that a freely willing person should somehow attain 'good' ends because in accepting responsibility for one's actions a motivation to plan and improve one's circumstances supposedly arises.

Thus, the fact that the person lives from psychological birth within an increasingly complex network of dialectical patternings such as *self–other, me–not me, do–don't, happy–sad, like–dislike, right–wrong, good–bad, obey–disobey, anxious–relaxed,* and so forth suggests that the fundamental stance of the human condition is that of 'taking a position on' the possibilities open to one for knowing, doing, adapting, and so forth, the sum of which we refer to as *behavior*. It is never simply a question of what 'happens' to the person as life carries on. The meaning of what takes place demands that the individual place something akin to a co-ordinate of axes, a personal construction (Kelly, 1955) on events which frames them meaningfully. There is a basic relativity in all our lives, a "stationary state" (Bohr) that we freely choose to affirm albeit on grounds which we take to be compelling.[5] Euclidian *fixity* has as much relevance for psychology as does Cartesian *fluidity*. Is it conceptually tenable for psychology in its present state of development to frame and investigate a teleological explanation of behavior? What is needed, and how well can such a formulation meet the emerging facts of our discipline? We move next to the closing section of this paper, in which my efforts to establish a rigorously telic explanation of behavior will be reviewed.

5. Logical Learning Theory

In selecting the word *logical* to characterize human learning we return to the Heraclitean attribution (in the true sense of the word; refer above) of a *logos*, that is, a rationale, a patterned hence lawful order to the universe. Aristotle unquestionably drew on the Heraclitean precedent when he defined the formal cause. Heraclitus was referring to the patterns which we see in nature and which are not repeating themselves over time so much as simply 'being there' within the flux and change of passing events. The human being is thus a logician ("logos identifier") who, relying on *both* dialectical and demonstrative meanings, 'comes at' the world taking a vis-à-vis stance in relation to experience which is never completely independent of the person's phenomenal structuring. Logical learning theory is thus an introspective formulation, a *logical*

[5] It is important to appreciate that the 'freedom' in free will refers to the dialectical alteration of grounds for the sake of which one is then (psychically) determined. There is no contradiction between free will and determinism. The 'will' or 'will power' aspect refers to the fact that once a ground is taken, a premise is affirmed, the person is (finally) caused to behave in a determinate fashion. The determination is telic, not efficiently caused, but still *predictable* if we but know the grounding premises under affirmation by the organism preliminary to and in the course of behavior.

phenomenology in contrast to the traditional *sensory* phenomenologies of the gestalt psychology tradition (Rychlak, 1981b, pp. 763–765).

If the human being is confronted with a logical order, including dialectical relations that are 'illogical' based on demonstrative assumptions (where *left* never means "right," etc.), then a need to choose or affirm one or the other of such conflicting alternatives is fundamental to all knowing. That is, the human is called on to order (pattern) such inconsistencies in coming to know (learn) things. Life experience presents itself as patternings in various stages of completion (i.e., 'possibilities') and the individual takes a stance in relation to such alternative patterns *not* as efficiently caused 'effects' but as a predicating, premising 'cause' of what is to follow.[6] Logical learning theory even accepts the likelihood that people sometimes affirm *both* poles of a dialectical meaning relation—people do expect to have their cake and eat it too. To a logical learning theorist, the notorious Zen koan, "What is the sound of one hand clapping?" makes dialectical sense even as it is demonstratively illogical. But the more usual process of what we call 'rational' thought is to affirm one meaning—the true, correct, preferred, right, best, most popular, and so forth alternative—over the other(s) possible in the continually arising patterns of lived experience.

Another reflection of human dialectical reasoning is seen in what logical learning theory terms an *affective assessment*. This is taken to be a basic reflection of the human being's evaluative capacity, to align into premises linguistic and/or sensory patterns which *also* bear a positive or negative quality of meaningfulness to the reasoner concerned. Emotions are bodily discharges which occur as feelings (sensory patterns) in response to certain situations of life. There is a cognitive aspect (linguistic pattern) to emotion, of course, but the person is not rendering a psychological assessment in experiencing an emotion. The 'evaluation' in emotion is biologically connected and not exclusively psychological. Emotions are notoriously difficult to identify, and when they occur they press their biological evaluations on the person more as 'effects' than 'causes.' When we refer to affection (affective assessment) we are thinking of a purely psychological capacity to assess that which is under conceptualization, including emotions, and align a premise (opinion, bias, etc.) accordingly. It is possible for any one person to assess an emotion (e.g., sexual arousal) positively in one life context (e.g., on a

[6] There are limitations to such personal causation, of course. The human cannot 'will' himself or herself through walls. All we wish to point out here is that agency is never itself 'shaped' by experience. Agency begins with and is an essential aspect of experience. As stated in note 5 above, we are not free of the need to take a position on (premise, predicate) life.

dance floor) yet consider it affectively negative in another life context (e.g., during a religious rite). Affection is on the side of mentation (formal/final causation predominant), whereas emotion is held to be more on the side of a biological discharge (material/efficient causation predominant).

The reason affection is so important to an understanding of mentation is because it is assumed that the individual *from birth* affirms premises and extends their meanings based upon the 'choice' resulting from such estimates of worth, positive or negative qualities, and so forth, of what is being learned. Recent surveys of the literature in infant learning (see Sameroff & Cavanagh, 1979) do not contradict the view that infants exercise considerable influence over what they will 'learn' or 'be conditioned to.' An interest factor along the lines of affective assessment is readily adapted to the observed facts, for infants do appear to show preferences and are not simply 'associating' responses to contingent stimuli. But how is it possible for infants to be described as affectively assessing and thereby premising (cognizing, 'encoding') their existence preferentially? Is not the child too primitively organized to conduct such a complex task as rendering an assessment? Mediation theory would have such abilities 'shaped into' the child thanks to environmental conditionings or cybernetic 'inputtings' (encodings) of various types. Logical learning theory, which holds that there is no such thing as a shaping or inputting/encoding which is free of affective assessment, presents at this point a basic theoretical term which rests upon the meaning of final causation. We have never had a final-cause construct in psychological learning theories, wherein the stimulus–response and its first-generation offspring, the input–output conception, have held sway in the best traditions of efficient causation.

The term proposed by logical learning theorists is *telosponse,* defined as the affirmation of a meaningful premise (e.g., as a visual image, language term, statement, or judgmental comparison) relating to a referent (point, end, goal, reason, etc.) that acts as a purpose for the sake of which behavior is then intended (performed, enacted, etc.) (Rychlak, 1977, p. 283). The term *purpose* is restricted to the 'meaning of a concept' and *intention* refers to the fact that an organism is behaving 'for the sake of' such meanings. Phrased in causal terminology, purpose focuses on formal causation whereas intention gets at final causation. Thus a pencil is a practical tool devised by humans for their use. The pencil serves a purpose (as a concept). But the pencil *qua* pencil knows no purpose. It is the human being, acting telosponsively, who behaves for the sake of this purposive meaning and thereby intends it to come about. Intentions can be of two types. In the case of an *action intention* the person picks up the pencil because he or she has a note to write; however, the person may frame an *understanding intention* by simply noting that the pencil is at hand for possible use if a need should arise (pp. 283–284). As an introspective theoretical formulation, logical learning theory does not assume that learning is always observable in the literal sense of overt movements. As Bandura (1978) has amply demonstrated, people can learn by observing others, and they can enact such learning or not depending upon circumstances as *they* view and affectively assess them.

Since observed motions are held to be instrumentalities, carried out as action intentions (effects) rather than as responses or outputs, logical learning theory has jettisoned psychology's traditional reliance on frequency/contiguity principles. At least, such principles are discounted in an extraspective sense. From the introspective perspective, contiguity is viewed as a reflection of the importance of patterning (logos) to understanding. Obviously, patterning into form is more readily possible if events occur close together. It is difficult for the learner to get the 'connection' between two events if several hours separate them. A puzzle 'falls into place' more readily if we have an overview of the parts all at once, in context. In this sense, each learning is akin to Bohr's individualized steady state. We confront the task 'as it is' in its present (semi-)form. And as for frequency, practice makes perfect because with each 'trial' the person is mentally enriching the 'routine' being enacted, and perfecting thereby the overt movements within the pattern. But just as the Cartesian motion does not create the pattern, motion does not create the habitual routine. People continually shift their premises in learning tasks, improving on them, making errors, trying a new strategy, and so forth. Indeed, the richer the understanding, the less likely it is that a learner will perform it in the rote fashion which frequency implies.

In place of frequency/contiguity principles, logical learning theory relies upon the principle of tautology to explain how premises are brought forward as understanding and action intentions. The tautology is itself a *relation of identity*, but this formal-cause conception can be construed in either an extraspective or an introspective manner. Thus, framed in the more familiar extraspective manner we would have '*A* is *A*' or '*A* = *A*.' It is this interpretation of tautology which bears the onus of being "redundant information." However, a tautology can be framed introspectively in the sense of 'If *A* then *A*.' By thinking of the first *A* in the latter formulation as a premised meaning (pattern), and the second *A* as the extension of this meaning tautologically into the person's ongoing thoughts, behaviors, and so forth, we can see how telosponsivity is made possible. For example, an individual may premise, "The door is over there and I want to leave the room" (symbolized by the first *A*) and extend this meaning tautologically by leaving the room (symbolized by the second *A*). Moving feet and a door opening may track the 'If *A* then *A*' course of events, but such instrumentalities do *not* explain the pattern of behavior anymore than they did in the Cartesian mathematical space. To explain the behavior we must know the framing premises for the sake of which the person enacts the motions observed. There may be reflexive (totally automatic, formal/final causal) aspects to motor behavior, but the essence of psychological activity is 'bringing to bear' rather than 'responding'.

If learning involves tautologizing 'redundancies', does this mean that there is no creative innovation possible? Actually, logical learning theory holds that we must ground new knowledge in old knowledge, because just as we noted in physics, a 'set' ground must be in place in order for the telic intelligence to have a 'that' for the sake of which meaning-extensions are carried forward as inductions and deductions.[7] But it so happens that the analogy is a limited form of tautology in which there is a relation of *partial* identity. Analogy brings a dialectical note into the patternings of tautologies, for now we can speak of the analogy and the disanalogy (see Oppenheimer, 1956). It is through analogical reasoning or, as Royce (1964) has noted, through metaphorical reasoning that highly intuitive cognitive processes are made possible. Here, then, is the source of creativity in human reason. Bertrand Russell (1919) has noted how the tautological principle is fundamental to creative reasoning in mathematics (pp. 204–205). It is fascinating to observe in Einstein's theoretical development a definite series of tautological extensions, identifying matter as energy (Kondo, 1969, p. 45), inertia as gravity (pp. 69–70), and gravity as curved space (p. 76). We therefore see no problem with creative innovation based on tautological principles of explanation.

It must be appreciated that the telosponsive process of human mentation need not be thought of as taking place 'over time.' Tautological explanation removes the necessity of relying on time's passage to account for learning and behavior. There are no antecedents thrusting, impelling, or cuing consequences along in telosponsivity. To ask for an antecedent to telosponsivity is like asking the modern physicist to 'reduce' formal causation to efficient causation. We have already seen that this is impossible. How, then, are we to speak about the 'flow' of extension in meanings being brought forward as inductions and deduc-

[7] We sometimes forget that *both* induction and deduction require a framing ground in order for them to take place. Deductive grounds are likely to be called principles or assumptions and inductive grounds are likely to be called hypotheses or generalizations. Both forms of grounds act as 'thats' for the sake of which meaning is under extension telosponsively.

tions? At this point logical learning theory relies on what is essentially a progressing or unfolding order (logos), as in the logical steps of a syllogism. The major premise (all men are mortal) combined with the further affirmation of the minor premise (this is a man) leads determinately to the conclusion (this man is mortal). The premises in this instance act as *precedents,* which when combined into a pattern of meaning *sequaciously* (i.e., in a logically necessary manner) bring about the conclusion. There is a 'flow' of meaning from precedents to conclusion, but only in a metaphorical sense. Nothing moves. No time passes.

Hence, in logical learning theory, rather than speaking of antecedents impelling or cuing consequents, we speak of precedent and sequacious meanings. A precedent meaning is thus one that goes before others in a logical order or arrangement (*sans* time considerations) to establish through tautological extension the nature of the meanings that follow it. A sequacious meaning is one that follows logical sequence, that flows from the meanings of precedents (*sans* time considerations), extending these patterns in a logically necessary (i.e., determinate) fashion according to the principle of tautology. It is possible to affirm erroneous premises and to align the premised meanings incorrectly; therefore there is no assurance that a precedent-sequacious line of reasoning is correct, or even rational, for that matter. As noted above, it is also possible to affirm and extend dialectically opposite lines of thought at the same time. Unconscious understanding intentions are assumed to arise when the 'other side' of a consciously affirmed alternative is still entertained by the person during sleep, and so forth.

It is important to stress that we are not proposing that a homunculus exists within the person, doing the affirming, deciding, assessing, and choosing for the telic organism. Mechanistic theories, which describe the person extraspectively, require such an introspectively framed homunculus to direct the apparatus 'from the inside' when they attempt to capture telic description. Logical learning theory *begins* its account from the introspective theoretical perspective and, thanks to dialectical reasoning, construes the person as a self-reflexive identity capable of directing thought from the very outset of existence. In place of Descartes' classic "I think, therefore I am," logical learning theory advances "I think, and realize through dialectical reasoning that I could be thinking otherwise; therefore *I* exist as an agent of my thought." So important is the dialectic to telic theorizing that the logical learning theorist is prepared to say: *No dialectic, no telosponsivity!* Cybernetic, information-processing machines do *not* reason dialectically. They are exclusively demonstrative 'reasoners,' taking the programmed assumptions they are given as 'primary and true' realities not open to question. Human nature is different from the machine's nature precisely because it balances a dialectical with a demonstrative reasoning capacity.

How well does logical learning theory meet the facts of brain structure and brain process? Eccles (Popper & Eccles, 1977) has proposed a detailed theory of brain function which employs experimental data and is remarkably comparable to the tenets of logical learning theory. First of all, there are several suggestions of a dialectical organization in the brain's functioning. There are basically two kinds of neurons in operation, one which forms excitatory synapses and one which forms inhibitory synapses (p. 232). The corpus callosum has fibers joining brain halves which are in a mirror-image relationship with each other (p. 241; see also Sperry, 1977). The prefrontal lobes are in a reciprocal relationship with the limbic system (Popper & Eccles, 1977, p. 349). The brain is anything but a model of cybernetic parsimony. Indeed, the fundamental organization of nerve fibers is modular, in which up to 10,000 nerve cells are locked together by mutual connectives. Each of these modules takes electrical power from its neighbor, given only the chance to be activated. Thus, Eccles opines: "We think the nervous system always works by conflict—in this case by conflict between each module and the adjacent modules" (p. 243). Finally, Eccles suggests that there is a two-way communication between certain modules of the brain and the self-conscious mind (p. 285). Thus, he theorizes in a telic vein as follows: "The self-conscious mind is always as it were working backwards and forwards, and we could even say that in all of its perceptual processes it is moulding or modifying the modular activities in the brain in order to get back from them what it wants [memories, etc.]" (p. 514).

Penfield's (1975) dualistic theory is equally compatible with the tenets of logical learning theory, holding that there are two brain mechanisms, a higher and a lower. The highest brain mechanism has direct contact with the temporal lobes and the prefrontal areas of the cerebral cortex. These areas evolved more recently than the older motor and sensory areas of the diencephalon. It is this older cortex which has a cybernetic, computer-like quality about its functioning. This is where information gleaned from past life is most probably stored. But the interpretation given to such stored information as knowledge is framed by the higher brain mechanism, which is directed by a totally different energy source—the mind! The mind acts independently of the brain in the same way that a human programmer acts independently of the computer he or she uses to organize data and extract information from. The mind has no memory, relying instead on the computerlike brain. Summing up, Penfield (1975) says: "A man's mind, one might say, is the person [note: this is *not* a homunculus!]. He walks about the world, depending always upon his private computer, which he programs continuously to suit his ever-changing purposes and interest" (p.61).

There are some remarkable examples of dialectical reasoning and understanding intentions in the clinical reports of Penfield, in which he employed the method of electrical cortical stimulation. In one case, a young South African patient lying on the operating table in Montreal, Canada, reported that he was *also* at that very moment laughing with his cousins on a farm in South Africa (Penfield, 1975, p. 55). He was conscious of being in two places at once. In other instances, after Penfield had caused a patient to move his hand through electrical stimulation of a motor area he was told by the patient: "I didn't do that. You did." Indeed, the patient may reach over with his other hand and *oppose* the involuntary movement. Finally, when patients are made to vocalize by stimulating their speech center they are likely to say afterward: "I didn't make that sound. You [Penfield] pulled it out of me" (p. 76). All of these examples suggest a dialectical capacity to violate demonstrative unipolarities such as not being in two places at once or saying what was not actually said, or doing what was not actually done. These patients were obviously in a vis-à-vis relationship with their ongoing mentation. They were cognizant of when it was that they as agents were intending to say or do things, and when they were not. If it is possible to transcend the flow of patterned neural activity in this manner, then it is possible to redirect the course of such activity in ways that logical learning theory suggests.

We do not have to deny that mechanisms of the nervous system and physical body in general *do* exist and function. Penfield's electrical stimulation was obviously efficiently and materially causing such mechanisms to come into play. But what the evidence further suggests is that, consistent with the tenets of logical learning theory, such mechanisms are *instrumentalities* of something else—a higher brain center or self-conscious mind. Since neither Eccles nor Penfield relied upon a dialectical formulation, the resultant theories of these eminent brain scientists continue to press quasi-cybernetic terminology on the one hand and an uncertain realm of direction on the other. Eccles emphasizes that the question "where is the self-conscious mind located? is unanswerable in principle" (Popper & Eccles, 1977, p. 376). Could it be that a dialectical formulation suggesting that patterns noted in brain processes serve as possible meanings for the individual, who must *as an aspect of this very process* contribute to it by 'taking a position' on what meaning will be known, understood, extended, enacted, and so on? Would such a theoretical conception negate the dualism now required by a completely demonstrative formulation? It is not for us to propose brain theories, but surely it must be evident to the uncommitted reader that nothing currently in vogue among brain specialists would contradict the style of explanation proposed by logical learning theory.

Some mention should be made of the empirical support for logical learning theory, which has been duly put to test for roughly 20 years to date. The major thrust of this research has been to demonstrate that affective assessment—according to the theory, an *unlearned* human capacity—influences learnings of various types, and that it cannot be 'accounted for' by the principles of frequency and contiguity. When we operationalize affective assessment, we speak of the resultant measure as *reinforcement value* (thereby keeping a clear distinction between our theory and the method of proof used in its support; see Rychlak, 1977, on this point: pp. 168–172; 228–229). Subjects in experiments are asked to rate verbal or pictorial items for reinforcement value by judging them as being liked (much, slightly) or disliked (much, slightly) on two occasions (pp. 327–329). Only those experimental items (words, CVC trigrams, pictures of faces, IQ subtests, etc.) which a subject rates as reliably liked or disliked (preferably "much") on both pretesting occasions are then used in an experiment testing the role of affection in learning, IQ performance, recognition of faces, modeling performance, operant reinforcement, and so forth. Considerable research has been done to establish that affective preference is not reducible to the frequency/contiguity measures of association value, familiarity, pronounceability, and the like (see Rychlak, 1977, Chapters 9 and 10, for an overview of this research).

The weight of evidence in over 60 studies to date established that when subjects premise a task, the ambience of the experimental situation, or their own competence as a person in a *positive* way, they extend meaning more readily along the liked than the disliked course of learning. That is, they learn what they like about the materials more readily than what they dislike. In entirely methodological terms (i.e., without claiming the definitive theoretical account of the experimental observation), this has been termed the "positive reinforcement-value effect." On the other hand, if subjects premise the learning task, the ambience of the experimental situation, or their personal competence *negatively*, they extend meaning more readily along the disliked than the liked dimension and thereby either collapse the positive reinforcement-value effect into insignificance or reverse it entirely as a "negative reinforcement-value effect." These effects have been shown to play a role in personality, intelligence, interpersonal relations, projective and objective testing, and many other aspects of human behavior (for an overview, see Rychlak, 1981d). More recent work on logical learning theory has taken the form of demonstrating dialectical aspects of free recall, paired-associates learning, and impression formation.

The final point to be made concerning traditional learning explanations is that at present conditioning principles are being severely challenged by the ubiquitous necessity of subject awareness in order for either classical or operant conditioning to 'work' (see Brewer, 1974). It is a continuing puzzle to the logical learning theorist how traditionalists can go on pretending somehow to 'control' the behavior of subjects when it is 90% certain—i.e., 90% of the studies on awareness find—that the actual cause of behavior is a willingness on the part of subjects to take the premise offered by the experimenter as an experimental design and to enact it telosponsively. Of course, it can also be shown that some subjects are perverse enough dialectically to negate the intentions of this experimental design and thereby intentionally subvert the predicted outcome (see Page, 1972). Whatever conditioning *is* in human behavior it most certainly *is not* what psychology's traditional, efficient-cause explanations of behavior had claimed it was. We are in the midst of a Kuhnian revolution in learning theory, but not enough psychologists seem to realize this fact, or realizing it they are unwilling to cast off the outworn terminology even as they seem to be changing these meanings in ways more compatible with theories of agency (e.g., Bandura, 1978).

6. Conclusion

Though psychology is at a crossroads, the direction to be taken is clearly marked for those who are willing to read the signs along the way to a more distinctive science in the future. We require a teleological theory of behavior to match the developments taking place in the other sciences. We have not taken up each of these sciences in turn, such as biology, chemistry, astronomy, and so forth, but changes are occurring in the theories of these sciences similar to those we have seen taking place in modern physics. Material and efficient causation will simply not cover the explanations which are emerging. And even more important, we as psychologists cannot in our wildest dreams expect to account for the *people* who work in these sciences and concoct such theories, based on the nineteenth-century premises of a Newtonian science. We must begin to think of behavior with formal and final causation added to the description of events, employing thereby *all four* of the original Aristotelian meanings. Hopefully we will have the confidence to propose constructs which are distinctively human—that is, teleological. Rather than turning Watson's model-T machine into an IBM computing machine ready to run, let us see if it is not possible to think of human beings as telosponding organisms, ready to reason for the sake of purposive meanings in an intentional fashion.

7. References

Aristotle. *Posterior analytics* and *Topics*. In R. M. Hutchins (Ed.), *Great books of the western world* (Vol. 8). Chicago: Encyclopedia Britannica, 1952.

Bacon, F. *Advancement of learning*. In R. M. Hutchins (Ed.), *Great books of the western world* (Vol. 30). Chicago: Encyclopedia Britannica, 1952.

Bandura, A. The self system in reciprocal determinism. *American Psychologist*, 1978, *33*, 344–358.

Bandura, A. Self-referent mechanisms in social learning theory. *American Psychologist*, 1979, *34*, 439–440.

Blumenthal, A. L. The founding father we never knew. *Contemporary Psychology*, 1979, *24*, 547–550.

Bohm, D. *Causality and chance in modern physics*. Philadelphia: University of Philadelphia Press, 1957.

Bohr, N. *Atomic theory and the description of nature*. Cambridge: Cambridge University Press, 1934.

Boring, E. G. *A history of experimental psychology*. New York: Appleton-Century-Crofts, 1950.

Bradley, J. *Mach's philosophy of science*. London: Athlone Press, 1971.

Brewer, W. F. There is no convincing evidence for operant or classical conditioning in adult humans. In W. B. Weimer & D. S. Palermo (Eds.), *Cognition and the symbolic processes*. Hillsdale, N.J.: Lawrence Erlbaum, 1974.

Burtt, E. A. *The metaphysical foundations of modern physical science* (rev. ed.). Garden City, N.Y.: Doubleday, 1955.

Cassirer, E. *The problem of knowledge*. New Haven: Yale University Press, 1950.

de Broglie, L. A general survey of the scientific work of Albert Einstein. In P. Schilpp (Ed.), *Albert Einstein, philosopher scientist* (Vol. 1). New York: Harper & Row, 1949.

Dollard, J., & Miller, N. E. *Personality and psychotherapy: An analysis in terms of learning, thinking, and culture*. New York: McGraw-Hill, 1950.

Easterbrook, J. A. *The determinants of free will*. New York: Academic Press, 1978.

Einstein, A. *Essays in science*. New York: Philosophical Library, 1934.

Farrington, B. *Francis Bacon: Philosopher of industrial science*. New York: Henry Schuman, 1949.

Feuer, L. S. *Einstein and the generations of science*. New York: Basic Books, 1974.

Frank, P. *Philosophy of science*. Englewood Cliffs, N.J.: Prentice-Hall, 1957.

Heider, F. *The psychology of interpersonal relations*. New York: Wiley, 1958.

Hull, C. L. *Principles of behavior*. New York: Appleton-Century-Crofts, 1943.

Jonçich, G. *The sane positivist: A biography of Edward L. Thorndike*. Middletown, Conn.: Wesleyan University Press, 1968.

Kant, I. *The critique of pure reason*. In R. M. Hutchins (Ed.), *Great books of the western world* (Vol. 42). Chicago: Encyclopedia Britannica, 1952.

Kelley, H. H. The processes of causal attribution. *American Psychologist*, 1973, *28*, 107–128.

Kelly, G. A. *The psychology of personal constructs* (2 vols.). New York: Norton, 1955.

Kondo, H. *Albert Einstein and the theory of relativity*. New York: Franklin Watts, 1969.

Kuhn, T. S. *The structure of scientific revolutions* (2nd ed.). Chicago: University of Chicago Press, 1970 [1st edition, 1962].

Kuhn, T. S. *The essential tension: Selected studies in scientific tradition and change.* Chicago: University of Chicago Press, 1977.

Marrow, A. J. *The practical theorist: The life and work of Kurt Lewin.* New York: Basic Books, 1969.

Mountcastle, V. B. The view from within: Pathways to the study of perception. *Johns Hopkins Medical Journal,* 1975, *136,* 109–131.

Mussen, P. *The psychological development of the child* (3rd. ed.). Englewood Cliffs, N.J.: Prentice-Hall, 1979.

Neisser, U. *Cognitive psychology.* New York: Appleton-Century-Crofts, 1967.

Oppenheimer, R. Analogy in science. *American Psychologist,* 1956, *11,* 127–135.

Page, M. M. Demand characteristics and the verbal operant conditioning experiment. *Journal of Personality and Social Psychology,* 1972, *23,* 304–308.

Penfield, W. *The mystery of the mind.* Princeton: Princeton University Press, 1975.

Popper, K. R., & Eccles, J. C. *The self and its brain.* New York: Springer-Verlag, 1977.

Randall, J. H., Jr. The development of scientific method in the school of Padua. *Journal of the History of Ideas,* 1940, *1,* 177–206.

Rosenblueth, A., Wiener, N., & Bigelow, J. Behavior, purpose and teleology. *Philosophy of Science,* 1943, *10,* 18–24.

Royce, J. R. *The encapsulated man: An interdisciplinary essay on the search for meaning.* Princeton, N.J.: Van Nostrand, 1964.

Russell, B. *Introduction to mathematical philosophy.* London: Allen & Unwin, 1919.

Rychlak, J. F. (Ed.), *Dialectic: Humanistic rationale for behavior and development.* Basel, Switzerland: Karger, 1976.

Rychlak, J. F. *The psychology of rigorous humanism.* New York: Wiley-Interscience, 1977.

Rychlak, J. F. *Discovering free will and personal responsibility.* New York: Oxford University Press, 1979.

Rychlak, J. F. The false promise of falsification. *The Journal of Mind and Behavior,* 1980, *1,* 183–195.

Rychlak, J. F. *A philosophy of science for personality theory* (2nd ed.). Malabar, Florida: Krieger, 1981. (a)

Rychlak, J. F. *Introduction to personality and psychotherapy: A theory-construction approach* (2nd ed.). Boston: Houghton Mifflin, 1981. (b)

Rychlak, J. F. Freud's confrontation wtih the telic mind. *Journal of the History of the Behavioral Sciences,* 1981, *17,* 176–183. (c)

Rychlak, J. F. Logical learning theory: Propositions, corollaries, and research evidence. *Journal of Personality and Social Psychology,* 1981, *40,* 731–749. (d)

Sameroff, A. J., & Cavanagh, P. J. Learning in infancy: A developmental perspective. In J. D. Osofsky (Ed.), *Handbook of infant development.* New York: Wiley, 1979.

Schrödinger, E. Discussions of probability relations between separated systems. *Proceedings of the Cambridge Philosophical Society* (Vol. 31). Cambridge, England, 1935.

Simon, Y. *The great dialogue of nature and space* [Edited by G. J. Dalcourt]. Albany, N.Y.: Magi Books, 1970.

Slife, B. D. Psychology's reliance on linear time: A reformulation. *Journal of Mind and Behavior,* 1981, *2,* 27–46.

Spence, K. W. *Behavior theory and conditioning.* New Haven: Yale University Press, 1956.

Sperry, R. W. Bridging science and values: A unifying view of mind and brain. *American Psychologist,* 1977, *32,* 237–245.

Taylor, R. Purposeful and non-purposeful behavior: A rejoinder. *Philosophy of Science,* 1950, *17,* 327–332.

Tolman, E. C. *Purposive behavior in animals and men*. New York: Appleton-Century-Crofts, 1967.

Virgil. *The aeneid*. In R. M. Hutchins (Ed.), *Great books of the western world* (Vol. 13). Chicago: Encyclopedia Britannica, 1952.

Watson, J. B. *Behaviorism*. New York: Norton, 1924.

Weimer, W. B., & Palermo, D. S. (Eds.). *Cognition and the symbolic processes*. Hillsdale, N.J.: Lawrence Erlbaum, 1974.

Weizenbaum, J. *Computer power and human reason: From judgment to calculation*. San Francisco: Freeman, 1976.

Wiener, N. *The human use of human beings*. Boston: Houghton Mifflin, 1954.

Wightman, W. P. D. *The growth of scientific ideas*. New Haven: Yale University Press, 1951.

Zilsel, E. The origins of Gilbert's scientific method. In P. P. Wiener & A. Noland (Eds.), *Roots of scientific thought*. New York: Basic Books, 1957.

Zukav, G. *The dancing wu li masters: An overview of the new physics*. New York: Bantam Books, 1979.

3

Teleology Is Secondary to Theoretical Understanding in the Moral Realm

Walter B. Weimer

Rychlak's discussion ranges over numerous topics in philosophical psychology, centering on such issues as internal versus external perspective in observation, the theory of causality, the measurement problem in quantum physics, the relational nature of the mental order, and many more. Given such diversity, it is impossible to discuss more than a sampling of his claims, and I will confine my remarks to two classes of points: those on which I feel other authors have preceded Rychlak and have been more systematic and coherent and points on which Rychlak's analysis is incorrect. The critical nature of my remarks should not be allowed to obscure fundamental areas of agreement, such as the centrality of intentional and teleological theory in the moral sciences, the relational nature of the psychological order, and the importance of understanding the epistemology of physical theory. It is because I agree strongly on such issues that I am concerned to make the case as convincing as possible in their favor.

1. Alternative Concepts of Key Points

Rychlak argues that final-cause analyses are necessary for psychology, and claims that modern physics has abandoned material and efficient causes in favor of purely formal analysis of patterns. He assumes

Walter B. Weimer • Department of Psychology, Pennsylvania State University, University Park, Pennsylvania 16802.

that physics is taking a further step toward final causes in quantum theory and that such a shift legitimates comparable theorizing in psychology. However, the epistemological muddle surrounding the measurement problem in quantum physics has nothing to do with causality and is instead focused on the nature of conceptual understanding and the problem of meaning.

Modern physics has no final causes. Physical theory attempts to understand, to rationalize, the events and processes in our universe. Causality, in any of Aristotle's senses, is not found in the universe at all: Since the time of Hume (1739/1888) we have known that causality is a conceptual rather than a real entity and is thus present or absent from our theories depending on how, and for what purpose, we formulate them. Any theory that specifies principles according to which events are 'related' or covary in specifiable fashion is 'causal.' Thus even mathematics and logic are causal in the formal-cause sense. Aristotle's material and efficient causes are particular types of theoretical relation that applied to commonsense interactions in classical physics. That they do not apply in the physics of the very small or very hot or the universe as a whole is an indication that they are not adequate theoretical explanatory constructs for those domains. The measurement problems in quantum physics have *nothing* to do with causality, but rather with the relation of epistemology to ontology, and descriptive analysis to meaningful theoretical analysis. Bohr's and Heisenberg's Copenhagen interpretation is a positivistic stratagem treating all theories as instruments of description rather than explanatory conjectures about reality and is thus a 'noncausal' or descriptive analysis since it denies the possibility of theoretical understanding for patterns of events in the quantum realm. In contrast to such subjectivism and instrumentalism, philosophical realists, such as Einstein and Popper (e.g., 1972), search for 'causal' or theoretical accounts of the quantum descriptions. Indeed David Bohm proposed a 'causal' interpretation of quantum phenomena in opposition to positivistic descriptivism in 1952 (see Bohm, 1978; Bohm & Hiley, 1975). Those of us who are realists are faced with the task of specifying how our knowledge-gathering capacities can be understood so that we can know how epistemology and ontology are related, how our knowledge relates to the real (see Weimer, 1982).

The facile assertion of Bohr (accepted by Rychlak at face value) that quantum phenomena have 'free will' and the loose formulation that particles 'know' what others are doing are found not in serious physical science but only in sensationalistic popularization. To accept such irresponsible accounts as what quantum physics requires is as silly as to accept the contents of *Psychology Today* as what psychology is. The Copenhagen account is purely formal, and if it can have a causal interpretation at all it is only as the denial of the possibility of any of Aristotle's categories other than the formal; that is, to understand that positivistic instrumentalism denies material and efficient causality is to understand that it denies final causality as well. Particles are not agents; agents are psychological subjects understood by analogy to our minds and actions. Physics does not legitimate teleology in psychology. Psychological analysis must stand on its own feet rather than the clay of the false idol.

The social realm is teleological because it is known by analogy. All the moral sciences, including psychology, are teleological. This is so because the domain of analysis is specified in terms of (human) action and its relation to intentional agency. This has been known for centuries in other areas, such as economics. As Ludwig Mises put it in *Human Action* (1949): "For the comprehension of action there is but one scheme of interpretation and analysis available, namely, that provided by the cognition and analysis of our own purposeful behavior" (p. 26). We understand the behavior of other agents by analogy to the workings of our own minds. The data on which psychology, sociology, economics, and so forth are based are not physical but rather functional and intentional. This is due to the shift in perspective that Rychlak emphasizes: All the data are introspective in the moral realm—the extraspective is limited to the purely physical. As Hayek (1948) said,

> The position of man, midway between the natural and social phenomena—of the one of which he is an effect and of the other a cause—brings it about that the essential basic facts which we need for the explanation of social phenomena are part of common experience, part of the stuff of our thinking. (p. 126)

Thus we know the intentional and teleological by analogy:

> We invariably interpret their [other agents'] action on the analogy of our own mind: That is, we group their actions, and the objects of their actions, into classes or categories which we know solely from the knowledge of our own mind. (p. 63)

We do not need to ape physics to be 'scientific,' and indeed if we do we will have no science at all, but rather the extraspective correlates to psychology. Put another way, physics has no final causes because its domain is nothing but the extraspective, not because it is efficient or formal causal, or because its patterns of events are somehow of a particular character. All the moral sciences deal with teleological concepts instead of physical ones, and that has been well known (at least outside of behavioristic psychology) at least since Hume's *Treatise* in 1739.

The moral sciences study relational orders. Rychlak discusses relations as central to teleological theory. But in this matter as well the classic moral science authors are clearer and more coherent. One example is Hayek's discussion in *The Sensory Order* (1952) of why the sensory and other mental orders must be relational rather than physical. Another is the catallactic order of market exchange, discussed by Adam Smith (1776) as ably as most contemporary theorists. One way to see the relational character of both realms is to note that there are no absolute mental or sensory qualities, nor any absolute values underlying money prices. Both the mental and market orders consist of solely relational structures in which knowing any particular is a matter of rank ordering or ordinally scaling it in relation to the entire order. There are neither absolute mental qualities nor independent values for money and goods. Consider marginal utility or subjective use of scarce resources (economic goods). As Mises (1949) summarized:

> Action does not measure utility or value; it chooses betweeen alternatives. There is no abstract problem of total utility or total value. There is no ratiocinative operation which could lead from the valuation of a definite quantity or number of things to the determination of the value of a greater or smaller quantity or number. There is no means of calculating the total value of a supply if only the values of its parts are known. There is no means of establishing the value of a part of a supply if only the value of the total supply is known. There are in the sphere of values and valuations no arithmetical operations; there is no such thing as a calculation of values. (p. 121–22)

Thus the price of goods in the market order is relational only, determined by demand and supply within the order as a whole. There is no real value of money independent of the order: Prices are rank orderings relevant to consumer preference—they can never reflect a ratio scale.

The analogous conclusion holds for the mental order, as Hayek (1952) indicated:

> It seems thus impossible that any question about the nature or character of particular sensory qualities should ever arise which is not a question about the differences from (or the relations to) other sensory qualities; and the extent to which the effects of its occurrence differ from the effects of the occurrence of any other qualities determines the whole of its character. (p. 35)

If these examples seem unfamiliar (or worse) to psychologists, the conclusion I would prefer one to draw is that their provincialism has prevented psychologists from either understanding or taking advantage of the conclusions of other investigations in the moral sciences. There is a wealth of theory and research in related areas that can save us much needless repetition of inventing the wheel—all we need do is abandon positivistic prescriptions in philosophy of science and the tendency slavishly to imitate physics as an ideal of science (see Weimer, 1980).

2. Errors and Inaccuracies

Rychlak makes several claims that cannot withstand critical scrutiny. Consider some that can have serious consequences for his account.

Not all cognitive theories are mediational. Rychlak cites Weimer and Palermo (1974) as indicative of the claim that cognitive theory is mediational. Although this may be true of traditional sensory information-processing approaches (as I have argued; see Weimer, 1977), it is not true of my work, nor the majority of theorists in Rychlak's cited source, nor the more recent Weimer and Palermo (1982). Those theories are structural and formal rather than mediational (or temporal), and Rychlak's discussion is simply incorrect on this point.

All relations are equally relational. The distinction between unipolar and oppositional relations that Rychlak maintains is also incorrect. All relations are 'bipolar' or 'oppositional.' No part of a relational order can 'stand alone' or be absolute, as the discussion above should indicate.

Teleology is essentially conceptual and only incidentally human. Rychlak's conclusion identifies the teleological with the distinctly human (note that this contradicts the earlier, Bohr-inspired assertion that particles have purposes). This misassimilates teleology, which is a property of conceptual systems, to those entities which, in this region of the universe, have conceptual systems. The conceptual need not be restricted to the human, and a logical learning theory should not be so restricted either.

Creativity cannot be tautological except in explicit reconstruction. Rychlak appears not to comprehend the major message of Chomsky's revolution in linguistics: That creativity is rule-governed productivity, the ability to make infinite use of finite means. Creativity requires the distinction between surface and deep, or tacit and explicit, levels of analysis and recursive principles that generate surface particulars from underlying abstract structures. Logical learning theory fails to make the surface-deep distinction in any principled manner and thus fails to address genuine novelty. On this point Rychlak's formulation is as deficient as the mediational theories he is concerned to reject.

Formal analysis does not occur in time. Rychlak emphasizes that his theory speaks of precedent and sequacious meanings rather than temporal antecedents and consequents. But that absence of temporal "push and pull" is a property of all formal conceptual analysis rather than unique to logical learning theory. Logic and mathematics, to say nothing of the pure logic of choice (e.g., Mises' discipline of praxeology), transformational linguistics, structural analysis in psychology (see Weimer, 1984), and all other explanatory theories of complex phenomena are atemporal in this sense. There is no magical process, called 'dialectical' reasoning, that transcends temporal determination; all conceptualization in formal terms does so as a matter of course. Theoretical explanation is instantaneous, even if the theory is one of time-bounded processes that unfold in physical clock time.

3. References

Bohm, D. *Wholeness and the implicate order.* London: Routledge & Kegan Paul, 1978.

Bohm, D., & Hiley, B. J. On the intuitive understanding of nonlocality as implied by quantum theory. *Foundations of Physics,* 1975, *5,* 93–109.

Hayek, F. A. *Individualism and economic order.* Chicago: University of Chicago Press, 1948.

Hayek, F. A. *The sensory order.* Chicago: University of Chicago Press, 1952.

Hume, D. *A treatise of human nature.* L. A. Selby-Bigge (Ed.). Oxford: Oxford University Press, 1888. (Originally published 1739.)

Mises, L. *Human action.* New Haven: Yale University Press, 1949.

Popper, K. *Objective knowledge.* Oxford: Oxford University Press, 1972.

Smith, A. *The wealth of nations.* London: J. M. Dent and Sons, 1776.

Weimer, W. B. A conceptual framework for cognitive psychology: Motor theories of the mind. In R. Shaw & J. D. Bransford (Eds.), *Perceiving, acting, and knowing.* Hillsdale, N.J.: Lawrence Erlbaum, 1977.

Weimer, W. B. Psychotherapy and philosophy of science: Examples of a two-way street in search of traffic. In M. Mahoney (Ed.), *Psychotherapy process.* New York: Plenum Press, 1980.

Weimer, W. B. Ambiguity and the future of psychology: Meditations Leibniziennes. In W. B. Weimer & D. S. Palermo (Eds.), *Cognition and the symbolic processes* (Vol. 2). Hillsdale, N.J.: Lawrence Erlbaum, 1982.

Weimer, W. B. Limitations of the dispositional analysis of behavior. In J. R. Royce & L. P. Mos (Eds.), *Annals of theoretical psychology* (Vol. 1). New York: Plenum Press, 1984.

Weimer, W. B., & Palermo, D. S. (Eds.). *Cognition and the symbolic processes.* Hillsdale, N.J.: Lawrence Erlbaum, 1974.

Weimer, W. B., & Palermo, D. S. (Eds.). *Cognition and the symbolic processes* (Vol. 2). Hillsdale, N.J.: Lawrence Erlbaum, 1982.

3

On Reasons and Causes

Daniel N. Robinson

Having defended the distinction between reasons and causes often and having argued that psychological explanations cannot ignore the former (Robinson, 1976, 1978, 1979), I can only commend Rychlak for recognizing the distinction and for developing certain theoretical implications arising from it. Before examining his version of the distinction and the theoretical uses to which he puts it, however, a few words must be devoted to the historical passages in his essay.

Impelled by several secondary sources of questionable perspicuity, Rychlak repeats a number of traditional libels suffered by Aristotle at least since the thirteenth century. Thus we learn early in Rychlak's essay that, "Bones do not have it as their 'aim' to hold up the muscles . . . any more than leaves have it as an 'aim' to shade fruit." Aristotle's accounts, we are told, "were heavily teleologized for he viewed motion as the 'act of a thing in potency'. Aristotle always related the 'nature' of anything to its ends." The fact, of course, is that Aristotle explicitly rejected the notion of *bones* having aims or psychological attributes of any kind. The separate parts of the body are, "wholly insensitive and consequently not perceptive even of objects earthy like themselves (Aristotle, *On the Soul*, 410^b).[1] Moreover, in the realm of purely material transactions, he left ample room for the operation of chance and accident, thus rejecting rigid determinism even in physics. The 'nature' of anything, accordingly, is not invariably "related . . . to its ends." Even in treating motion as the *logical* mediator between potentiality (only that which *can* be moved will move) and actuality (the empirical fact of its

[1] All references to Aristotle are taken from *The basic works of Aristotle,* edited by Richard McKeon (New York: Random House, 1941).

Daniel N. Robinson • Department of Psychology, Georgetown University, Washington, D. C. 20057.

movement), he was concerned not to offer a theory of motion but a definition of it. Thus, there is nothing eccentric, even under modern lights, about the claim that motion "is the act of a thing in potency." In most ordinary settings, the observable correlate of the transition from potential to kinetic energy is *motion*.

That Aristotle subscribed to the view of nature having "ends" is unarguable, but it is important to recognize that 'nature' here refers to the total order (form) of the conceived universe. Nature, in this sense, is a kind of unseen intelligence whose only directly knowable parallel is *mind*—not bones or teeth. In defending teleology against the essentially Darwinian theory advanced by Empedocles, Aristotle was careful to divorce his position from excessive versions of it:

> If a man's crop is spoiled on the threshing floor the rain did not fall for the sake of this—in order that the crop might be spoiled—but that result just followed. (Aristotle, *Physics*, 198^b)

Intention and design were proposed only to account for things and events which could not arise from the mere accidents and collisions of physical entities. The fact that such intentions were not visible scarcely worried Aristotle:

> It is absurd to suppose that purpose is not present because we do not observe the agent deliberating. Art does not deliberate. If the ship-building art were in the wood, it would produce the same results *by nature*. If, therefore, purpose is present in art, it is present also in nature. (199^b)

It is also important to recall that Aristotle did not depreciate the role of explanations grounded in efficient causation. He was very much the mechanist in his scientific accounts of physiological processes. *Metaphysics* arises precisely because the mechanistic theories of physics are insufficient to establish the very principles upon which physics must depend. What divides the metaphysical and the physical realms is the difference between *temporal* order and the *logical* order:

> In order of time, then, the material and the generative process must necessarily be anterior to the being that is generated; but in logical order the definitive character and form of each being precedes the material. (*On the Parts of the Animals*, 645^b)

When Rychlak moves to the modern period, still other confusions (or opportunities for confusion) arise. He discusses Helmholtz's influence on Wundt through which the latter is said to have believed that "a true scientist opposes telic description in *any* form." This is partially true of the somewhat fictional Wundt bequeathed by E. G. Boring, but it is surely not true of the Wundt of the *Völkerpsychologie*, who insists that human *character* is determinative of significant human actions and is *not* part of the causal order treated by the natural sciences. It is not widely known that Wundt's productive career included a book on Leibniz. Nor is it generally recognized by historians of psychology—or, at least, by the loyal disciples of Boring—that Wundt's *voluntarism* was grounded in philosophically acute analyses beyond the scholarship of many of his later critics.

Turning now to the body of Rychlak's argument there appear to be two principal paths leading the author to recommend 'teleological' theories in psychology. The first is analogical, the second analytical. It is the second that deserves particular attention, but some comments on the first will not be out of order.

Read under a certain light, there would seem to be an unintended irony in the competing—even the contrary—uses to which Rychlak puts the old and the new physics. We are warned throughout the early pages of his essay against patterning psychology on the Newtonian model. We are called on (correctly, I think) to eschew the easy temptation of equating psychology with physics. But then—all of a sudden—we are introduced to a new physics wherein subatomic phenomena appear to allow the indeterministic and formal-cause properties which the older physics sought to proscribe. But if we were mistaken in adopting the mechanistic physics that flowered in the seventeenth century, why are we now to derive solace from a newer physics that happens to be less constraining? To state the case briefly, if bluntly, why should it matter *one iota* what passes for the received truths of physics in *any* epoch if our concern is with human psychology? A deterministic physics does not establish psychological determinism unless, of course, radical psychological materialism is first taken for granted. But on the radically materialistic reduction, we do not have a new explanation of psychological events; we have their utter elimination. That is, on this sort of reduction, psychology is not "like" physics; it simply *is* physics. It seems apparent, however, that Rychlak would not accept this reduction, which is to say that he accords psychological events existential status. To enjoy this status, there must be something about events of this kind that permits them to be distinguished from purely physical events. And, to the extent that this is so, it is entirely unclear how a science of the latter can serve either as a model of or an explanation for the former. If the old physics was somehow inapplicable, it must be for reasons other than the fact that it was flawed or incomplete. Rychlak surely is not saying that what was wrong with the notion of a Newtonian psychology is that Newton's concept of mass fails to respect how mass changes at hypervelocities. Indeed, what Rychlak seems to be saying is that the sort of Newtonian psychology Locke attempted to create was *all wrong*—was based on a deep conceptual mistake—because there happen to be bona fide psychological realities that simply cannot be reduced to or modelled by or translationally identified with merely material (mechanical) transactions. Again, if this is Rychlak's position, it would seem to make modern physics, *qua* physics, as beside the point as classical physics. To claim, as Rychlak does, that "the fact that formal causation has emerged as the *central cause* in modern science enables psychology to begin postulating theories of a suitably telic nature" is to claim that psychology derives its theoretical licenses from physics. Before we can even dispute such a claim we must be instructed as to its rationale. One would have thought that the license to theorize was granted by the dictates of reason in company with a set of reliable observations. In the realm of 'phenomena', scarcely any can claim to be more directly and frequently observed than our immediate awareness of our purposes. Long before there was physics there were persons equipped with this awareness and, presumably, prepared to account for it. Indeed, the connection between an actor's intentions and the actions themselves will survive any and every physical theory, in much the same way as a toothache survives every theory of its causation. We do not become 'enabled' to have toothaches as a result of discoveries in neurophysiology, and we certainly did not have to wait for modern physics to grant us the right to "begin postulating theories of a suitably telic nature."

In this same connection, we learn from Rychlak that, since "there *is no* efficient-cause substrate to reality," psychologists must come to grips with the "theoretical necessity that there *is no* efficient-cause substrate to behavior." This is a fallacious deduction since the ontological status of the subject-terms is different in the two cases and since the two cannot be given logically kindred definitions. (*Behavior* can be defined ostensively but *reality* cannot. The reality of modern physics is, as Rychlak notes, largely formal. This is surely not the status of behavior.) But apart from the logical incoherence of the maxim, there seems to be something unarguably wrong about it as a statement of fact. The usual sense of 'efficient cause', at least since the time of Hume, is an event or condition sufficient to bring about another event or condition. In such a sequence, one event reliably precedes the other; the two are 'constantly conjoined'. These properties are fully preserved in countless neural-behavioral sequences such that there is nothing odd about the claim that, for example, "Stimulation of motor cortex is the 'efficient cause' of the movement of the contralateral limb." Let me not digress here to examine the nomological shifts engaged as we move from molar to molecular to atomic and finally to subatomic levels of observation and theory. Clearly, at that observational level at which the term *behavior* is psychologically informing, the traditional meaning of *efficient causation* is applicable—*no matter what is going on in the sanctuaries of theoretical physics*.

On the analytical side of the argument, Rychlak leaves at least one reader utterly confused as to the special advantages of "logical learning theory" and the (allegedly) sound support it receives from Rychlak's own experiments. There is the now usual brandishing of quotations from "eminent brain scientists," but nowhere in this essay do we find a specific prediction of "logical learning theory" supported by the available evidence of the neural sciences. We are told that computers do not engage in dialectical reasoning (which is no surprise, since they are not *engaged* in any psychological sense at all), but that (somehow) the brain—by virtue of excitatory and inhibitory synapses—might be. The problem here is that if, *in principle*, any of the stuff of the nervous system could reason dialectically then, *in principle*, there is no reason why a computer cannot. Rychlak cannot have it both ways. If there is something about dialectical reasoning that precludes the very possibility of computers doing it, the same characteristics must preclude the possibility of brains doing it.

This, of course, is the subject of much speculation in the fields of computer science, philosophy of mind, artificial intelligence, and cognitive psychology. What is beyond speculation, however, is that *we* reason dialectically (at least sometimes). But although this is true, Rychlak has done little more than note it. His essay is punctuated in the final pages by a number of Hegelian positings, contrapositings, etc., but it does not rise higher than the level of *reportage*. The impatient reader is inclined finally to mutter, "Yes, but *so what*?" It surely does not follow, for example, from the fact that infants display pattern-preferences, that they enter the world possessed of dialectical competence. As it happens, kittens enter the world with similar 'preferences' and, for all we know, so do frogs and mosquitos. In such research, the term 'preference' is statistical, not affective in its reference. The infant 'prefers' in the sense of gazing longer at one pattern than at other and different ones, or gazing at it more frequently, or habituating to it only after relatively long exposures. The only psychological theory embarrassed by findings of this sort is the radical *tabula rasa* theory to which no one (and certainly not Locke) ever seriously subscribed. The data we now have make it quite obvious that the advanced species arrive on the scene with a fair share of hard-wired processors selectively responsive to environmental features of this or that sort. Some of the hard-wiring seems to be the gift of evolutionary pressures, for example, auditory cells that respond selectively to the vocal range of the same species, visual cells that respond selectively to hand-shaped stimuli. But there is nothing in this *perceptual* equipment that establishes, implies, or logically entails a corresponding *rational* (dialectical or otherwise) process. The defects of traditional associationism and its modern behavioristic progeny are well known and have been often recounted. Rychlak, however, misleads the reader if he is proposing that these defects somehow count as evidence in favor of his own formulations. *Ex nihilo nihil venit*.

The tautologizing redundancies that Rychlak would install as the principle of learning finally seem to be no more than awkwardly formalized expressions of the Plain Man's account of why he does what he does. In the illustration, "If *A* then *A*," what are we told other than the fact that, to understand Smith's getting up and walking out the door it is important to posit that he intended to leave the room? It is generally true that "to explain the behavior we must know the framing premises for the sake of which the person enacts the motions observed"; this is why the court of law seeks to establish *opportunity and motive*. The entire institution of justice arises out of the common knowledge of the human race that some actions could have been otherwise, that the actor performed by choice, that the motions observed were not regulated by the causal determinations of physics, and that what happened would not have happened had the actor intended otherwise. It would seem to be late in the season of thought to invent a new psychology able to uncover the obvious. My point is not that psychology can safely ignore human reasons; nor is it that human reasons can in principle be absorbed into the causal nexus of purely material transactions. The point, instead, is that psychology as an *experimental* affair is not likely to achieve more than an absurdly truant awareness of what every nonpsychologist has known for certain since our species jumped to the ground. Rychlak is right, of course, in demanding that a developed psychology be able to incorporate the fact of human rationality (intentionality, volition), but his essay does not indicate how this incorporation is to proceed. It is my own sense that the more refined accounts will, indeed, come from the armchair rather than the busy laboratory. What counts in such matters is not the furniture but the intellectual power of those sitting in it. What, after all, was Einstein's achievement if not armchair physics?

Finally, given the format of the *Annals*—a format I applaud—my comments will be subjected to remarks and rebuttals to which I will not be able to respond. It will not be out of place, therefore, for me to anticipate at least one complaint Rychlak may have with what I have expressed all too briefly here. My statements could be construed as suggesting that teleological theories are unnecessary because everyone is aware of his own purposes. And the rebuttal, of course, would take the form of alerting me to the fact that psychologists of a certain persuasion have doubted just these purposes and psychology itself is notably lacking in explanatory systems of the telic type. Let me be clear, then, in stating that my reservations regarding Rychlak's essay arise from the judgment that *his* teleological theory is not in fact a formal theory but a restatement in formal terms of the Plain Man's historical understanding of his own behavior. It is, therefore, a restatement of just that account which twentieth-century psychology judged to be unscientific, unparsimonious, unverifiable, mentalistic, and folksy. As I have been at pains to show in many articles and books, this judgment was not only presumptuous, but it did not lead to a "better" or a scientific account of human behavior. Instead, it led to the abandonment of most of the issues that warranted the very existence of psychology in the first place. My criticism is not with Rychlak's critique of the recent past, but with his implicit belief that the *same* mentalism once rejected will now succeed where the old one failed. To state the case tersely, the old one failed because its major tenets simply could not be incorporated into the framework of an observational, quantitative, laboratory science—the science psychology longed to be. I find very little in Rychlak's argument that will suffer such incorporation except in a very trivial way.

There are irreconcilable differences between reasons and causes. That the new physics is prepared to deal with a causeless universe is not to say that it is prepared to find *reasons* in it. *We* have all the reasons and they do not now surrender to scientific explanation. To dress them up in the raiment of tautologizing redundancies is but to conceal their directly but privately witnessed being; their irreducibly *psychological* being. They may arise *from* processes in the brain—although I cannot imagine how!—but they do not arise *as* processes in the brain. Life is literally pointless without them, but psychology has not the slightest notion of what to do with them. I wish Rychlak had given us a clue, but he has not, and neither has anyone else.

1. References

Robinson, D. N. *An intellectual history of psychology*. New York: Macmillan, 1976.

Robinson, D. N. Psychological explanation: Reasons or causes? Paper presented at the annual convention of the American Psychological Association, Toronto, 1978.

Robinson, D. N. *Systems of modern psychology: A critical sketch*. New York: Columbus University Press, 1979.

3

Ours Is to Reason Why

William J. Baker

When Rychlak first raises the distinction between efficient and final causes in the context of modern psychology, the reader is led to expect a new and important distinction necessitated by the nature of the phenomena which are specific to the purview of the science of psychology. It is both startling and disappointing, then, to have the chapter conclude by saying, in effect, if only psychology would be more like physics, then possibly it might achieve true scientific status. This rather distinct echo of turn-of-the-century American psychologizing is updated by Rychlak in his suggestion that this early belief was, in principle, correct but that it was tied to a too primitive concept of what physics was. In essence, then, he says that if psychologists will simply grasp the scientific basis for *modern* physics, and then follow the appropriate rubrics, we will finally achieve scientific respectability.

This frustrating confounding of a number of quite important issues deserves to be sorted out for many different reasons, but primarily because the original issue is an important one for modern psychology. Let us dispense with the *ad hominem* side of the problem first since it is more annoying than important. After a century of scientific research under the aegis of psychology, there seems little need left to justify its place among the other sciences, especially since that period included a great deal of serious effort to reduce its domain to other sciences, or to define it away; but these efforts have clearly and consistently failed. There is something there for which a separate science is needed. One still occasionally encounters ninteenth-century minds for whom there are no sciences other than physics and chemistry, and possibly biology, but they seem to be a vanishing breed, so let us consider more salient issues.

William J. Baker • Center for Advanced Study in Theoretical Psychology, University of Alberta, Edmonton, Alberta, Canada, T6G 2E9.

Now that psychology is emerging from the intellectual straitjacket of strict behaviorism, we can afford to look back seriously and more sympathetically at what Wundt was really trying to do 100 years ago. He proposed the need for a *new* science. His suggestion, that we begin by employing the methods of the old sciences, did not imply that we should try to physicalize or physiologize the psychological, but only that we should begin with the successful methods of the established sciences and expand or modify these as the situation within the domain of psychology might require. What he saw, then, was the existence of a set of phenomena which did not fall within the domain of any of the currently existing sciences, the "facts of consciousness,"[1] which deserved to be studied in accordance with acceptable scientific principles.

Note that what links the various sciences into a common whole is the acceptance of a set of scientific principles, not the acceptance of any specific method as such. The various branches of science are distinguished from each other primarily in terms of the sets of phenomena each is intended to study. The distinctions, then, are substantive. Methodological differences, which clearly exist, are simply a consequence of the nature of the phenomena under study rather than a basis for essential similarities or differences. Note too that these scientific principles (assuming, for the moment, that we could agree on what these are) act as a set of constraints in terms of which we can specify what would qualify as "doing science." These evolved primarily to distinguish "doing science" from "philosophizing."

In the heyday of logical positivism, where the prevailing motive was to place as much distance as possible between science and philosophy, the attempt was made to develop a set of scientific principles that would so constrain what qualified as scientific activity that the result would be value-free and totally objective. Pushed to an extreme, this is precisely the view that created the emphasis which Rychlak noted on descriptions in terms of efficient causes much to the detriment of explanations in terms of the nature of things, or the 'reasons why'.

Rychlak is quite correct when he sees this as a characteristic of science in general at the turn of the century, and he is also correct in suggesting that this was the philosophy *(sic)* underlying the evolution of behaviorism as the prevailing paradigm within psychology. It has taken some time for it to become evident, but it now seems generally recognized that the pursuit of total objectivity has not only failed, but that it was, in principle, wrong as an ideal to be pursued. No human activity can be value-free or totally objective, nor should it be.

[1]Wundt, W. *An introduction to psychology* (R. Pintner, trans.). London: G. Allen, 1912. (2nd German ed., originally published, 1902.)

Of course, the problem with this admission is that it blurs the line between science and philosophy, and even that between either of these and theology. But can there be a clear distinction among these when each pursues knowledge of the same set of facts? These, unlike the various branches of science, can address exactly the same set of phenomena. They differ in terms of the principles each adopts to define how to proceed with that pursuit (although only scientists have made a fetish out of disowning the other procedures).

When Rychlak suggests that observable behavior can be interpreted as teleological, that is, as goal-oriented, he lets loose the *bête noire* of behaviorists by suggesting serious consideration of an unobservable cause. Here the meaning of *unobservable* is, of course, not "publicly observable," which generally ignores the unique problem of psychological phenomena which are obviously observable for the conscious experiencer even though they are not for anyone else. Rychlak argues that unobservables, in the public sense, should be allowed (or, more strongly, required) in the science of psychology just as they are in the science of physics, but his analogy obscures more than it clarifies.

Can the concept of final cause be applied equally and unequivocally to both animate and inanimate entities? Is it unequivocal for all types of animate entities? When the leaf fails to shade the fruit, do we blame it for its failure? What, then, is the relation between agency or responsibility, especially as applied to human behavior, and the concept of final cause? Can entities function in a manner contrary to or simply different from what full knowledge of their publicly observable properties would lead us to predict? These issues seem to be uniquely associated with the nature of the phenomena within the domain of psychology whereas the physicist, in restricting his domain to the inanimate, has a decidedly simpler problem.

We can be presented with a description of the propagation of light as a function of the inverse square law and put this to effective use in many applied settings without considering why light should function in the manner observed. We would consider this a useful description even though it is limited to the level of efficient cause. However, Rychlak correctly suggests that few serious physicists would be satisfied with it. They would raise questions about the nature of light in order to explain why the facts are as observed. There is inevitably a gradual shading here from science to cosmology that ought to be perfectly acceptable since there will always be a philosophical position implicit in any scientific view.

In any case, we would accept the premise that light could not behave other than as it does, that its functions are fully determined by its nature and that of its environment. The issue of self-determination or responsibility is irrelevant; effective, reasonably deterministic laws could be written and unequivocally related to the nature of the phenomenon. But can it be ignored in the life sciences? Clearly, we could conceivably write an efficient-cause type of description that would represent *how* John went to work this morning. Although this might be interesting at some level, what relation would this bear to the question, *Why* did John go to work this morning? Is this one question, or several?

An analysis of the function of light-scattering will yield a useful basis for speculation about constraints on the nature of light. Similarly, analysis of the physical and physiological aspects of John's activities will enable us to speculate about constraints on his physiological nature. But what of the other *why*, his motive for going to work? Would that be discoverable in the same manner? If it were, then I would see no reason for a science of psychology; it should be reduced to physics and physiology. This same duality of *why* is badly confounded in modern psychology, recurring over and over in the areas of motivation, learning, perception, memory, or any other area of cognition. It is the basis for the vaguely sensed dissatisfaction that emerged with the constraints of S–R conceptualizations and the basis for the current emergence of serious concern with conceptions of *mental* 'states' and 'processes'. But even this latter is only a step, albeit in the proper direction, toward concern with the facts of consciousness or, more properly, with those conscious activities which define the set of phenomena psychology was meant to address.

Efficient-cause types of explanations within psychology, when restricted to publicly available observables, are, by that very restriction, limited to the physical and the physiological. Explanations which attempt to go beyond these, to infer constraints as a function of the possibly broader nature of the behaving organism, but which still use the analogy of physics, will seek to provide basically deterministic accounts since, for the inanimate, things cannot 'behave' in any manner not predictable from a knowledge of their publicly observable properties. In a quite serious sense, inanimate objects cannot 'behave' at all. It is this simple fact which makes their functions describable in a lawful or deterministic sense.

This neat determinism is perturbed by animism. Strong predictability seems to rapidly fade as we move through the various forms of living creatures and seems to vanish almost entirely in dealing with some aspects of human behavior. If one's concept of "doing science" is tied to strict determinism, then science would be limited to physics and chemistry. The life sciences, of course, force us to consider science from a nomothetic view based more on general expectations rather than inviolate laws.

Rychlak unfortunately fails to address any of these crucial issues which emerge in the life sciences in general, or psychology in particular, but it is precisely in these sciences that we are faced with a shift or a polysemy in the notion or scope of the concept of final cause. The living organism begins to be able to act "for its own sake" in a manner no longer neatly predictable in terms of the publicly observable features of that organism. There is a shift from being determined to self-determination. For psychology, this latter is the first of all observable (though not public) events because it is through a sense of this that an individual first differentiates the self from the not-self and he begins to discover that realm of phenomena which is, in fact, the unique domain of psychology, those "facts of consciousness" that Wundt sought to explain and which modern cognitive psychology is busy rediscovering.

3

Precedents and Professors—The Struggle Over Common Ground

Reply to Commentators

Joseph F. Rychlak

If I understand the outlook of Professors Baker, Robinson, and Weimer correctly, all three are either proteleology or at least willing to look at this possibility in psychological theorizing. They just do not care for what I am doing in this regard. I had asked the editors of this journal to seek comments from colleagues on the other side of the question, hoping to engage them in debate that might have some influence on the thinking of my opponents, as well as on my own. As one oriented to dialectical human reasoning, I believe that much is to be gained from such oppositional exchanges (see, e.g., Rychlak, 1972). Apparently, we were unable to draw the other side into such discussion. But never mind: I am up to my neck in debate with my own kind. My present critics have found my ideas and empirical tests seriously wanting. My presentation also seems to have led to misunderstandings. I hope that my own shortcomings will not prejudice the reader against further consideration of telic explanation in human behavior. I thank my critical colleagues for taking the time to read my paper and to comment as they have done. I hope that this opportunity for a rejoinder will reassure the reader that my scholarship and my way of reaching for the telic human image are not so deficient as my colleagues make them out to be.

I am surprised to find Professor Weimer claiming that I have taken literally Bohr's reference to the 'free choice' of an atom in moving from one stationary state to another. When I introduce Bohr's quotation, I say that he "startles us" in using such final-cause phraseology. A few

Joseph F. Rychlak • Department of Psychology, Loyola University of Chicago, Chicago, Illinois 60626.

paragraphs later I refer to the "decisions" of Bohr's atom to rearrange its particulate organization in quotation marks. And then in the next section of my paper I suggest that when Bohr spoke of the free choice of an atom to select its steady state condition, "he was framing, *however seriously,* [italics added] an extraspective teleology." It apparently slipped Weimer's mind that in the third paragraph of my paper I specifically state, "I contend that telic theory demands an introspective perspective." I could therefore not take an extraspective theory of atomic choices as my own in this obviously naive manner. Yet Weimer persists in ascribing such a position to me, based upon my supposed literal acceptance of Bohr's off-hand suggestion. What I actually say is that in his willingness to use telic terminology to refer to the atomic realm, Bohr was analogizing to the human realm—suggesting to me that he believed in some such concept for the human being.

By analogizing to the telic nature of human beings in his formal writings, Bohr set an example for us in the 1920s that we psychologists have never modeled in our formal theoretical accounts. The supplanting of efficient by formal causation gives me hope that we will be more ready to picture the human being in a teleological fashion in the future. I am *not* saying that atoms have free will and therefore it follows that people have free will. I *am* saying that physicists accept telic phraseology in conceptualizing people and hence it follows that psychologists might employ telic phraseology in their descriptions of human behavior without suffering in scientific status. Finally, I am suggesting that those psychologists who cling to efficient-cause reductionism have no reason to do so, since the history of science is proving that there is no such substrate to all things. That is, theories presuming this form of efficient-cause substrate simply fail to meet the observations required to validate them.

This takes us to the question of why I use arguments relating to the evolution of physics as a science. All three of my critics faulted me for this practice, suggesting that psychology must stand on its own feet and not "ape physics" as Weimer put it. Why do I take the tack that I do? Because I have researched the question for some 25 years now and concluded that psychology did *in fact* come to the nontelic position of today by patterning itself on the natural sciences. And a Newtonian form of physics undoubtedly played the leading role as model for our fledgling science. Again and again, the leading academic psychologists of the 1920s, 1930s, and 1940s—the ones who were doing experimental research—were patterning themselves after what they took to be *the* science of physics. Unfortunately, no one seemed to be reading in the new physics at the time. When Bergmann and Spence (1941) were spelling out what it took to do science correctly they even presumed to speak for the way in which a "primitive physicist" (p. 3) conducted science, in contrast to the lesser efforts of most psychologists.

I took these arguments seriously. I honestly believe today that the *reason* psychologists of a rigorous persuasion dismiss all talk of teleology in behavior is because of their historical traditions (which they fail to examine properly). They are, in good faith, conforming to a precedent meaning of *science* which they sequaciously extend to their work without personalizing the resultant practice (although any individual psychologist may have his or her personal reasons for modeling Newtonian physics). I have always taken the debate over scientific practice to be a purely technical problem, of what it means to theorize and what it means to validate one's claims on knowledge. Teleology suffers in the analysis because of the historical issues my paper delineated. On the other hand, my critics seem to advance *ad hominem* arguments to account for the lack of telic description in psychology today. Baker seems to assign differences in scientific outlook to "breeds" of men, Robinson believes that Wundtian voluntarism was "beyond the scholarship" of his critics, and Weimer tells us that it was the provincialism of psychologists which kept them from understanding or taking advantage of the conclusions of the moral sciences, where telic description was ready and waiting to be modeled.

I prefer to believe that psychology affirmed an ideology, nurtured by people like Titchener, Thorndike, then Watson, Hull, Spence, Bergmann, Skinner, and others—leading figures in the sorts of experimental efforts they designed and carried out. It was based upon a conception of scientific explanation that was about to be challenged, and I for one now think that it *has been* sufficiently challenged to clarify what is changing. It is easy to forget that Hull (1937, p. 2) analogized human behavior to the actions of a raindrop, claiming that the same 'laws' enter into human behavior that determine the 'behavior' of the raindrop, and in the process pointing out that human intentionality was therefore an illusion. These so-called laws were undoubtedly conceived as efficient-cause regularities. My colleagues may be too young to recall or even realize that as recently as 25 years ago it was common to hear in psychology that 'lower-level' laws presumably entered into or generated 'higher-level' laws, and hence in time all of behavior will be subject to basic physical lawfulness of this sort. I nurture the—perhaps naive—belief that if we can kick the struts of efficient causation out from underneath such outmoded, Newtonian remnants, if we show that once and for all there are no such efficient causes at the basis of reality, then we might make it possible for a more humanized account of behavior to be framed in the future.

Incidentally, it must not be overlooked that I think of people as telosponders, as agents of their behavior, whereas Weimer defends the cybernetic model as satisfactory for human characterization. I view people as having a transcendent, reflexive mental capacity to judge and align self-determined preferences, whereas Robinson believes that they are "hard-wired" into such stimulus-preferences at birth. Given these contrasting terminological usages, the reader might well wonder who among the disputants is really aping physics.

Though Professor Baker was in general more kind to me than the others, he did have a line which jolted me out of my seat. I am referring to the opening phrase of his third paragraph: "Now that psychology is emerging from the intellectual straitjacket of strict behaviorism. . . ." I do not believe this for a moment! Much of my writing has been devoted to an examination of the fallacy of this "we are changing" point of view, and I try to touch on this in the present paper where I talk about the nonchanges of modern cognitive psychology. Professor Weimer too finds these arguments without merit. All I can do at this point is refer the reader to papers in which I have taken up the supposed advances of behavioristic, mediational theorizing represented by Neisser (Rychlak, 1977, p. 208), who speaks of people constructing their environment; Irwin (Rychlak, 1977, p. 206), who speaks of act-outcomes; Mischel (Rychlak, 1976a), who speaks of people as active, aware problem solvers; and Bandura (Rychlak, 1979), who speaks of reflective thought.

All of these theoretical accounts probably strike Baker and Weimer as freeing psychology from the straitjacket of strict behaviorism. But except for possibly a loosening of the collar, I see no basic changes in these accounts from the traditional behavioristic explanations. They are all mediation theories, relying in the final analysis on efficient causation as viewed extraspectively. Weimer claims that he is not a mediation theorist and that I am quite wrong in my assertion that mediation theory is at the heart of all modern explanations of learning or cognitive processing. I stick by my original claim but state here and now that I am ready to be educated by Weimer on this point in the future. If he has indeed escaped mediational explanations in his theory, whereby initial [efficiently caused] shapings from the environment are the ultimate determinants of human mentation, then I applaud his efforts and hope to share the work of advancing teleological explanation with him in the future. But I remain skeptical for the present, since he denies my claims concerning the current situation in psychology.

Baker is disappointed that I failed to propose a unique approach to science for psychology to take. I have not done so because in my view what unites the sciences is the manner of proof they accept, *not* the uniformity of their theories or the taxonomies that might be devised over subject matter. The term *science* is derived from a Latin root meaning "knowledge" or "to know" (*scire*, to know) and the *scientific* usage adds "to make" or to demonstrate in the sense of doing something to show that one has knowledge (*facere*, to make, do). As scientists we place emphasis on our demonstrative reasoning capacities and attempt to *validate* that which we claim to know. This is a very involved issue, one I cannot go into given the restrictions of space placed upon me. I would just like to point out to Baker that I have called for a middle ground revolution in the science of psychology in which we open up our theoretical explanations to alternatives encompassing telic description even as we retain our present method of controlling and predicting to an empirical criterion (see Rychlak, 1977, p. 219). I am sorry to disappoint a colleague, but Baker should realize that I have addressed this question, as well as the other questions he raises, in my previous writings.

Weimer claims, "The measurement problems in quantum physics have *nothing* to do with causality, but rather with the relation of epistemology to ontology and descriptive analysis to meaningful theoretical analysis." In referring to "measurement" problems instead of "conceptual" problems Weimer may be thinking that the problems encountered in the subatomic realm may one day be cleared up when "better" instruments are built for measuring what is presumed to be taking place independent of the physicist's intellect. This may not be his view, but let me just emphasize the point that the unique occurrences and indeterminacy taking place at the subatomic level are *not* problems in measurement, but call for an altered view of reality. In rejecting causality the way he does in the above quotation, Weimer is merely expressing a preference in terminology over my own. Epistemology, ontology, descriptive analysis, and meaningful theoretical analyses can be subsumed by the formal and final causal meanings, as I have shown in an extensive survey of such ideas across history (Rychlak, 1977, pp. 8–31). I choose to retain the meanings of the causes because they do provide a common ground for discussion. They have a rich history and are easily understood. I consider them highly abstract, theory-construction constructs which subsume literally anything in experience through various combinations of their meanings. Of course, it is also true that simply by enumerating the meanings of the causes in our usages we shall not be able to complete the program of clarification that Weimer calls for. But to say that causality is not involved *at all* is merely to state a semantic preference.

The same applies to Robinson, who tells us, "There are irreconcilable differences between reasons and causes." In the *Gorgias* dialogue, Socrates (Plato, 1952, pp. 262–263) analyzes the reasons why men do things—for example, why a businessman should take a dangerous sea voyage, or a sick man take medicine. I count at least a dozen instances in which, rather than referring to the reasons for doing such things, Socrates speaks of such behaviors as done "for the sake of" ends—wealth and health as the case may be. What is so irreconcilable here between *reasons* and *formal/final causes?* Surely this is more a question of quibbling over terms than one of irreconcilable differences. I suspect that if Robinson really had something important to say on such differences, he would have done so.

Instead, Robinson found considerable space in which to, if not impugn, then at least slight my level of scholarship. I supposedly use questionable secondary sources. I am drawn to simplistic, plain-man, folksy accounts. He takes umbrage at my use of the examples stating "Bones do not have it as their 'aim' to hold up the muscles . . . any more than leaves have it as an 'aim' to shade fruit." He says that Aristotle specifically rejected the notion that anatomical structures like bones have psychological attributes of any kind. There are two points to be made here. First, the cited examples were not "thought up" by me. The bones and muscle example is taken from Bacon (1952, p. 45), who was in the process of criticizing Aristotle at the time; and the leaves shading fruit example is taken from Aristotle (1952), who says, "Leaves, for example, grow to provide shade for fruit" (p. 276). Robinson is correct in suggesting that it is 'nature' as a whole—and not specific items within nature like a leaf—that has the intention to fulfill ends. This is what I said in my paper: "Aristotle had suggested that nature operated for a purpose." Indeed, according to Aristotle, all things natural move through a genesis to fulfill their potential in the actuality of their ends.

Second, I do not see where this business about bones holding up muscles entitles Robinson to claim that I am attributing 'psychological' characteristics to Aristotelian descriptions of anatomical structures. I have noted in my prevous writings (e.g., Rychlak, 1981b, p. 277) that there seem to be at least three forms of teleologies—human, natural, and deity. I would never assign psychological characteristics to a natural teleology. I would do so only to a human teleology and expect theologians to speculate about the deity's ends.

A more serious problem arises in Robinson's use of the example of a man's crop being spoiled on the threshing-floor to support his claim that in Aristotelian philosophy "Intention and design were proposed *only* [italics added] to account for things and events which could not arise from the mere accidents and collisions of physical entities." I *strenuously object* to this highly erroneous characterization of Aristotelian philosophy, since it is obvious to anyone who spends conscientious time on the peripatetic philosopher's writings that teleology was fundamental to his world view. Accidents and collisions occurred but they were quite secondary! Of course, it is easy to draw Robinson's conclusion if we use the arguments of Aristotle's opponents as if they were his own. This is what Robinson's scholarship has accomplished.

The threshing-floor quotation cited by Robinson is taken from Book II, Chapter 8 of the *Physics* (Aristotle, 1952, p. 276). In Chapter 8, Aristotle sets out to explain (1) why nature belongs to the "class of causes which act for the sake of something [final causation]" and (2) the place of *necessity* in nature. He even grumbles a bit about how other writers on natural occurrences like rain rarely give proper consideration to point (1), preferring to stress what I would call the mechanistic explanations of point (2). He goes on to pose a number of questions that a mechanist might raise against the teleologist, such as: "Why should not nature work, not for the sake of something [final cause], nor because it is better so, but just as the sky rains, not in order to make the corn grow, but of necessity?" (p. 275). In other words, why should we posit natural ends? Things just happen of necessity.

To make his opponent's attitudes against teleology even more telling, Aristotle then uses the lines quoted by Robinson about the grain being ruined on the floor of the threshing room. Surely rain did not 'intend' that the ultimate end of a crop's destruction come about. Everything just happened necessarily—from the initial rainfall to the raising of the crops to the final destruction. Aristotle then discusses the formation of our teeth, which seem so beautifully organized (formal causation) in the mouth as supposedly taking shape merely by necessity and without an end being reflected. He then steps back from such arguments of the nonteleologist and says:

> Such are the arguments (and others of the kind) which may cause difficulty on this point [i.e., the point of natural purpose]. Yet it is impossible that this should be the true view. For teeth and all other natural things either invariably or normally come about in a given way [implying intelligent design, etc.]. (p. 276)

Aristotle is clearly *rejecting* the view that Robinson has him embracing. Aristotle devotes the remainder of Chapter 8 to a vigorous defense of telic descriptions of natural items, ending with the conclusion: "It is plain then that nature is a cause, a cause that operates for a purpose" (p. 277). The point here, as in all Aristotelian explanations, is that there is a genesis in natural objects which move from potentiality to actuality in an end by fulfilling the formal-cause patterns which they supposedly have within their natures from the outset (recall, for example, the old saw about the tree being presaged in the acorn). This was where he drew his evidence for an intelligent action in nature, and this is why I would find it impossible to call Aristotle a mechanist in *any* of his theorizing—as Robinson is so willing to do. Aristotle has teleology at the very heart of his world view.

There is no better demonstration of this world view than in his interpretation of a chance occurrence. Robinson tells us that Aristotle left "ample room for the operation of chance" in his explanations. Well, this is true enough, but this glib allusion masks the crucial fact that what Aristotle took as the meaning of chance was *not* what we mean today. Today we are likely to calculate the chances of literally any event taking place with no thought of purpose in the calculation. Yet for Aristotle a chance event occurred only occasionally, appearing to be the outcome of a rational or natural purpose, but actually taking place by accident. It was a sort of off-shoot of telic action. Thus Aristotle says, "Chance is an incidental cause in the sphere of those actions for the sake of something which involve purpose" (p. 273). An example he gives of a chance occurrence is that of a man going to a certain location for one purpose but finding to his joy that he collects money from debtors that he happens to meet there "by chance." Now, he could have gone to this location through intelligent reflection, but he did not. His purpose was not to collect money but something else. Even so: "Intelligent reflection . . . and chance are in the same sphere, for purpose as an ingredient of chance implies intelligent reflection" (p. 273).

It is, of course, true that Aristotle had no need of employing all four causal meanings in *every* description of something. The genetic principle reflected in the growth of a bone in our leg or a tooth in our mouth was framed extraspectively in the supposed purpose of nature as a whole, but acting through individual objects and events. The resultant natural teleology did not mean that each item within nature was intending its end in a self-reflexive, psychological manner—as Robinson seems to have taken my comments to imply. Finally, I cannot accept Robinson's easy paralleling of kinetic energy with Aristotle's views on movement or motion. Aristotle clearly places the *end* attained (as presaged in the potentiality of a natural form) over the *process* of motion or movement, whether this takes place in organic development or in the motions of human behavior. The concept of kinetic energy relates to the process of motion, and ends are not considered relevant to the descriptive account.

I am also aware of the reinterpretation we are getting these past few decades on Wilhelm Wundt and therefore cited Blumenthal (1979), one of the leading scholars in this regard. But, as Robinson admits, it was Titchener's image of Wundtian psychology that was incorporated in America. And, furthermore, it is my present belief that Wundt may have invited the reductionism that Titchener attributed to him through an early acceptance of Helmholtz's views on the nature of scientific explanation. Wundt's voluntarism is not an easy conception to understand, if we mean by this a true intention (final cause), for as he noted: "In giving an account of the particular causes which determine volition, we shall only recognize as *determinate* motives those which give it a definite direction, and which act like simple forces, incapable of further analysis" (Wundt, 1907, p. 231). The phrase "simple forces" was Helmholtz's, by which he meant an efficient-cause substrate moving things along by means of the law of conservation of energy (Cassirer, 1950, p. 86). Wundt's own conception of an adequate scientific description spelled out the nature of these simple forces even more clearly: "We must trace every change back to the only conceivable one in which an object remains identical: motion" (Wundt, 1907, p. 88). We might, then, be a bit more understanding of poor Titchener on this matter of precisely what Wundtian psychological explanation was about. Volition which relies on underlying efficient-cause forces is just an early manifestation of the sorts of mediation models (volition as a middle term) that I claim predominate in psychology today. Wundt surely gave at least two messages to his students to act on—one sounding telic, but one sounding decidedly reductive and efficiently causal in nature.

I wonder how many of our colleagues in psychobiology, sociobiology, ethology, medical psychology, and the like agree with Weimer's claim that "the data on which psychology, sociology, economics, and so forth are based are not physical but rather functional and intentional." Howard Kendler (1981) is only the most recent of a series of distinguished psychological experimenters who have claimed precisely the opposite—that psychology's best hope is in drawing data from a biophysical source. Surely 'functional' data might involve biophysical conceptions. Angell (1907, p. 72) argued this way in his structuralism versus functionalism debate with Titchener. Several of the points made by Weimer appear to me to support logical learning theory rather than detracting from its basic thrust. *Of course* moral sciences study relational orders. *Of course* economics deals with patternings of preference. *Of course* formal analysis does not require the passage of time for its insights to be determined. Weimer seems irritated because logical learning theory can subsume such teleological formulations. The real question is: Can the instrumental (nontelic) learning and cognitive theories of today do justice to such humanistic formulations? I agree with Weizenbaum (1976) that they cannot.

The reaction of my colleagues to the concept of dialectical reasoning is both interesting and disappointing. Baker ignores the concept, Weimer calls it a myth, and Robinson says that it is beyond speculation that human beings reason dialectically—"at least sometimes." I think Robinson should be informed that someone besides Hegel took an interest in the dialectic. To say that I am positing in an Hegelian manner could not be more incorrect. Hegel's use of the dialectic was extraspective, as a world principle. I employ it more in the introspective sense of transcendence and self-reflexivity (Rychlak, 1976b). Dialectical tracts can be traced to the sacred writings of the Vedic culture in India (c.1500 B. C.; Raju, 1967, p. 44), wherein dialectic was given a "many and one" interpretation in which opposites defined totalities (Rychlak, 1976b, p. 11). Hinduism and Buddhism then carried this style of thinking forward into the Eastern philosophies which prove so popular today. Some of the earliest languages known to humanity reflect dialectical meanings in key words. Mo Ti (c.470–391 B. C.) in China and Socrates (c.470–399 B. C.) in Greece both headed schools of dialectic with absolutely no cultural contact. In 1326, Adam de Brome founded the college of Oriel at Oxford University with the expressed goal of training scholars in sacred theology and the "art of dialectic." I personally favor Jung's (1975) view that Hegel had projected his unconscious psychic processes onto the universe. And, since he was a human being, it happened that these intellectual machinations took on a dialectical formulation.

When Weimer says that all relations are oppositional he is abstracting the concept of relation and thinking of the two ends as opposite, one to the other. He is thinking of the relation ('this versus that' end) and not the unipolar items related, which is, of course, the point I have in mind. When we relate the words 'red' and 'barn' together in an essentially stereotypical manner, so that 'red barn' brings to mind an image of the structure we expect to see on our weekly drive through the country, we are surely relating unipolar concepts. The color and the structure are not united relationally until *we* unite or pattern them in thought. On the other hand, a tie of 'dominance' to 'submission' is intrinsically bound together even though we might use other bipolar words in the conceptualization (rough–weak, pushy–shy, etc.). Whereas 'red' can be affixed (patterned) to items other than 'barn', dominance can *never* be separated from submission no matter how it is employed otherwise in relational ties of meaning. Surely this difference between demonstrative and dialectical meaning relations is obvious.

I have never suggested that dialectical reasoning is impossible to conceive in a cybernetic machine of the future. Quite to the contrary, I suggested precisely this possibility in the 1968 edition of one of my books, which is now in a second printing and carries a page discussing this very point (see Rychlak, 1981a, p. 363). I hope that Robinson is correct in suggesting that computer analysts are looking into dialectical reasoning today. I know that I have been trying to get some of them interested in this topic for the past few years, with no real success. The problem is, a machine which can reason like a human being will be just as fallible as a human, and there is very little interest in a self-doubting, impetuous, or opinionated machine. In any event, I and my colleagues in logical learning theory are presently designing experiments which tap what we take to be dialectical reasoning in human beings. Weimer and Robinson apparently believe that an organism which can reason to the opposite of inputs, many times over, is not capable of true creativity and freedom from the unidirectional control under which all machines operate. I saw no arguments why this should be the case in their critiques. Weimer would like logical learning theory to have a tie to deep-structure theorizing. As I view Chomsky to be in the same Kantian vein that enriches logical learning theory, this would certainly be a fine union to arrange. Logical learning theory is open to growth in the future. We are only getting under way and invite others to help us become more comprehensive.

The evidence I present from brain research as well as the adult and infant conditioning research is not meant to be definitive. Such experiments will never in themselves establish the validity of dialectical reasoning in human behavior. Indeed, I cannot think of a way in which I can establish this position if it is not accepted as a precedent possibility in the first place. But surely we can see in the conflict theory of brain functioning, the seemingly 'illogical' activities of the brain stimulation research, and the dubious evidence that infants can be conditioned in everything they do that there is room for a theory of agency resting upon a dialectical conception of human reason. Robinson says that the data collected to date "have made it quite obvious that the advanced species arrive on the scene with a fair share of 'hard-wired' processors selectively responsive to environmental features of this or that sort." Infants are 'prepared' (Seligman, 1970) by natural endowment to react to certain stimuli, but (a) how this enters into their behavior—whether in a telic or machine-like manner—is still under debate, and (b) the fair-share range of such behaviors is not so great as Robinson's phraseology implies (see Sameroff & Cavanagh, 1979).

Children are (in a sense) "hard-wired" to curl their fingers inward to the palm of their hand instead of outward, and in a sample of normal infants there is *no* variance in this activity. But in any study of selective reactions to stimuli we always find a range of preference—as to the infant's fixing longer on colored than on noncolored stimuli. We always find some infants who show no preference and some who look to the noncolored items significantly longer. It does not seem to occur to Robinson that such a group preferential tendency might be freely arrived at. If 90% of the infants prefer stimulus A over stimulus B why does it necessarily follow that the processes involved in their uniform preference are machine-like? Could there not simply be an aesthetic quality involved, bringing the infants to exercise telic (final-cause) judgment in a consistent manner? Why do we insist on pressing material-efficient cause explanation in this unexamined, impulsive fashion, which ultimately stems from our admiration for the physical sciences in any case (refer above)? It is the *patterning* of the wiring into meaning that counts, and some experts in child study believe that nature does some of the patterning but that given the manner of this patterning it allows a contribution to be made by the child at the very outset of life. I am suggesting that some of this patterning is bipolar and a natural capacity to reason dialectically enables the developing human being to transcend even nature's hard wiring to generate alternatives that were *never* encompassed in this natural patterning to begin with.

In closing, I would like to call the reader's attention to what has taken place here. We have witnessed four individuals providing the grounds for the sake of which they *would* and *do* believe in the *same* thing! There is a certain amount of intellectual muscle-flexing taking place, but apart from these academic frailties I think we have witnessed a demonstration of the telic human nature at its best. Logical learning theory contends that we can only know what we know, that we must extend in tautological fashion assumptions that frame our understanding to endow sequaciously that which we are taking up with meaning. Each of the participants of the present exchange has a unique set of assumptions concerning the topic under consideration—the question of teleology in human behavior. I presented my assumptions and the empirical work that flowed from these precedents. But my critics found these precedents, if not totally wrong, then at least flawed in some irreparable manner. Or they faulted me for not considering other precedents which they believed were essential to any true capturing of the telic conceptualization. Surely they did not bargain to accept *my* terms. Dialectic is nonsense, or reducible to wiring in the organism. Telosponsivity is a figment of my imagination and no more. Causes do not exist. I find all of this a marvelous reflection of our humanity. It is this style of behavioral description—in which people act as logicians, aligning their case to the opposite as an opponent is making his or hers (Rychlak, 1972)—that I am trying to capture in my logical learning theory. It would be my hope that a small number of readers, some of whom have to this point remained in their armchairs, would begin to frame explanations along this line and submit them to experimental test. In time, I feel certain that the teleologists in psychology will effect a major Kuhnian revolution in the human image.

1. References

Angell, J. R. The province of functional psychology. *Psychological Review*, 1907, *2*, 61–91.

Aristotle. *Physics*. In R. M. Hutchins (Ed.), *Great books of the western world* (Vol. 8). Chicago: Encyclopedia Britannica, 1952.

Bacon, F. *Advancement of learning*. In R. M. Hutchins (Ed.), *Great books of the western world* (Vol. 30). Chicago: Encyclopedia Britannica, 1952.

Bergmann, G., & Spence, K. Operationism and theory in psychology. *Psychological Review*, 1941, *48*, 1–14.

Blumenthal, A. L. The founding father we never knew. *Contemporary Psychology*, 1979, *24*, 547–550.

Cassirer, E. *The problem of knowledge*. New Haven: Yale University Press, 1950.

Hull, C. L. Mind, mechanism, and adaptive behavior. *Psychological Review*, 1937, *44*, 1–32.

Jung, C. G. Letter to Joseph F. Rychlak, 27 April 1959. In G. Adler & A. Jaffe (Eds.), *C. G. Jung letters* (Vol. 2). Princeton, N. J.: Princeton University Press, 1975.

Kendler, H. H. *Psychology: A science in conflict*. New York: Oxford University Press, 1981.

Plato, *Gorgias*. In R. M. Hutchins (Ed.), *Great books of the western world* (Vol. 7). Chicago: Encyclopedia Britannica, 1952.

Raju, P. T. Metaphysical theories in Indian philosophy. In C. A. Moore (Ed.), *The Indian mind: Essentials of Indian philosophy and culture*. Honolulu: University of Hawaii Press, 1967.

Rychlak, J. F. Communication in human concordance: Possibilities and impossibilities. In J. H. Masserman & J. J. Schwab (Eds.), *Man for humanity: On concordance vs. discord in human behavior*. Springfield, Illinois: Charles C Thomas, 1972.

Rychlak, J. F. Comments on "The self as the person." In A. Wandersman, P. Poppen, & D. Ricks (Ed.), *Humanism and behaviorism: Dialogue and growth*. New York: Pergamon Press, 1976. (a)

Rychlak, J. F. The multiple meanings of "dialectic." In J. F. Rychlak (Ed.), *Dialectic: Humanistic rationale for behavior and development*. Basel, Switzerland: Karger, 1976. (b)

Rychlak, J. F. *The psychology of rigorous humanism*. New York: Wiley-Interscience, 1977.

Rychlak, J. F. A non-telic teleology? *American Psychologist*, 1979, *34*, 438–439.

Rychlak, J. F. *A philosophy of science for personality theory* (2nd ed.). Malabar, Florida: Krieger, 1981. (a)

Rychlak, J. F. *Introduction to personality and psychotherapy: A theory-construction approach* (2nd ed.). Boston: Houghton Mifflin, 1981. (b)

Sameroff, A. J., & Cavanagh, P. J. Learning in infancy: A developmental perspective. In J. D. Osofsky (Ed.), *Handbook of infant development*. New York: Wiley, 1979.

Seligman, M. E. P. On the generality of the laws of learning. *Psychological Review*, 1970, *77*, 406–418.

Weizenbaum, J. *Computer power and human reason: From judgment to calculation*. San Francisco: Freeman, 1976.

Wundt, W. *Lectures on human and animal psychology* (4th ed.). New York: Macmillan, 1907. (1st ed., 1894)

4

The Hypotheses Quotient

A Quantitative Estimation of the Testability of a Theory

K. B. Madsen

Abstract. This paper discusses the various interpretations of testability as a criterion for distinguishing between scientific theories and other theories. A method for a quantitative estimation of the testability of scientific theories is introduced. This method is called the *hypotheses quotient* (HQ). The procedure for calculation of the HQ is presented and discussed, and the method is demonstrated by detailed examples: analyses of Freud's topographical theory and his theory of anxiety. In connection with the analyses of Freud's theories, some concepts are introduced from the author's comparative metatheory, which is called *systematology*.

This paper presents a method for the quantitative estimation of the testability of a scientific theory. The method consists of the calculation of the *hypotheses quotient* for a given theory. We shall present the hypotheses quotient in detail later, after a historical introduction of the development of the testability criterion.

1. The Testability Criterion

This section presents a very brief survey of the historical development of the testability criterion for scientific theories. For a deeper analysis of this problem, the reader is referred to some of the major works in the philosophy of science (Bunge, 1967; Kuhn, 1962; Popper, 1934; Radnitzky, 1968; Törnebohm, 1975).

K. B. Madsen • Royal Danish School of Educational Studies, Copenhagen N. V., Denmark.

History shows many different philosophical positions regarding the testability of a hypothesis. The followers of logical empiricism stated that scientific statements and hypotheses, as well as empirical generalizations, should be *verifiable*—that is, that their real truth should be supportable through empirical methods. They went so far as to state that scientific statements were meaningless if they were not verifiable. Later, they became aware that these stringent demands not only excluded metaphysical (ontological) statements from science but that they also excluded many statements that were an already acknowledged part of science. Therefore, they had to modify their demands so that hypotheses and other statements should only be verifiable in principle. Furthermore, even though a hypothesis had been verified by all experimenters and other observers, this did not exclude the possibility that in the future this hypothesis might be falsified. This more moderate demand on scientific utterance has often been expressed, so that hypotheses and other statements must be *confirmed* or supported.

On the basis of dissatisfaction with the logical empiricists' demands for verification, Popper formulated in *Logik der Forschung* (1934) another principle of scientific theory. Scientific hypotheses and other statements should be *falsifiable;* that is, their real truth should be tested through empirical methods. As soon as a hypothesis is falsified, according to Popper's principle, it should be rejected as a scientific statement. Scientific theory, then, should consist only of falsifiable but not yet rejected hypotheses. This principle was just as powerful as the demands of the logical empiricists for verification when it was used on ontological and other philosophical hypotheses. Popper thus rejected both Marxism and psychoanalysis as scientific theories, because in his opinion they were not falsifiable.

Since the Second World War, there have been many philosophers of science who have criticized Popper's demands for falsifiability. Both Bunge (1967) and Kuhn (1962) stated that demands for falsifiability are unrealistic because they exclude a number of well-acknowledged hypotheses and theories from the sciences. Bunge has especially stated that stochastic hypotheses cannot, or can only with great difficulty, be falsified, whereas they can be supported or confirmed easily in a moderate form. As we know, stochastic hypotheses play a great role in modern science; for example, in nuclear physics, which it would be absurd to exclude from the sciences.

Kuhn has stated that the history of science shows many cases of falsified theories' having been maintained by researchers until better theories have been developed. New theories become accepted when they can explain the same things as old ones can, as well as when they can explain things the old theories could not explain. Kuhn does not deal so much with single hypotheses as with theories that form systems of hypotheses. He states that it is possible to maintain a theory as a whole even though it has been falsified by single experiments or other observations. One can simply modify one or another of the theory's hypotheses. The theory as a whole can be maintained if it still proves useful for explanations, predictions, and interpretations of the phenomena within its research area. Theories will only be rejected when all of their important hypotheses must be revised and/or when a new and better theory arises.

Kuhn's theory of science has the advantage of a better agreement with the factual development of the sciences than those of Popper and the logical empiricists. Therefore, we will use it for the time being. According to this liberal principle, scientific hypotheses must only be confirmable and/or falsifiable. However, no test of a hypothesis can be considered absolute any more than a falsification can. A test in itself can be shown to be wrong. It is, then, not possible to make absolute demands for scientific hypotheses and theories to be real truths. It is only possible to regard scientific knowledge as preliminarily acceptable hypotheses.

2. The Hypotheses Quotient

2.1. Definition of the HQ

While working on a comparative metatheoretical study of psychological theories (Madsen, 1959), I found that the 20 theories analyzed were different in many ways. One difference was the relative number of two kinds of hypotheses: *theoretical* hypotheses and partly *empirical* hypotheses. Before describing this difference, we must present our definition of the term *hypothesis*.

A hypothesis is a general proposition that formulates the relationship between two or more terms (variables), among which at least one term is hypothetical or transempirical (i.e., referring to an unobserved or unobservable intervening variable or to an explanatory construct).

In accordance with this definition, a *theoretical hypothesis* is defined as a hypothesis formulating the relationship between two or more hypothetical (transempirical) terms. If we introduce the abbreviation *H* for hypothetical (transempirical) term, then we can abbreviate theoretical hypothesis to *H–H hypothesis*.

A partly *empirical hypothesis* is defined as a hypothesis formulating the relationship between at least one descriptive (empirical) term and one or more hypothetical terms. Thus, (partly) empirical hypotheses establish relationships between intervening variables (or hypothetical constructs) on the one hand and empirical (observable) variables on the other. The empirical variables can be classified into *independent* variables and *dependent* variables. In psychology, independent variables are often defined as *stimuli* or *S-variables,* and the dependent variables are often defined as *behavior* or *R-variables.* (This last class of variables may be defined so broadly that it also includes *phenomenological variables,* i.e., the private, conscious experience of the experimental subject, as expressed in public verbal communications.) Using the conventional terminology, we can abbreviate the two subclasses of empirical hypotheses as *S–H hypotheses* and *H–R hypotheses.*

With the introduction of this conventional terminology, we can express the relationship—or relative number of theoretical hypotheses and (partly) empirical hypotheses—as a ratio in the following formula for the hypotheses quotient (HQ).

$$HQ = \frac{\Sigma\,(H\text{–}H)}{\Sigma\,[(S\text{–}H) + (H\text{–}R)]}$$

The formula is constructed in such a way that the lower the HQ, the higher the testability of the theory.

Before we present the various interpretations of this formula, we must prevent misunderstandings by pointing to a consequence of our definition of hypothesis. Only two categories of hypotheses can exist:

1. Theoretical hypotheses (*H–H* hypotheses)
2. *Partly* empirical hypotheses, including *S–H* hypotheses and *H–R* hypotheses

There can be no *completely* empirical hypotheses (symbolized as *S–R* relations) because we have defined *hypothesis* as a general proposition containing at least one transempirical (hypothetical) term.

Therefore, we shall introduce a new term, *datathesis,* as a name for a proposition describing either a single observed variable or a relationship between two or more empirical variables. Datatheses that describe relationships between empirical variables include:

1. The *functional relationship* between an independent (*S*) variable and a dependent (*R*) variable (thus, this category of datatheses includes the so-called *S–R* laws)
2. The *correlation* between two or more dependent (R) variables

These two categories of datatheses (*S–R* theses and *R–R* theses) are not hypotheses according to our definitions. Even in the case in which a scientist predicts a not yet observed relationship (between *S*-variables and *R*-variables), the thesis should not be called a hypothesis but a *provisional (or tentative) datathesis*. Furthermore, such a datathesis must not be included in the calculation of the HQ as we have defined it, at least not if it is to be in accordance with our present interpretation of the HQ, which is the result of a development that we present in the next section.

2.2. Interpretation of the HQ

In our first comparative metatheoretical study of psychological theories (Madsen, 1959), we also used the term *theory-empiri-ratio* for the HQ. This term (t/e-ratio) was used to express our first interpretation of the HQ, namely, as "a measure of the hypothetical abstraction (versus empiricism) of the theories" (Madsen, 1959, p. 335). In a later work (Madsen, 1974), we introduced the term *explanatory power* as a better expression of our interpretation of the HQ. We thought that the HQ was an estimation of what Miller has called "the ratio of facts to assumptions" (Miller, 1959). At that time, we thought we could just as well include purely empirical formulations (*S–R* laws) in the calculation of the HQ. (In that case, the *S–R* laws should be added to the *S–H* and the *H–R* hypotheses.) However, we did not include the *S–R* laws, because only a few of the theories analyzed had any explicit formulations of *S–R* laws. In order to make the theories more comparable, we therefore decided to exclude the *S–R* laws from all the HQs.

Later we came to doubt if it was correct (or, rather, if it served any purpose) to include *S–R* laws in the calculation of HQs. The reason for this was that in the case of living psychologists—and most of our subjects are living—there might still be an increment in the number of *S–R* laws to be predicted, or found empirically and explained from the set of hypotheses in their theories. Even in the case of no longer living psychologists, an increase could occur in the number of *S–R* laws produced by followers using a theory in predictions and explanations.

If we do not include *S–R* laws in the calculation of the HQ, then we must reinterpret the hypotheses quotient. In this case, it is not an estimation of the explanatory power of a theory, but of the *potential* explanatory power. In other words: the HQ is an estimation of how broadly hypothetical terms are anchored to descriptive terms by means of postulated functional relationships.

This last interpretation of the HQ, as an estimation of *potential explanatory power*, is rather close to a formulation of Popper's, to be found in his intellectual autobiography *Unended Quest* (1976), where he writes (p. 86):

> The better theories are those with the greater content and the greater *explanatory power* (both relative to the problems we are trying to solve). And these, I showed, are also *the better testable theories;* and—if they stand up to test—the better tested theories. (Italics added.)

Thus, Popper relates *explanatory power to testability.* That inspired us to our last interpretation of the HQ. The formula is a quantitative estimation of the testability of a theory.

In order to demonstrate the plausibility of this interpretation, the next section of this paper will present the details of the method of calculation of the HQ, with an example.

3. Calculation of the HQ

3.1. The Procedure of Calculation

It is obvious from the constitution of the HQ formula that a prerequisite for calculation of the HQ of a certain theory is that there exist a *classification* of the hypotheses of the theory into two categories—*theoretical* hypotheses (*H–H* hypotheses), and partly *empirical* hypotheses (*S–H* and *H–R* hypotheses). In order to make this classification, the theory has to be formulated with such a degree of precision that the classification can be done while reading the original text.

This is rather easily done with systematic and exact theories, such as Hull's theory. But, unfortunately, many (perhaps a majority of) psychological theories are not written in such a formal, systematic, and precise way that their hypotheses are easy to classify directly. There are even theories—such as Maslow's theory—that are written so that the main hypotheses are *implicit* in the text.

In such a case, it is necessary to first make an *explicit formulation* of the hypotheses implicit in the theory. The steps in the most difficult cases are:

1. Explication
2. Classification
3. Calculation

But it is obvious that the explication of the implicit hypotheses is a more or less subjective interpretation of the text. This introduces a source of error. In order to reduce the influence of subjective interpretation, I have proceeded as follows:

1. The process of explication is made public; that is, it is presented in the metatheoretical text so that the reader can follow and check it. First, large parts of the analyzed text are quoted. Second, the content of the quotations is concentrated in an explicit formulation of the hypotheses, which are contained in the quotation according to the interpretation of the metascientist in question (in this case, the present author). Third, the explicit verbal formulation is translated into a partly symbolic formulation, which makes it easy to classify the hypotheses into the theoretical and the partly empirical.

The symbolic formulation uses an *S–H–R* notation, but the application of our metatheoretical method is *not* limited to so-called *S–R* theories or behavioristic theories. The *S–H–R* sequence can be generalized to an *independent-intervening-dependent variable sequence* (or input–transformation–output sequence), which is applicable to *all theories:* physical, biological, psychological, and social.[1]

2. The interpretation is checked by the author of the theory. The results of the explication and systematic reconstruction of the theory are presented to the author of the theoretical text, so that he can approve the interpretation of his own theory. This is, of course, only possible in cases of the theories of contemporary psychologists.

This procedure was followed in the comparative metatheoretical study of 22 contemporary psychological theories (Madsen, 1974). However, there were two exceptions: One of the authors died before the analysis started (Woodworth), and one died while the analysis was going on (Maslow). As Maslow's theory was one of the least explicit and systematic theories, we have greater uncertainty about its HQ (0.13) than about any other, which is indicated by a question mark in Table 1 (see Table 1).[2]

[1] Thus, the HQ formula could be reformulated into a version applicable *to all* sciences:

$$\text{HQ} = \frac{\Sigma\,(H\text{–}H)}{\Sigma\,[(E_{in}\text{–}H) + (H\text{–}E_{dep})]}$$

where E_{in} is equal to "independent, empirical variable" and E_{dep} is equal to "dependent, empirical variable." Bunge (1967) introduced a similar formula for explanatory power.

[2] The unexpectedly low HQ of Maslow's theory may be explained by the following:

1. Maslow's theory of motivation is the most precise and systematically organized part of Maslow's whole production. Only this motivational theory is analyzed for calculation of the HQ.
2. I had to reconstruct Maslow's theory to such an extent that it could be claimed that the HQ is valid only for my version of Maslow's theory and not for the original version.

Table 1. The Hypotheses Quotients (HQ) of some Psychological Theories Analyzed in Madsen (1974) and Madsen (1959) Respectively

Modern theories	HQ	Earlier theories	HQ
Catell	0.09	Tinbergen	0.11
Maslow	0.13?	Hebb	0.13
Duffy	0.14	McClelland	0.14
Miller (I)	0.20	Hull	0.36
Pribram	0.29	McDougall	0.43
Bindra	0.30	Lewin	0.50
Atkinson and Birch	0.33	Murray	0.71
Berlyne	0.38	Young	0.82
Brown	0.38	Allport	1.00
Konorski	0.54	Tolman	1.43
Woodworth	0.57		
Miller (II)	0.60		
Festinger	0.84		
Atkinson	0.86		

3.2. An Example: The Case of Freud's Theory

In order to demonstrate the method of calculation of the HQ, we will analyze a text and show how we arrive at an HQ for it. For the purpose of demonstration, we have selected a part of Chapter 7 of *The Interpretation of Dreams* (1900/1953, pp. 535–543). The reasons for the selection of this text are:

1. Freud's theory is the most important theory in the history of psychology.
2. The selected text introduces in relatively few pages the well-known topographical theory and the so-called *lenzemodel,* which is among the earliest explanatory models in psychology.
3. Freud's theory is one of the theories that Popper criticized as being untestable and therefore—according to his criterion—unscientific.
4. Freud's text is a well-suited object for demonstration because it belongs to the majority of nonformalized theories in psychology but is rather clear and precise in presentation (e.g., more so than Maslow's text).

The precision of the presentation is facilitated by the construction of a graphic model (which is lacking in the case of Maslow). Thus, Freud may be conceived to be in the middle of a dimension ranging from the most precise and systematically formalized theories (such as Hull's, Lewin's, Cattell's, and Atkinson's) to the least precise and systematic theories (such as Maslow's).[3]

As guidance for the selection of the more or less implicit hypotheses in the text, we use the graphic model. According to this, we have found the following (implicit) hypotheses, which we quote in the order in which they appear in the text.

1. *Quotation:* "The first thing that strikes us is that this apparatus, compounded of Ψ-systems, has a sense of direction. All our physical activity starts from stimuli (whether internal or external) and ends in innervations." (p. 537)
2. *Quotation:* "We shall suppose that a system in the very front of the apparatus receives the perceptual stimuli but retains no trace of them and thus has no memory, while behind it lies a second system which transforms the momentary excitations of the first system into permanent traces." (p. 538)
3. *Quotation:* "The first of these *Mnem.* systems will naturally contain the record of association in respect to *simultaneity in time;* while the same perceptual material will be arranged in the later systems in respect to other kinds of coincidence, so that one of these later systems, for instance, will record relations of similarity, and so on with the others." (p. 539)
4. *Quotation:* "But when we consider the dream-wish, we shall find that the motive force for producing dreams is supplied by the *Ucs;* and owing to this later factor we shall take the unconscious system as the starting point of dream formation." (pp. 541–42)
5. *Quotation:* "The only way in which we can describe what happens in hallucinatory dreams is by saying that the excitation moves in a *backward* direction. Instead of being transmitted toward the *motor* end of the apparatus it moves toward the *sensory* end and finally reaches the perceptual system." (p. 542)
6. *Quotation:* "According to our schematic picture, these relations are contained not in the *first Mnem.* systems but in later ones; and in case of regression they would necessarily lose any means of expression except in perceptual images. *In regression the fabric of the dream-thoughts is resolved into its raw material.*" (p. 543)

On the basis of the quoted passages, we can formulate some explicit

[3] In spite of these formal drawbacks, I regard Maslow's theory as very original and thought-provoking (see Madsen, 1981).

hypotheses.[4] These explicit formulations are made in a partly symbolic way. The letters *S, H,* and *R* are used to indicate that the term in parenthesis after the letter is a term for an independent *S*-variable, a hypothetical variable, or a dependent *R*-variable. Arrows are used as symbols for functional relationships. Thus, we have the following hypotheses:

Implicit in the first quotation is:
1. *Hypothesis:* S (External) → H (Perception)
Implicit in the second quotation is:
2. *Hypothesis:* H (Perception) → H (Memory)
Implicit in the third quotation is:
3. *Hypothesis:* H (Memory_1) → H (Memory_2)
Also implicit in the *first* quotation is:
4. *Hypothesis:* S (Internal) → H (Unconscious motives)
Implicit in the fourth quotation is:
5. *Hypothesis:* H (Ucs motives) → H (Preconscious system)
What is implicit in the fifth quotation is:
6. *Hypothesis:* H (Pcs repression) → H (Memory_2)
Also implicit in the *fifth* quotation is:
7. *Hypothesis:* H (Memory_2) → H (Memory_1)
Implicit in the sixth quotation is:
8. *Hypothesis:* H (Memory_1) → H (Hallucinatory pcpt.)
And finally, implicit in the *first* quotation is:
9. *Hypothesis:* H (Perception) → R (verb. rep.)

On the basis of these *explicit formulations,* we can make the following *classification* of the hypotheses:

1. *H–H hypotheses:* Hypotheses nos. 2, 3, 5, 6, 7, and 8. In sum: 6 hypotheses.
2. *S–H hypotheses:* Hypotheses nos. 1 and 4. In sum: 2 hypotheses.
3. *H–R hypotheses:* Hypothesis no. 9. In sum: 1 hypothesis.

If we calculate the HQ on the basis of this classification, we have:

$$\text{HQ} = \frac{\Sigma\,(H\text{–}H)}{\Sigma\,[(S\text{–}H) + (H\text{–}R)]} = \frac{6}{2 + 1} = 2.0$$

This is the highest HQ we have found. And, as you will remember, the

[4] In addition to the quotations, we are (as already mentioned) guided by the *graphic model,* and we have the full text as *a context for understanding.*

higher the HQ, the lower the testability of the theory. While evaluating this HQ, we must keep in mind that the text we have used as the basis for the calculation is only one of Freud's many theories. Furthermore, the topographical theory—as well as its later substitute, the structural theory—is the most abstract among the whole hierarchy of theories in Freud's theoretical system. Lower in the hierarchy, we find Freud's theories about instinctual drives, forms of mental energy, and anxiety. Still lower in the theoretical hierarchy—closer to the empirical foundation—are the theories about dreams, slips of the tongue, and neurotic symptoms.

3.3. Another Example from Freud

In order to demonstrate that the other theories are lower in the hierarchy of abstraction and therefore supposed to have a lower HQ (and higher testability), we have reconstructed Freud's theory of anxiety (presented in Freud's *Problems of Anxiety,* 1926). The reconstruction is presented in the form of a graphic model (see Figure 1).

In addition to the model, we have also made the following explicit formulation of Freud's hypotheses.[5]

1. *S* (Trauma) → *H* (Anxiety
2. *S* (Danger) → *H* (Perception)
3. *H* (Perception) → *H* (Expectation)
4. *H* (Expectation) → *H* (Anxiety)
5. *H* (Anxiety) → *H* (Defense)
6. *H* (Defense) → *H* (Change of Pcpt.)
7. *H* (Drives) → *H* (Perception)
8. *H* (Super-Ego aggression) → *H* (Perception)
9. *H* (Anxiety → *R* (Flight)
10. *H* (Anxiety) → *R* (Conscious Experience)
11. *H* (Anxiety) → *R* (Organic Processes)

These hypotheses can be classified as follows:

1. *Theoretical (H–H) hypotheses:* Nos. 3, 4, 5, 6, 7, and 8. In sum: 6 hypotheses.
2. *Partly empirical hypotheses:*
   a. *S–H* hypotheses: nos. 1 and 2. In sum: 2 hypotheses.
   b. *H–R* hypotheses: nos. 9, 10, and 11. In sum: 3 hypotheses.

[5] In this case, we have omitted the text for practical reasons.

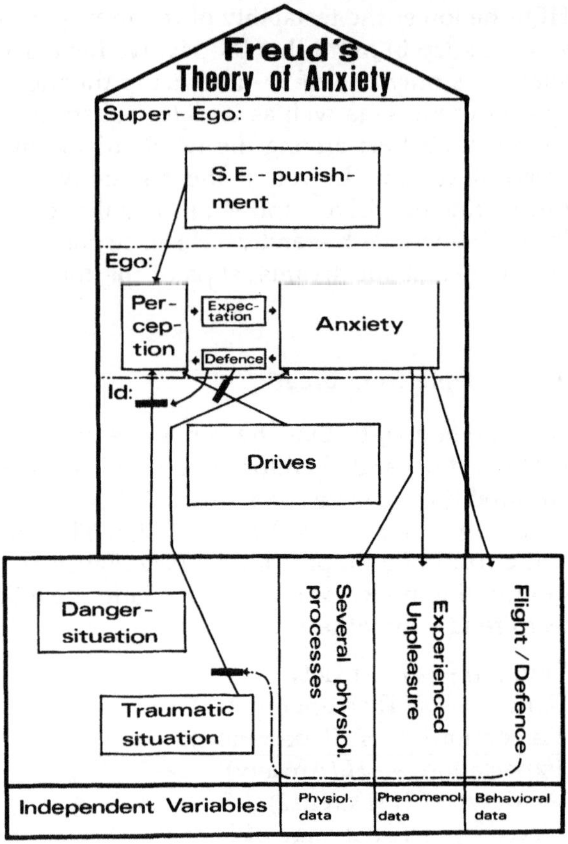

Figure 1. A graphic model of our reconstruction of Freud's theory of anxiety.

On the basis of this classification, we can calculate the HQ:

$$HQ = \frac{\Sigma\,(H\text{–}H)}{\Sigma\,[(S\text{–}H) + (H\text{–}R)]} = \frac{6}{2 + 3} = 1.20$$

Thus, the testability of Freud's theory of anxiety *is* higher (the HQ lower) than the testability of his topographical theory. However, even a HQ of 1.20 is among the highest we have found (see Table 1). Before leaving the testability of Freud's theories, we should mention that some of them probably have still higher testability (and lower HQ). Unfortunately, we have not yet analyzed these.[6]

[6] Later I have calculated the HQ for the theories presented in *Introductory Lectures to Psychoanalysis* (1917) about faulty actions (HQ = 0.33), dreams (HQ = 0.60), and neuroses (HQ = 0.71).

4. Some Metatheoretical Reflections

4.1. Half-formal Theories

We should like to emphasize that testability and fruitfulness are not identical. There may be theories with a low degree of testability (high HQ) that have been of great heuristic value by inspiring the creation of many theories with higher testability (and lower HQ). We think that many psychologists will agree with our evaluation of Freud's theories as some of the most fruitful in the development of psychology.

We were inspired to this conception of the relationship between testability and fruitfulness by Bunge (1967). He criticized Popper's testability criterion and drew attention to theories in physics that are not directly testable but nevertheless are important to the development of physics. Bunge especially mentions classical mechanics and cybernetics. He regards these theories as *half-formal* theories, with an epistemological or metatheoretical status between the *formal* theories of logic and mathematics and the *nonformal* (empirical) theories of physics and other nonformal sciences investigating the real world. Such half-formal theories are not directly falsifiable, but they can be indirectly verified through their use in relation to the empirical (and directly testable) theories of physics.

4.2. Research Programs

A similar distinction is made by Popper, who distinguishes between testable scientific theories and nontestable *metaphysical research programs*. In Popper (1976), this concept is defined as "*a possible framework for testable scientific theories*" (p. 168).[7]

As an example of metaphysical research programs, Karl Popper mentions Darwin's theory of evolution by natural selection. This example demonstrates that a metaphysical research program may be very fruitful for the development of scientific knowledge, even when it is not a testable scientific theory itself. I think that Popper could (and should) have mentioned the theories of Marx and Freud as examples of metascientific research programs. He has earlier described these theories as

[7] Thus, the concept of metaphysical research programs is very similar to what Kuhn (1962) has called *paradigms*. The paradigms include the steering philosophy of the world and of science. Furthermore, they include some main hypotheses and empirical research work, which has assumed the role of an ideal for scientists in a whole field (e.g., Copernicus, Newton, Einstein, and in psychology, Freud, Pavlov, and Piaget).

not testable (and not scientific). Furthermore, Freud made a similar distinction by introducing the term *metapsychology*.

4.3. Freud's Metapsychology

Freud sometimes used the term *metapsychology*. The first mention of it appears in a letter (no. 41) to Wilhelm Fliess (1896), in which it is not defined. In a later letter (no. 84) to Fliess (1898), it was used again, this time with some suggestion of a definition.

> It seems to me as if the wish-fulfillment theory gives only the psychological and not the biological or rather metapsychological explanation. (Incidentally I am going to ask you seriously whether I should use the term 'metapsychology' for my psychology which leads behind consciousness.)

Freud uses the term sometimes in his later published works. The most explicit definition we have found is in *Das Unbewuste*, from 1915, where he writes (p. 140 in *Studienaufgabe, band III*, which is equal to Standard Edition, vol. 14, p. 181):

> Ich schlage vor, dass es eine *metapsychologische* Darstellung genammt werden soll, wenn es uns gelingt, einen psychischen Vorgang nach seinen *dynamischen, topischen,* und ökonomischen Beziehungen zu beschreiben.

It is obvious from these two proposals for definitions that Freud uses the term *metapsychology* for those explanations that are *transempirical* ("leads behind consciousness") and uses hypothetical constructs (of the dynamic, topographical-structural, and economic categories) to explain the functioning of the "psychical apparatus."

In our metatheoretical terminology, this is equal to the following: metapsychological explanations are explanations made (mainly) by means of *theoretical* hypotheses.

One of the most important metapsychological texts is the part of Chapter 7 of *The Interpretation of Dreams* analyzed in the preceding pages. Although Freud did not call this a metapsychological text, it is obvious from the later use of the term *metapsychology* that this text belongs to that category. As we have already seen, Freud used *theoretical* hypotheses in this text mainly to explain the function of the psychical apparatus. Thus, it is a *topographical* (structural) theory. (It is later supplemented by dynamic and economic hypotheses.) It is also obvious that the explanation "leads behind consciousness." This text, then, has all the defining characteristics required for a metapsychological text.

If we look at the text again, it is clear that the passages containing the implicit hypotheses are only a small part thereof. The main part of the rest of the text is about the construction of the explanatory model and the graphic presentation of the *lenzemodel*. Thus, the main part of the text belongs to the same level of abstraction as the hypotheses. We call it the *hypothetical level*.

In addition, there are parts of the text that belong to a higher level of abstraction. This level contains the *philosophical and metatheoretical presuppositions* of the theory construction and empirical research. Therefore, we have named this abstract level the *metalevel* and the theses constituting this part of the text are called *metatheses*. These theses may be roughly classified into two main categories: metatheses containing the presupposed philosophy of the world, or *ontology*, and metatheses containing the philosophy of science (including epistemological, metatheoretical, and methodological presuppositions).

Freud's philosophy of the world is presented—or at least hinted at—on page 536, from line 7 to line 20. There is a short presentation of his philosophy of the mind/body problem. This is a version of the so-called *identity-theory* or *double-aspect theory*, which is incorporated in this model of the psychical apparatus.

Freud's philosophy of science is presented—or at least hinted at—on the same page 536, from line 20 to line 35. This philosophy of science—or, rather, *metatheory* in the more precise sense—is a version of an *instrumentalist metatheory*, as Freud presents arguments for using analogies in the form of explanatory models. There are other metatheses spread throughout the text—the largest number on page 543, where he also points to the fruitfulness of using explanatory models.

Thus, we may conclude this part of our analysis of Freud's metapsychology in the following way: A *metapsychological text* is a text that contains some metatheses, as well as some hypotheses. The latter are mainly theoretical hypotheses.

Table 2. An Overview and Comparison of Metatheoretical Concepts

Levels of abstraction	Bunge's half-formal theories	Popper's research program	Freud's metapsychology
Metalevel	No (or few) metatheses	Some metatheses	Some metatheses
Hypothetical level	Mainly theoretical (and perhaps a few empirical) hypotheses	Mainly theoretical (and a few empirical) hypotheses	Mainly theoretical (and a few empirical) hypotheses
Datalevel	No datatheses	No datatheses	No datatheses

4.4. Comparison

It is obvious from this analysis of Freud's text as a metapsychological text that his concept of metapsychology is rather similar to Popper's concept of a metaphysical research program. Both terms signify texts including metatheses and hypotheses, but not datatheses. It is also clear that these two concepts have some similarity to Bunge's concept of a half-formal theory, which designates a text including mainly theoretical hypotheses, no datatheses, and perhaps a few implicit metatheses. To clarify the comparison further, we have made a comparative matrix (see Table 2).

From an inspection of Table 2 it is clear that there is enough similarity between Bunge's half-formal theories and Freud's metapsychology to make the following conclusion possible: Although metapsychological texts lack empirical content (no datatheses and no or only a few partly empirical hypotheses), they may be just as heuristically fruitful for psychology as the half-formal theories have been for physical science. Thus, *the HQ is a quantitative estimation of testability, but not of heuristic value.*

5. Concluding Remarks

The Hypotheses Quotient (HQ) is presented here as a quantitative estimation of the testability of a theory. The procedure of calculating the HQ is presented in detail and discussed critically. We have demonstrated the method by an analysis of two of Freud's theories and calculated the HQ of these theories.

In connection with our analysis of Freud's theories, we have applied some concepts from our comparative metatheory called *systematology*. A summary of this metatheory is organized as a graphic model (see Figure 2).

Some of the problems in the calculation of the HQ can be avoided by using a computer program developed for the purpose. In addition, we have developed an alternative ratio, called the *M–H–D* ratio. However, the presentation of these supplementary methods requires more space than this chapter allows. Therefore, we refer the reader to two books (Madsen, 1974, 1975).

ACKNOWLEDGMENTS

The author gratefully acknowledges the assistance of Linda Kalish, presently at the University of Copenhagen, for her English language corrections and for typing the manuscript.

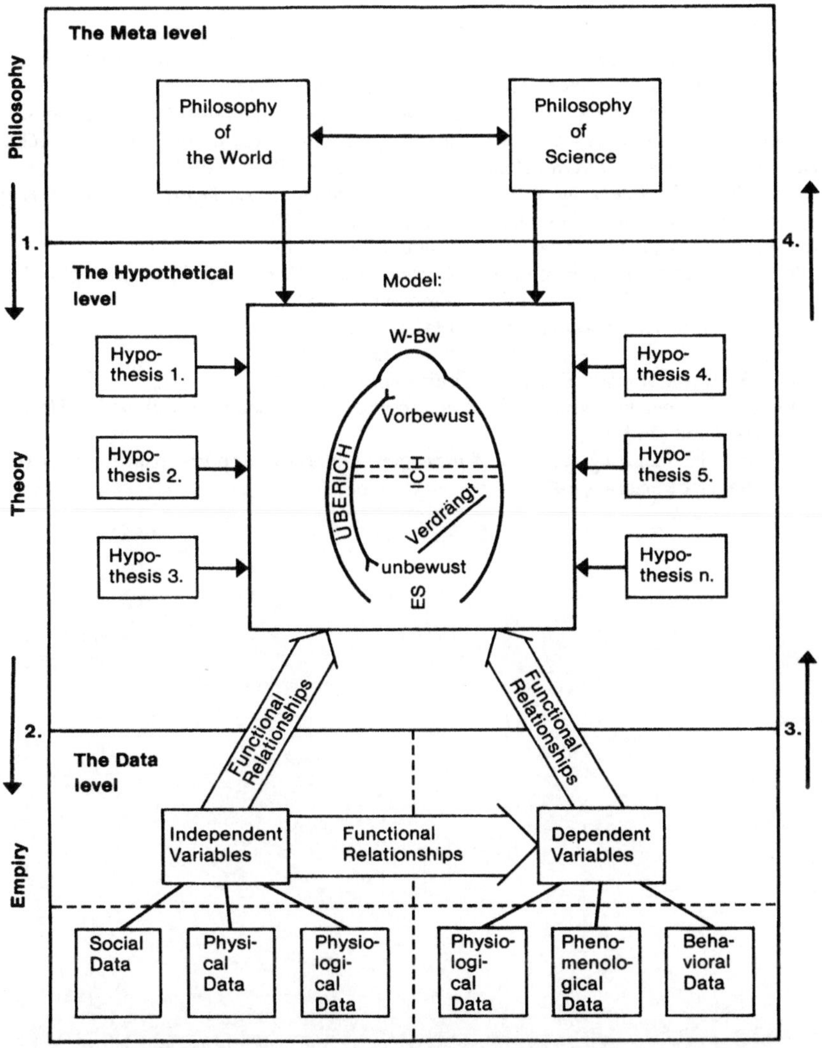

Figure 2. A graphic model of our metatheory—called *systematology*. This metatheory conceives a scientific text as consisting of three levels of abstraction: the *metalevel*, containing metatheses (i.e., propositions about the philosophy of the world and the philosophy of science); the *hypothetical level*, containing hypotheses and the explanatory model (in this case, Freud's structural model); and the *datalevel*, containing datatheses (i.e., general functional relationships and specific descriptive propositions).

6. References

Bunge, M. *Scientific research. I and II*. New York: Springer-Verlag, 1967.

Freud, S. *The standard edition of the complete psychological works of Sigmund Freud* (J. Strachey, Ed. and Trans.). London: The Hogarth Press, 1953–1974.

Kuhn, T. S. *The structure of scientific revolutions*. Chicago: University of Chicago Press, 1962.

Madsen, K. B. *Theories of motivation*. Copenhagen: Munksgaard, 1959 (4th ed., 1968).

Madsen, K. B. *Modern theories of motivation*. New York: Wiley, 1974.

Madsen, K. B. *Systematology: A comparative metatheory for psychologists*. (in Danish; English ed. in preparation). Copenhagen: Munksgaard, 1975.

Madsen, K. B. *Abraham Maslow*. Copenhagen: Forum, 1981.

Miller, N. E. Liberalization of basic S–R concepts. In S. Koch (Ed.), *Psychology—A study of a science (Vol. II)*. New York: McGraw-Hill, 1959.

Popper, K. *Logik der Forschung*. Wien: 1934.

Popper, K. *Logic of discovery*. London: Hutchinson, 1959.

Popper, K. *Unended quest*. Glasgow: Collins/Fontana, 1976.

Radnitzky, G. *Contemporary schools of metascience*. Copenhagen: Scandinavian University Books, 1968 (2nd ed., 1970).

Törnebohm, H. *Inquiring systems and paradigms*. Göteborg: Reports from the Department of Theory of Science, 1975.

4

Logic and Psycho-logic of Science

Lewis Wolfgang Brandt

In my comments I shall follow as closely as possible the structure of Madsen's chapter on the Hypotheses Quotient in order to facilitate comparison. Since Madsen states that he bases his work on the philosophy of science by Kuhn, I shall do the same in my comments wherever I consider it appropriate. My discussion from the point of view of a logic of science will be followed by considerations from the perspective of a psycho-logic of science. I say *a* and not *the* logic of science because there exist various logics of science (cf. Radnitzky, 1968). The reader will see how much Madsen's and mine overlap.

1. The Testability Criterion

Although he refers to later works, Madsen's historical introduction ends with Kuhn (1970). If Kuhn (1970) may have agreed with Madsen's testability criterion, Kuhn (1970), at least according to my reading, did not. I shall return to this discrepancy later.

2. The Hypotheses Quotient

Madsen distinguishes between *hypotheses* and *datatheses*. His definition of a datathesis indicates that not only the two variables included

Lewis Wolfgang Brandt • Department of Psychology, University of Regina, Regina, Saskatchewan, Canada, S4S 0A2

but also their relationship is an empirical one. He defines a *hypothesis* as "a general proposition that formulates the relationship between two or more terms (variables), among which at least one term is hypothetical or transempirical," but he does not say anything about the relationship. He disregards propositions in which the relationship is hypothetical or transempirical. For example, a 'causal' relationship is hypothetical so that the proposition '*A* caused *B*' is transempirical even if *A* and *B* are observable. However, '*A* preceded *B*' is an empirical proposition, if both *A* and *B* are observable and *B* was actually observed as following *A*.

In his definition of a hypothesis he further does not distinguish between "unobserved" and "unobservable" variables. Yet, it is exactly this distinction which, according to Holzkamp (1964), enables one to differentiate between the "true liability" of a theory and the merely "apparent liability" (*unechte Belastetheit*). The former cannot be further reduced, whereas the latter can. True liability is, according to Holzkamp, one of the criteria for the "empirical value" of a theory (cf. Brandt, 1967), which is the central issue in Madsen's article.

Madsen's definition of a "partly empirical hypothesis" assumes that "empirical (observable) variables" are not themselves always theory laden (Hanson, 1969). He considers stimuli and behavior to be empirical whereas Kuhn (1970a) stated that "people do not see stimuli; our knowledge of them is highly theoretical and abstract. . . . We *posit* the existence of stimuli to explain our perceptions of the world, and we posit their immutability to avoid both individual and social solipsism" (pp. 192–193; emphasis added).

Madsen excludes from his HQ formula what he calls datatheses. He justifies this exclusion on the basis that in his previous work "only a few of the theories analyzed had any explicit formulations of *S–R* laws. In order to make the theories more comparable, we therefore decided to exclude the *S–R* laws from all the HQs." It seems to me that the comparison of theories in terms of their "empirical value" would have been strengthened by the inclusion of *S–R* laws. Nor do I accept Madsen's second reason for excluding these laws: If theories are to be compared, they must be compared at any given moment in history. The argument that a present theory may include new *S–R* laws in the future makes "the critical discussion . . . [of] what may be called the 'science of the day' " (Popper, 1976, p. 86) impossible.

The reader may realize that my last quotation is from the same page from which Madsen quoted. As I interpret the passage quoted by Madsen, theories are to be compared in terms of *both* their scope *and* their falsifiability. However, Madsen disregards the scope of theories entirely in his comparisons.

3. Calculation of the HQ

Madsen outlines three steps in his translation of a text into his formulae. However, in his first example he omits the second step of concentrating "the content of the quotations . . . in an explicit formulation of the hypotheses." In his second example he presents only the last step. Hence the reader cannot closely compare Madsen's interpretation of the text with his or her own interpretation.

I find it unfortunate that Madsen (1) chose for his illustration a theory which its author completely revised in later years and (2) quotes from an English mistranslation (Brandt, 1961, 1966, 1977, 1980). The quotations themselves seem rather arbitrarily selected since on the one hand they do not correspond to single hypotheses either by Madsen's own or by my interpretation and since on the other hand some hypothetical constructs seem to appear in more than one quotation. Moreover, what I understand to be the hypothesis contained in the second quotation has been experimentally supported by the Bell Telephone Laboratories (1962) research into "short-term visual memory."

Taking, then, simply what Madsen lists as hypotheses without being able to discern how he culled these from the quotations, I am at a loss to understand how he arrives at considering perceptions, hallucinations, anxiety, and expectations to be hypothetical constructs after having stated earlier that he includes "private, conscious experience" under "*R*-variables." If Madsen, as Freud did, distinguishes perceptions from a perceptual system, he ought to state this. Nor is it clear to me whether he is aware that trauma and danger are also private experiences which may or may not be conscious.

Madsen's discussion of testability omits any reference to how a given theory is to be tested. He disregards the fact that Freud (1937) stated that his theory is to be tested on the basis of deductively derived interpretations. Freud spelled out clearly the criteria for acceptance or rejection of a "construction" (interpretation). I presented an example of such psychoanalytic hypothesis-testing elsewhere (Brandt, 1974).

In terms of his quotation from Popper (1976), Madsen ought to have compared various theories in terms of their hypothetical constructs such as drive, defence, and superego on the one hand and their scope on the other. Suggestions for criteria for such comparisons can be found in Holzkamp (1964).

4. Some Metatheoretical Reflections

Since Madsen's own metatheoretical reflections at least implicitly question the usefulness of his HQ, I shall not enter into them in any detail but rather present my own.[1]

Madsen recognized that, by definition, theories—Freud frequently referred to his own as fantasy or speculation—always contain ideas. Ideas cannot be experimentally investigated. Hypothetical constructs are constructions of a mind. This fact is usually overlooked by philosophers of science when

> they wish . . . to compare theories as representations of nature, as statements about 'what is really out there.' Granting that neither theory of a historical pair is true, they nonetheless seek a sense in which the later is a better approximation to the truth. I believe nothing of that sort can be found. (Kuhn, 1970b, p. 265)

From the point of view of such a metascience it is futile to compare various theories in terms of their respective "quantitative estimations of testability"—even assuming that consensus could be achieved on the criteria for such a measurement. Such quantification leads to what Bunge (1967) called *Dataism*. Madsen himself ends his article far removed from what its title promised. As his figures illustrate (reminding me of Dadaism), he has now completely abandoned testability for heuristic value. As I shall indicate in my next section, this means a switch from evaluating a theory in terms of its ability to stop curiosity to evaluating it in terms of its tendency to stimulate further curiosity. These are, however, not logical, that is, intellectual, but affective, psycho-logical criteria.

[1] The space allotted to me for my comments does not permit me to go into more detail with respect to differences between Madsen's and my own interpretation of Freud. For example, I would have chosen as a statement of "Freud's philosophy of the world":

> Das Unbewusste ist das eigentlich reale Psychische, *uns nach seiner inneren Natur so unbekannt wie das Reale der Aussenwelt, und uns durch die Daten des Bewusstseins ebenso unvollständig gegeben wie die Aussenwelt durch die Angaben unserer Sinnesorgane.* [The unconscious is the truly real psychic, *as unknown to us in its inner nature as the real of the external world, and given to us through the data of consciousness just as incompletely as the external world through the indications of our sense organs.*] *Die Traumdeutung,* p. 617f. Freud's emphasis.

5. Psycho-logic of Science

As Kuhn demonstrated already in the work referred to by Madsen (Kuhn, 1962), scientists do not abandon a theory because of its low testability. The scientist–philosopher Lichtenberg wrote two centuries ago, "Most apostles do not defend their statements because they are convinced of their truth but because they once asserted their truth." Einstein did not reject quantum mechanics on the basis of low testability but with the *theo-logical* argument, "*der* würfelt nicht" (letter to Max Born dated Dec. 4, 1926), Freud did not revise his theories constantly because the earlier ones had low testability, but because he could not explain later observations to his own satisfaction on the basis of his earlier theories. James Deese did not reject behaviorism after having espoused it and worked within its framework for many years because of its low testability, but because it no longer made sense to him.

Mutatis mutandis, the psychologists who stick to any one of the many, frequently incommensurable theories presently on the market do so not because of its high testability but because they once espoused it for whatever personal, psychological reasons. They will, therefore, not abandon it because of its low testability—by whatever criteria.

Theories are explanations for one's experiences. In Einstein's words: "The object of all science, whether natural science or psychology, is to co-ordinate our experiences and to bring them into a logical system" (1923, p. 1). (His own "logical system" included, as the above statement to Born indicates, even theo-logic). An explanation is, according to Braithwaite (1961), that "which provides any sort of intellectual satisfaction," that is, which satisfies one's curiosity and stops one from asking further questions. In other words, psycho-logic lies at the root of all scientific inquiry and theorizing.

This has been acknowledged by P. W. Bridgman, to whose *The Logic of Modern Physics* (1927) psychologists often refer without having closely read it and who wrote in its introduction:

> It is of course the merest truism that all our experimental knowledge and our understanding of nature is impossible and non-existent apart from our own mental processes, so that strictly speaking no aspect of psychology . . . is without pertinence. (pp. x–xi)

One generation later, he wrote in the preface to his *The Way Things Are* (1959): " 'proof,' without which no science is possible, is entirely an affair of the individual and is therefore private, with the result that any creative science is of necessity private rather than public" (p. v).

The role played in science by what I call psycho-logic is also recognized by Kuhn (1970) when he discusses beliefs and values as important aspects of a "disciplinary matrix," namely, that which binds the members of any community of scientists together.

It is, therefore, not useful to differentiate theories in terms of their testability. Neither psychologists nor other scientists will abandon a theory in which they have some kind of investment on the basis of some logical criteria if they have no psycho-logical reasons to abandon it. Nor will they prefer the simpler (more parsimonious) of two theories on any but psycho-logical grounds. For even the principle of parsimony cannot be justified logically but only psycho-logically. There is no logical reason for explaining complex events in simple terms. For some people it is easier to reason in simple terms. Others enjoy complexities.

I have found it enjoyable to compare theories on psycho-logical grounds—without any expectation of inducing anyone to give up a theory presently adhered to. As a matter of fact, my psycho-logical classification of psychological theories implies that any given psychologist adheres to a given theory for psycho-logical reasons and hence will continue to support that theory as long as he or she does not change psychologically.

My method of comparison does not attribute to any one theory "better approximation to the truth." Instead, I compared a number of psychological theories in terms of their respective correspondence to various developmental levels (Brandt, 1982). My conclusions are "objective" by Bridgman's (1959) definition in so far as "there is more than one . . . method of getting to the same terminus," that is, in so far as several of my classifications coincide.

By relating various psychological theories to different levels of intellectual functioning in terms of Piaget and of object relations in terms of Freud, I found these theories to belong to corresponding developmental levels. For example, a theory relying on operational definitions uses thinking on Piaget's level of concrete operations, since operational definitions are concrete (Brandt, 1983). And a theory admitting only quantitative data belongs to Freud's anal stage on which measurable amounts are of concern.

6. Conclusion

As the history of science and the present situation in psychology demonstrate, scientists do not give up a theory they have once accepted for another one because of logical arguments. This is so because they do not espouse any theory in the first place for logical, but for psychological reasons.

Any comparison of theories which uses quantification does so on the basis of psycho-logical preferences, of valuing quantification higher than classification in terms of qualitative differences.

My own preference is for discerning qualitative differences among theories, among people, among experiences. I shall not succeed in inducing any psychologist to give up a theory he or she now holds. Neither will Madsen succeed in doing so.

7. References

Bell Telephone Laboratories. *Short-term visual memory* (film). 1962.

Braithwaite, R. B. Explanation. In *Encyclopaedia Britannica*. Chicago: Benton, 1961.

Brandt, L. W. Some notes on English Freudian terminology. *Journal of the American Psychoanalytic Association*, 1961, *9*, 331–339.

Brandt, L. W. Process or structure? *Psychoanalytic Review*, 1966, *53*, 374–378.

Brandt, L. W. Wanted: A representative experiment. Review of *Theorie und Experiment in der Psychologie* by K. Holzkamp. *Contemporary Psychology*, 1967, *12*, 192–193.

Brandt, L. W. Experiments in psychoanalysis. *Psychoanalytic Review*, 1974, *61*, 95–98.

Brandt, L. W. Psychoanalyse *versus* psychoanalysis: Traduttore, traditore. *Psyche*, 1977, *11*, 1045–51.

Brandt, L. W. Psicoanalisis no es Psychoanalyse. *Cuadernos Psicoanaliticos*, 1980, *2*, 25–34. (Spanish translation of an invited paper presented under the title "Psychoanalisis is not *Psychoanalyse*" at the meeting of the Canadian Society for Cultural and Intellectual History at Saskatoon in June, 1979.)

Brandt, L. W. *Psychologists caught: A psycho-logic of psychology*. Toronto: Toronto University Press, 1982.

Brandt, L. W. Gedanken über Gedankenlosigkeit einschliesslich "operationaler Definitionen."—Ein Ritt durch die Wüste. In G. Bittner (Ed.), *Personale Psychologie. Festschrift für Ludwig J. Pongratz*. Göttingen: C. J. Hogrefe, 1983.

Bridgman, P. W. *The logic of modern physics*. New York: Macmillan, 1927.

Bridgman, P. W. *The way things are*. New York: Viking, 1959.

Einstein, A. *The meaning of relativity*. Princeton: Princeton University Press, 1923.

Freud, S. (1937) Konstruktionen in der Analyse. In *Gesammelte Werke* (Vol. 16). London: Imago, 1945.

Hanson, N. R. *Perception and discovery*. San Francisco: Freeman, Cooper, 1969.

Holzkamp, K. *Theorie und Experiment in der Psychologie*. Berlin: de Gruyter, 1964.

Kuhn, T. S. *The structure of scientific revolutions*. Chicago: University of Chicago Press, 1970.

Kuhn, T. S. Reflections on my critics. In I. Lakatos & A. Musgrave (Eds.), *Criticism and the growth of knowledge*. Cambridge: Cambridge University Press, 1970.

Popper, K. *Unended quest*. Glasgow: Collins/Fontana, 1976.

did not produce any injury to the [illegible] place in the [illegible] of [illegible] connections [illegible]

[illegible] which [illegible] surgery [illegible]

[illegible]

[illegible] and [illegible] treatment [illegible]

References

[illegible]

Bj[illegible] (1982) [illegible]

[illegible] J [illegible] 6:221–[illegible]

[illegible]

[illegible] Mayo Clinic [illegible]

[illegible] (1981) [illegible]

[illegible]

[illegible] J Otol Rhinol Laryngol [illegible]

[illegible]

[illegible]

[illegible] (1960) [illegible]

[illegible]

[illegible] Otolaryngol [illegible]

4

Sound Theories and Theory Soundings

Mark F. Ettin

Psychoanalysis has been variously referred to as a metapsychology, an *ism*, an art, an esoteric cult, a doctrine, a theology, a demonology, and a therapeutic technique. It has been characterized as *un-*, *pre-*, *proto-*, *pseudo-*, *anti-*, and genuinely scientific, and its precepts have been given the status of laws, constructs, paradigms, beliefs, canons, hypotheses, or convenient fictions. Books, conferences, and even careers have been devoted to the study of the scientific merit and heuristic explanatory power of psychoanalysis.

Specific epistemological concerns have included the logic, reliability, validity, and evidence for the confirmation of Freudian constructs. Investigations and critical reviews focus on the semantic and syntactical rules linking hypothetical variables with one another and to the observable data, the clarity and specificity of various definitions, terms, and processes, and the nature of the analytic session as an investigatory setting. Madsen's paper is integrative and conciliatory in offering psychoanalysis a place at the table of scientific texts, while calculatingly seating it with the "guess list" of barely testable theses. Freud might interpret this as an ambivalent gesture.

Early critiques of Freudian theory often read like the love–hate relationships depicted in Victorian romantic novels; one wonders at the vehemence implied. Any ultimate consideration of psychoanalysis must take into account the vast amount of energy expended on its attack, defense, or elucidation. Freud has taught us this. Later expositions of Freudianism are handled more reasonably. Yet the complex and elusive

Mark F. Ettin • Department of Psychiatry, UMDNJ-Rutgers Medical School, Piscataway, New Jersey 08854.

nature of understanding ourselves and others can never be rendered in a merely logical manner. The matter at hand is simply not that simple. Years ago, I attended a lecture by a noted epistemologist speaking on "whether it was possible to know others existed." Throughout his entire talk and complex arguments, the speaker never gazed above his shoe tops. He failed to notice that half the bored audience was leaving the auditorium. Some correspondence between logic and reality is necessary to engage the common senses and engender belief and conviction. Freud has schooled us that we are emotional and often irrational beings, capable of great complexity and contradiction. Madsen demonstrates his own paradoxical capacities when seemingly compelled

> to emphasize that testability and fruitfulness are not identical. . . . There may be theories with a low degree of testability (high HQ) that have been of great heuristic value by inspiring the creation of many theories with higher testability (and lower HQ). We think that many psychologists will agree with our evaluation of Freud's theories as some of the most fruitful in the development of psychology.

This critique will specifically address the methodology of Madsen's HQ. It will then be argued that difficulties in Freudian theory arise not from any inherent structural or methodological flaw, but rather from the application by some of its adherents. Madsen's larger body of work, called *systematology,* will be used in comparing the processes of scientific theorizing with psychoanalytic psychotherapy. The author readily confesses to the use of psychoanalytic precepts and analysis throughout the critique with the conscious intention of demonstrating the range, utility, and commonplace acceptance of the theories.

1. Methodological Considerations

Madsen's HQ methodology effectively demonstrates the problems encountered in progressing from a speculative philosophy to a scientific empiric. In its current form, the HQ is more the product of the addition of a series of personalized judgments than it is a generalizable scaling method with which to evaluate the testability of scientific theories. Madsen's methodology is riddled with reliability problems. He confesses that his method is a "more or less subjective interpretation of the text . . . [which] introduces a source of error." Madsen's attempt at objectifying the procedure by "explication," "classification," and "calculation" is unconvincing and contrary to sound psychometric procedures. Let us critically examine the three steps in his HQ process.

Explication. Madsen's choice and explication of bits of Freud's theoretical texts are fraught with problems of *selection, inclusion,* and *rendition.* This reader was unable to discern the logic of Madsen's selection of which passages in Freud's text were to be rated. No criteria of selection are provided. Some statements were counted whereas other seemingly central and equally cogent propositions were passed over. What marks the door to such choices? One also puzzles over how a few statements or a chart can be a sufficient sample of Freud's writing on which to base generalizations. Hundreds of pages, including those filled with observations and datatheses, are excluded from the calculation. Perhaps Madsen means only to demonstrate his method by using material in an exemplary manner. If so, then the general numerical conclusions about the testability of Freud's topographical and anxiety theories are misleading.

Any evaluation of psychoanalytic precepts runs into problems of *rendition.* Madsen's methodology is heavily dependent upon the language with which the theory is stated. The actual written text is literally scored. The descriptive language is a variable independent of the concepts to be explained. A more noun-laden, metaphoric rendition may bias toward *H–H* ratings, whereas the identical concept described in a behavioral, phenomenological, or process language may encourage *S–H* or *H–R* scoring. Some theorists' genius pours forth in an abundance of metaphor and abstraction. Yet, their imagistic renderings can be tied down or operationalized. It is the work of disciples and technicians to turn smoke rings and lightning into nuts and bolts.

The first problem of rendition occurs when translating Freud from German into English. For Madsen's purposes, the translation must be particularly faithful to the structure of the language as well as the more generalized meaning. The second problem of rendition is in choosing which version of Freud's constantly evolving writing to evaluate. The third problem of rendition comes about through confusing Freudianism with psychoanalysis. Many writers, contemporaneous and subsequent to Freud, have reported or refined the originator's theory. There is no one fully accepted body of work called *psychoanalysis.* The organization and language utilized in depicting analysis is quite diverse. Rappaport (1960) has devoted much scholarship to systematizing psychoanalytic theory. Schafer (1976) has attempted to translate analytic concepts into a gerund language. Dollard and Miller (1950) present a rendition of psychoanalysis couched in Hullian language. To answer questions of explication, Madsen must clearly face problems of language, selection, inclusion, and rendition.

Classification. Madsen's classification and symbolic transformation procedures are also open to criticism. An independent rater, as this psychologist found, would be ill-equipped to follow Madsen's scoring and to duplicate his judgments. No specific or formal rules for rating and transforming the text are provided. An argument can also be made that even the most theoretical constructs can be operationally grounded. Madsen presents no studies or evidence that inter-rater reliability is possible. Thus his procedure remains an exercise in idiosyncratic philosophizing, not a potentially empirical procedure. Checking with original authors (if they happen to be alive, and Freud is not), provides no evidence for anything but, at best, a mutual consensus and, at worst, a conspiratory folly *à deux*. Without a reliable measure of testability, no valid, empirical comparisons of theories are possible.

Calculation. The last methodological criticism concerns the mathematical validity of the HQ formula. Madsen presents a simple summation method. His formula gives an even weighting to all the statements chosen. Surely, some of the analyzed propositions are more central or critical to the theory than others. Some hypotheses represent basic proponents, the axioms or postulates of the system. Other propositions may be corollaries derived by logical inference or may merely be less important to the theory as a whole. Any mathematical formula must distinguish centrality and assign a relative weight to various constructs. Madsen proceeds like the furniture mover who charges by the piece without respect to weight: a piano and a pencil are tallied equally.

Madsen's methodology of scoring each selected statement separately assumes that the constructs chosen are independent and, in fact, unrelated. Yet, the statements may in actuality depict the various elements of a larger proposition. How, then, should they be scored, as a network or as separate instances?

Another problem in scoring and weighting is using a simple dichotomous classification of hypotheses into either "partly empirical" or "theoretical" categories. No matter how carefully accomplished, such a rating is arbitrary since this summary process fails to account for the full range of possibilities. Clearly, all empirical hypotheses are not equally grounded in experience. Similarly, not all theoretical hypotheses are equally lofty and abstract. Some may require fewer transformations to reach their empirical root.

It should be noted that the HQ is an application of Madsen's larger body of work, *systematology*, described at the end of his article and graphically presented in his Figure 2. Madsen demonstrates the worthy pursuit of a theory of theories, or a "metascience." In fact, his larger text (Madsen, 1974) reads like a diagnostic manual or classification system descriptive of the various content and focal points of theories. Like Freud, he has chosen a tripartite topographical model based on levels of abstraction. The most abstract, or "metastratum," is concerned with the more philosophical propositions of a theory, including epistemological, methodological, ontological, and metatheoretical issues. The "hypothetical stratum" describes the explanatory terms or intervening variables of a theory. The connection between such constructs is called the syntax of the theory. The "descriptive stratum" is the most sensual level of a theory and is comprised of the raw data represented by phenomenological, behavioral, or physical descriptions of events and objects. The connection between the hypothetical and descriptive strata is referred to as the "semantics," "bridging principles" (Hempel, 1966), or "correspondence rules" (Boles, 1967) of the theory. The HQ is intended to be a measure of "how broadly the hypothetical terms are anchored to descriptive terms." Madsen's systematology also describes certain properties and functional relationships of the theory as a whole.

Madsen has, in effect, summarized and neatly organized a variety of basic principles of philosophical analysis and applied them to compare heretofore diverse motivational theories. This heuristically fruitful larger body of descriptive work provides a meaningful context in which to understand Madsen's attempt at quantification. His systematizing efforts hold up separately from the methodological difficulties of the HQ itself.

Let us now utilize some of Madsen's principles of analysis to investigate the logical and emotional allures, illusions, and allergic reactions to Freudian concepts and their application.

2. Freud as Philosopher and Scientist

Why does Freudian theory evoke such powerful responses? Freud himself was very well schooled in both science and the philosophy of science. In his theoretical and logical exposition, he was not given to drawing wild or doctrinaire conclusions, nor did he mistake hypotheses for concrete evidence.

> We are justified, in my view, in giving free rein to our speculations so long as we retain the coolness of our judgment and do not mistake the scaffolding for the building. (Freudian scientific methathesis identified by Madsen from *The Interpretation of Dreams*, p. 535.)

Another telling metaphilosophical statement is stated by Freud:

> The view is often defended that sciences should be built upon clear and sharply defined basal concepts. In actual fact, no science, not even the most exact, begins with such definitions. The true beginning of scientific activity consists rather in describing phenomena and then proceeding to group,

> classify and correlate them. Even at the stage of description it is not possible to avoid applying certain abstract ideas to the material in hand, ideas derived from various sources and certainly not the fruit of the new experience only. . . . They must at first necessarily possess some measure of uncertainty; there can be no question of any clear delimitation of their content. . . . Thus, strictly speaking, they are in the nature of conventions; although everything depends on their being chosen in no arbitrary manner, but determined by the important relations they have to the empirical material—relations that we seem to divine before we can clearly recognize and demonstrate them. It is only after more searching investigation of the field in question that we are able to formulate with increased clarity the scientific concepts underlying it, and progressively so to modify these concepts that they become widely applicable and at the same time consistent logically. Then, indeed, it may be time to immure them in definitions. (quoted in Rickman, 1957, pp. 70–71)

It is suggested that validating psychoanalytic propositions is not beyond the pale of science. Nothing inherent in Freud's methodology or formulations is counter to the scientific method. Freud based his theorizing on thousands of hours of clinical observation and a comprehensive study of literature and anthropology. His concepts are well grounded in data. However, it is true that Freud was not concerned with formally testing, or confirming, his concepts. It is suggested here that the proof or disproof of theoretical notions can occur during three different steps in the scientific process. And at each point, differing criteria are necessary to establish the validity of the hypotheses.

White (1970) has identified three classic theories of truth: coherence, pragmatism, and correspondence. Let us add a fourth model of truth seeking: the stochastic methods of modern empirical science. The mark of an acceptable descriptive theory is the parsimonious accounting for of heretofore unrelatable experiences. A descriptive theory such as Freud's psychoanalysis is data-based, compelling, coherent, and pragmatically valuable. A good descriptive theory has a certain aesthetic veracity. While the rules for transforming datatheses into hypothetical constructs must be clearly specified, the most that can be said about such a descriptive theory is that it meets the *coherent and pragmatic criteria of acceptance.*

Confirmation or verification requires that hypothesized constructs now be used not merely as descriptive *post hoc* tools. Rather, hypotheses must be thrown back on experience and either statistically predict future occurrences or correspond with some independent set of explanations.

Testing a theory involves structuring a methodology that formally investigates its hypothetical implications. A feedback loop is created wherein implied relationships are experimentally manipulated and statistically evaluated. The most that can be said about applying empirics to a theory is that currently it meets the *statistical probability criteria of confirmation.*

Verification is the most stringent and telling test of a theory. *To verify* implies a correspondence, consistency, or identity with the larger body of science and the nomothetical network of existing knowledge. Rose (1981) suggests that any phenomena may be studied at different levels of analysis. Thus, for example, a single human event occurs on a phenomenological, evolutionary, interpersonal, intrapsychic, behavioral, and biochemical level simultaneously. Rose suggests a hierarchy of explanation ranging from the more concrete, physical, and microscopic to the more abstract, phenomenological, and macroscopic. The physical theories have the advantage of greater specificity. The abstract theories have a wider domain and more relevance to the real, everyday world. Rose coaches against reductionistic thinking that equates higher levels of specificity with causes. Rather, epistemological levels are isomorphically related, as they seek to explain the same event. Translation between levels of analysis yields the most comprehensive *criteria of verification: the correspondence of a hypothesis with the larger body of existing knowledge.*

The relationships within and between fields with their various criteria of validity are graphically represented in Figure 1.

As Freud suggests, a theorizer may "divine" connections before

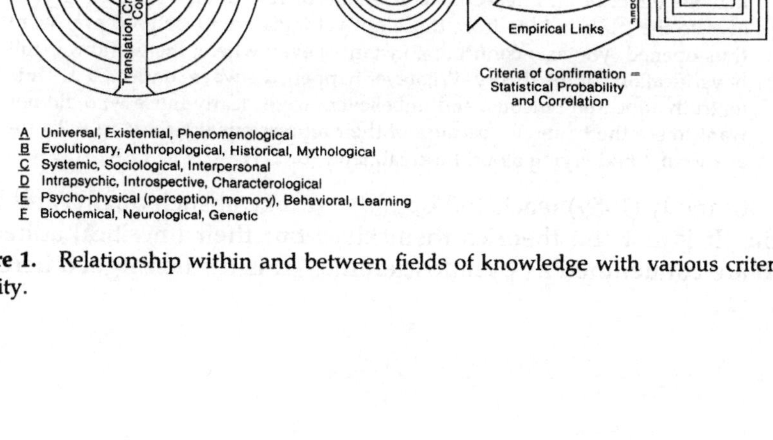

Figure 1. Relationship within and between fields of knowledge with various criteria of validity.

having the methodological sophistication to prove or demonstrate their validity. Additionally, intuitive, coherent theorizing may precede the criteria of verification made possible by technological advances in related fields. Hempel (1966) states that:

> The strength and support that a hypothesis receives from a given body of data should depend only on what the hypothesis asserts and what the data are: the question of whether the hypothesis or the data were presented first, being purely a historical matter, should not count as affecting the confirmation (or verification) of the hypothesis. (p. 38)

Many great thinkers were ahead of their time. Darwin awaited the evidence made possible by advances in microbiology that verified man and ape were related genetically. Einstein's relativity theory preceded the space travel that confirmed many of its propositions. Confirmation of Freud's concepts hinges on a methodology capable of formally testing its precepts. Verification awaits technological advances in such related fields as sleep and dreams, subliminal stimulations, perception, memory, neurology, and emotion. All knowledge is relative and, if useful, may be accepted for a time until confirming evidence increases or decreases the probability of its truth value. Verification occurs when a theory is borne out by independent inquiries into related matters. Falsification results when theoretical hypotheses are directly contradicted by findings at isomorphic, epistemological levels.

3. And Here We Go Again

Consistently fiery reactions to Freud belie the philosophically rooted and admittedly descriptive nature of his psychoanalytic contentions. Kennedy (1959) quotes an article by Popper (1957) giving a more emotional objection to Freudian theory. This theory appeared

> to be able to explain practically everything that happened. . . . [The] study had the effect of an intellectual conversion or revelation—of opening your eyes to the truth hidden from those not yet initiated. Once your eyes were thus opened, you saw confirming instances everywhere: the world was full of verifications of the theory. Whatever happened always confirmed it. Thus its truth appeared obvious; and unbelievers were clearly those who did not want to see the truth . . . because of their repressions which were still "unanalyzed," and crying aloud for treatment. (p. 27)

Kennedy (1959) reacts to Popper's outrage at such dogma and blind faith: "It is not the theories themselves but their uncritical adherents who are condemned as pseudoscientific." It is further argued here that the prime difficulty with psychoanalysis may be its depiction, application and the behavior of its adherents, rather than any specific philosophical difficulties with Freudian theory itself.

One of the limiting factors in the perception of adherents is that the principles previously postulated are adopted *a priori* and that some hard-core analysts believe that Freud's system must be accepted as a whole, without critical review of the concepts or its application to particular cases. Such a doctrinaire approach is merely poor practice for the scientist or the practitioner. It is true that previous experience with other individuals can lead to generalized notions about "human processes." Each individual therapy is an idiographic study from which certain universalities are discovered. Yet, in no science can individual cases be lawfully predicted from group data. Generalizations must gingerly serve as a rough map against which the specific and unique contours of the individual's real life are tested. We must be careful not to "shrink" people's animate experience down to the generalized principles of a dead theory.

While Freud's theoretical terms describe ongoing processes, his concepts are often mistakenly reified and frozen into a noun language. Hypotheses are mistaken for axiomatic truths, and nosological categories imposed on diverse phenomenological observations. Breathing life into explanatory constructs chokes their utility and suffocates the real people whom they aspire to describe.

Freud, himself, was the first and foremost adherent to psychoanalysis. The reader is referred to Jones (1955) for a detailed descriptive analysis of Freud's character and personality. His biographer describes the way Freud "fended off any influences from without, however apparently helpful, as if they were interfering distractions or even designed to lead him astray." Jastow (1932) calls psychoanalysis "a great discovery made by the wrong man." The generally hostile reaction with which Freud's revolutionary and risqué speculations were met apparently solidified his already rigid personality structure. The father of psychoanalysis was much more facile at negotiating with disciples than with peers. Diversions from the mainline of theory, no matter how creative, were reacted to as desertion, which Freud concretized by officially banning or "excommunicating" its proponents.

Because of Freud's personality and outside criticism, the psychoanalytic movement tended to become insular and self-protective, as did the excommunicated splinter groups. The long and vigorous training (including the training analysis) also tends to engender an elitist inbreeding and naive acceptance of ideas. The politics and indoctrination of analysts is powerfully described by Malcolm (1981).

It is also surprising that philosophers can get so charged up about a speculative theory such as Freud's. After all, the great works of philosophy are largely intuitive manifestos based on observation and logic. Reading some of the fiery denunciations of Freud leaves one with the impression that many philosophers know as little about psychoanalysis as some analysts do about philosophy of science. In this next section, let me make some suggestive attempts at integration.

4. A Clinician Looks at Theory Building

Individual psychotherapy is akin to theory building on a personal level and can be examined with respect to its functional relationships. The therapist's first task is semantic. The patient's words serve as minitheories or summary processes connoting the individual's phenomenological experiences. The therapist works with the individual in trying to clarify, understand, and more adequately describe the data of his or her life. In therapy, this clarification process is in itself ameliorative since, like the scientist, the individual gains more control and predictive power by developing a more adequate and exact depiction. The next task of the therapist is to look for patterns behind the empirics, regularities that occur in some predictable fashion. Hospers (1957), a philosopher of science, remarks that "knowledge in sciences begins with noticing regularities in the course of events . . . events occur over and over again in the same way." He goes on to talk about the function of theoretical explanation. "Amidst the constant diversity in our daily experience of nature . . . we trace the thin red vein of order through the flux of experience . . . we trace regularities so we can predict future occurrences and take precautions." Psychotherapists have talked about character structure in a similar fashion. Mann (1973) presents an image of themes recurring over time as a "thin red line that began in the past but remains active in the present."

Here Freudian theory has contributed by specifying certain of the correspondence rules or bridging principles between the data of personal experience and constructed reality. The following metatheoretical concepts are invoked:

1. Behavior is symbolic. (Royce, 1973, has defined this epistemologic stance as "metaphorism.")
2. Freudian theory expands on Hume's (Hendel, 1955) principle of "necessary connection," or "causal inference." Hume argues for a "native determination of the mind itself to find in the future the same pattern as we have witnessed in our past experi-

ence. . . . These mental ways are not rational; they are simply habits of mind, or dispositions of human nature. . . . We can detect three principles: the resemblance, the continuity in space and time, and the constant conjunction of any perceptions." Freud expanded this "gravitation of mind" to include emotional contiguity or an affect bridge that connects past phenomenological experience with current cognitive processes (semantic relationships: *S–H* and *H–R*), and various ideas, attitudes, memories, and intentions with other such mental phenomena (syntactical relationships: *H–H*). Freud argues for the primacy of "psychic reality" in establishing and maintaining these connections.

3. Freud borrowed the principles of ongoing development and closure from evolution and gestalt psychology, respectively. Behavior develops over time and seeks completion and successful conclusion.
4. Behavior is "overdetermined." A great deal of life experience goes into shaping our destinies. No one event or attitude can be said to cause our behavior. Rather, individuals are ongoing processes, and life events tend to reinforce or inhibit certain traits, dispositions, and patterns.

These metatheoretical principles combine to yield the transforma-

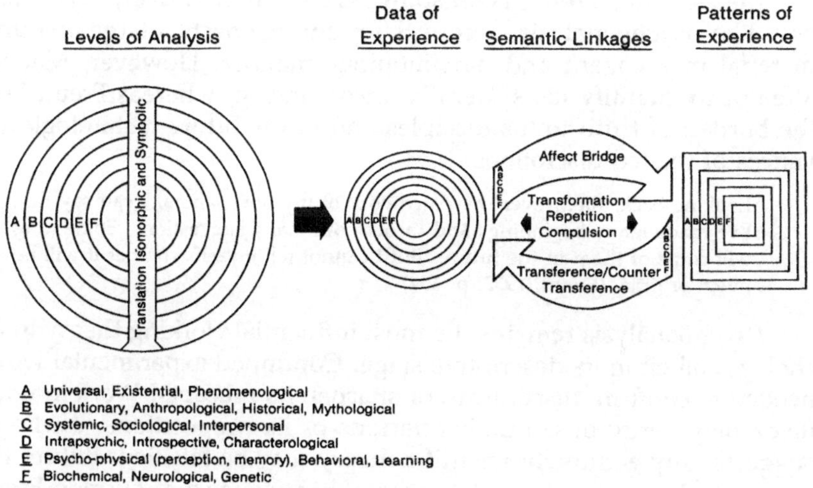

Figure 2. Transformation and translation of psychoanalytic psychotherapy.

tion principles of isomorphic transformation, repetition compulsion, and transference/countertransference. The bridging principles provide the therapist with the tools for understanding and affecting an individual's personal theoretical system. Repetition compulsion suggests that important experiences travel the affect bridge and are recreated or reenacted in patterns of behavior. Such a transformation seeks successful closure. Feelings or cognitions may be isomorphically symbolized at different levels of experience. Thus, for example, a heartache may manifest itself as an acute atrial fibrillation, a mournful mood, hurt feelings in an exchange, or a dream about losing something of importance. Psychoanalytically oriented therapy uses the transference/ countertransference relationship as the treating transformation, and the empirical measure of change. As old patterns are recreated in the therapy relationship, well-worn ways are tested for their accuracy, appropriateness, and utility. Psychoanalysis has it that as the transference/countertransference relationship changes, so does the new data of the patient's experience. The patient learns to perceive himself and the world afresh, as a freer, more satisfying, and accurate theory of life evolves. Figure 2 graphically represents the transformation and translation relationship in psychoanalytic psychotherapy.

5. Conclusion

Madsen, like Freud, is on sound ground descriptively. Systematology, like psychoanalysis, accounts for and connects a large amount of material in a cogent and parsimonious manner. However, Madsen's attempt to quantify his system is unconvincing, whereas Freud leaves the burden of truth to his disciples and to the future technological advances of related disciplines.

> Looking back, then, over the patchwork of my life's labors, I can say that I've made many beginnings and thrown out many suggestions. Something will come of them in the future. But I cannot tell myself whether it will be much or little. (Freud, 1952, p. 134)

Psychoanalysis remains the most influential working theory in psychology, albeit in its descriptive stage. Continued experimental work is needed to confirm, disconfirm, or sharpen its precepts. Work in related fields may verify or contradict various of its propositions. As Popper suggests, any genuinely scientific theory may be falsified. Future findings and advances in the wide sweep of knowledge may support Freud's wisdom or engender a paradigm shift (Kuhn, 1962). Surely, as the proverb suggests, "truth is the daughter of time."

6. References

Boles, R. C. *Theory of motivation*. New York: Harper & Row, 1967.

Dollard, J. & Miller, N. E. *Personality and psychotherapy*. New York, McGraw-Hill, 1950.

Freud, S. *An autobiographical study*. (Trans. J. Strachey.) New York: Norton, 1952.

Hempel, C. G. *Philosophy of natural science*. Englewood Cliffs, N.J.: Prentice-Hall, 1966.

Hendel, C. (Ed.) *Hume selections*. New York: Scribner's, 1955.

Hospers, J. *An introduction to philosophical analysis*. Englewood Cliffs, N.J.: Prentice-Hall, 1967.

Jastow, J. *Freud—His dream and sex theories*. New York: Pocket Books, 1932.

Jones, E. *The life and work of Sigmund Freud* (Vol. 2). New York: Basic Books, 1957.

Kennedy, G. Psychoanalysis: Protoscience and metapsychology. In S. Hook (Ed.), *Psychoanalysis, scientific method, and philosophy*. New York: New York University Press, 1959.

Kuhn, T. S. *The structure of scientific revolutions*. Chicago: University of Chicago Press, 1962.

Madsen, K. B. *Modern theories of motivation; a comparative metascientific study*. New York: Wiley, 1974.

Malcolm, J. *Psychoanalysis—The impossible profession*. New York: Knopf, 1981.

Mann, J. *Time-limited psychotherapy*. Cambridge, Mass.: Harvard University Press, 1973.

Mujeeb-ur-Rahmah, (Ed.) *The Freudian paradigm: Psychoanalysis and scientific thought*. Chicago: Nelson Hall, 1977.

Popper, K. Philosophy of science: A personal report. In C. A. Mace (Ed.), *British Philosophy in the mid-century*. London: Unwin, 1957.

Rappaport, D. *The structure of psychoanalytic theory: A systematizing attempt*. New York: International Universities Press, 1960.

Rickman, J. (Ed.) *A general selection from the works of Sigmund Freud*. Garden City, New York: Doubleday, 1957.

Rose, S. P. From causations to translations: What biochemists can contribute to the study of behavior. In P. P. G. Bateson & P. Klopfer (Eds.), *Perspectives in ethology*. New York: Plenum Press, 1981.

Royce, J. The present situation of theoretical psychology. In B. Wolman (Ed.), *Handbook of general psychology*. Englewood Cliffs, N.J.: Prentice Hall, 1973.

Schafer, R. *A new language for psychoanalysis*. New Haven: Yale University Press, 1976.

White, A. R. *Truth*. Garden City, New York: Anchor Books, 1970.

6. References

[illegible]

4

. . . But Discretion Were the Better Part of Valor

Dirk L. Schaeffer

If there is anything to be learned from Kuhn's (1962) trenchant analyses of science—or from the less exciting but more fruitful reflections of Campbell's "evolutionary epistemology" (e.g., 1977)—it is that science, even at its best, is mostly a series of noble mistakes. Theories and explanations are propounded, used, and ultimately discarded as the knowledge they generate inevitably reveals their fundamental shortcomings (Agnew, 1977). Consequently, it is difficult to dismiss Professor Madsen's Hypotheses Quotient (HQ) out of hand as fundamentally misguided; rather, it represents a genuinely noble quest for a metric that would allow fruitful comparison of theories to yield, one would hope, immense benefits in terms of the allocations of money, time, and intellect: a quest that is rarely undertaken, and that needs—despite Feyerabend (1970)—all the support it can get.

Unfortunately, however, this may be the best that can be said about Madsen's efforts. In practice, they seem to fall so far short of the desired goal as to hinder, rather than advance, our understanding of the nature of theory or theories. In attempting to document this statement, I shall address Professor Madsen's notions on three levels, roughly comparable to those of *data* (the specific hypotheses here identified, classified, and scored), *hypotheses* (the nature of the HQ generally), and *theory* (the ultimate rationale of the HQ). Under the last rubric I shall try to sketch out some alternative approaches to Madsen's goal which seem—at least on the basis of present ignorance—to involve somewhat fewer problems and counterintuitive assumptions than do Madsen's formulations.

Dirk L. Schaeffer • Psychological Assessment Consultation Evaluation (PACE), 10322A–121 Street, Edmonton, Alberta Canada T5M 1K8.

1. Data

Fundamental to Madsen's calculation of the HQ is the explication of the hypotheses contained in any given "theory" (the quotation marks are introduced only to allow the broadest possible treatment of this term), many of which, as he notes, may be "implicit" in the text. Madsen's major corrective for what he describes as the "more or less subjective interpretation" that this requires is by means of checking with the author. Even when this is possible, as it is not, of course, in the present Freudian instance, it may be apparent (see, e.g., Smedslund, 1978) that authors are perhaps the poorest judges of their own theories and that the method itself is, in consequence, highly questionable at this basic level. Consider, for example, the two Freudian theories which Madsen treats in the present paper. Schematically, the first may be graphed as follows:

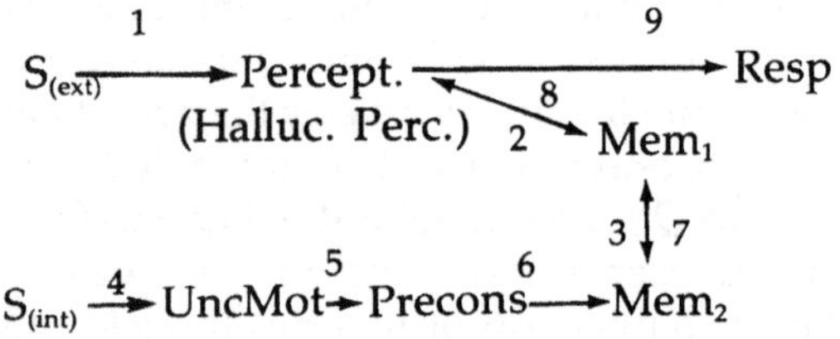

It is apparent here that there is a great deal happening on the bottom line of this schema that does not seem to be doing anything in terms of the rest of the theory: that is, it yields many *H–H* relations without any anchors on either the antecedent or consequent ends. How did this happen? Well, for one thing, Madsen reads the two sentence fragments: "All our physical activity starts from stimuli (whether internal or external)" and "we shall take the unconscious system as the starting point of dream formation" to yield the relations numbered 4 and 5. But is it not just as compelling to see "unconscious motives" simply as a part-definition or instantiation of "internal stimuli," collapsing these two steps into one? Similarly, Madsen reads the latter quotation as implying that there is a movement of these unconscious motives into the preconscious, prior to their having an effect on memories. This is explicitly denied, however, by the fifth quotation ("in hallucinatory dreams . . . the excitation moves in a *backward* direction") and in Freud (1900/1953, p. 541) which clearly locates the unconscious as *prior* to the preconscious, so that movement from the one system to the other would be in a *forward* direction. It is thus easily possible to simply eliminate "implicit" *H–H*

hypotheses 4 and 5 from Madsen's count, reducing the HQ ratio for this theory from 6/3 = 2 to 4/3 = 1.33.

A different, but related problem plagues Madsen's interpretation of Freud's theory of anxiety. The right half of a schematic of these hypotheses indicates, according to Madsen, that anxiety leads to three kinds of responses: flight, conscious experience, and organic processes (Hypotheses 9, 10, and 11). It does so, however, in three different kinds of people: phobics (who flee but do not experience anxiety consciously or in the form of organic symptoms); anxiety neurotics (who experience anxiety they can neither escape nor convert into symptoms); and hysterics (who show pseudosymptoms without experiencing anxiety and do not flee it). Thus, if one considers the *domain* of this theory, one must either collapse these three alternative and exclusive responses into one or argue that Freud has here offered three different theories of anxiety: phobic, neurotic, and hysteric. Either approach has the effect of eliminating two *H–R* hypotheses from Madsen's calculations, raising the HQ value for this theory from 6/5 = 1.20 to 6/3 = 2.00.

In combination, we have found that Freud's topographical theory, which Madsen terms "the most abstract among the whole hierarchy of theories," may score far lower on the HQ than the anxiety theory, which Madsen analyzes specifically to show that it falls below the topographical. Although I do not necessarily wish to argue here that Madsen's readings of Freud are wholly wrong or mine wholly right, it appears clear (1) that there is ample room for argument as to what is or is not "implicit" in such theories and (2) that even small arguments can have large effects on the resulting HQs.[1]

2. Hypotheses

Far-ranging as these problems are, they appear slight in comparison to Madsen's apparent conception of the nature of hypotheses within theories. We have already seen that he fails to distinguish the *domain* of a theory as a fundamental characteristic that must be taken into account

[1] Even leaving aside the problems of (1) the many other arguments that could be made on the level of the interpretation of Freud's text—the above examples are only those most easily exemplified—and (2) the nature of the calculation of the HQ, which excludes *S–R* hypotheses and is a ratio with fortunate asymmetrical properties, bounded by 0, 1, and infinity. The alternative: $HQ = \Sigma(H\text{–}H)/\Sigma(H\text{–}H) + (H\text{–}R) + (S\text{–}R)$ seems more appropriate.

in explicating hypotheses; closer reading of the construction of the HQ indicates that he fails to consider anything else of any importance either. Specifically, the HQ model functions entirely under the assumption that within theories all hypotheses are *equal, additive,* and *independent.* But surely no one would agree with any of these assertions: Theories are not mere lists of hypothetical statements, but rather interrelated networks of such statements; and it is precisely the interrelationships, and not the numbers, that matter in understanding, using, or evaluating a theory. At the simplest level, consider two brief theories. One, perhaps the basis of most learning and motivation theories, holds that stimuli lead to perception which leads to associations which lead to behavior. With one central *H–H* hypothesis balanced by *S–H* and *H–R* hypotheses at the antecedent and consequent ends, this comes as close to the ideal of positivistic psychology (the Bergmann–Spence– McCorquodale–Meehl model) as any two-stage theory can and must be regarded as a model of elegance. The second theory holds that the devil puts evil thoughts into our heads, which lead either to bad deeds or valiant prayer. Few of us would accept this as a useful model for psychology, or even a good theory (if only because its one *H–H* hypothesis is followed by two *H–R* hypotheses so that no antecedent anchor exists). But note that *both* these theories score .50 on the HQ.

Thus it appears that the HQ is so set up, on the basis of its assumption of the equality, additivity, and independence of hypotheses, that it can hardly tell us anything we might want to know.

3. Theories

If Madsen's view of theories, on the level of hypotheses, appears unconventional, his view of the overall significance of theory and the enterprise of theory construction appears extraordinarily conventional and thus difficult to criticize. Consequently, I do not now wish to give the impression of taking issue with anything Madsen suggests at this level; rather, I want only to suggest an alternative to the conventional view of theories which seems to me to be ultimately more fruitful than most present formulations, although immediately, perhaps, more painful.

Briefly, Madsen's and the conventional view suggest that theories are composed of propositions about (segments of) empirical reality which differ from anything you or I might casually say primarily in their "scientific" nature (again, the quotes are used only to allow for maximum breadth in the interpretation of this term). That is, they allow verification or falsification at some stage, they form a coherent network, they contain elements that are not purely empirical in nature, and they obey the common dictates of logic.

Let me try to oppose this viewpoint, particularly at the last step, by suggesting the following tentative definition of *theory* as: An interrelated network of quasi-tautological propositions about purported reality.

Only two words here will strike the reader as strange, and one is trivial. I say "purported" only because we do not yet quite know what *reality* is and thus have no good way of inferring whether any given statement is about "real reality" or not. "Purported" again allows maximum breadth.

But what can I mean by "quasi-tautological"? Essentially, that much of what goes on in theory construction and enunciation is more or less tautological in two specifiable ways. First, the propositions are tautological only within the restricted realm of *our present knowledge* of the phenomena of concern. And second, these tautologies commonly—although not invariably—articulate themselves as propositions which, when thought through, reveal themselves to be of such a nature that any alternative would be absurd, given our present state of knowledge. Thus they are not quite tautologies in the classical, logical sense.

Before probing some of the implications of this conceptualization of theory, it may be well to examine the evidence that can be adduced to support it.

It should be clear, for example, that this formulation helps account, perhaps better than any other, for the many anecdotes—familiar to any student of the history and philosophy of science, particularly the social sciences—of the extraordinary persistence of given theories, in the absence of any verifying data and even in the teeth of repeated falsification (Ross, 1977). Indeed, assuming that good theories are quasi-tautological in nature, it is difficult to see either the absence of verification or the pressure of a few discomfirming instances as sufficiently great to induce us to abandon a system that appears as intuitively compelling as an often only implicit web of tautology. Moreover, such blatant examples of theoretical fixation as the remark of Galileo's student, Torricelli (cited in Holzkamp, 1964)—"then if the balls of lead, of iron, of stone do not obey these laws, bad enough for them, we say then that we will not speak of them"—become, in this light, merely the proper exercise of scientific rigor, and—in their way—examples of science at its best.

Consider two further examples. Some years ago, when Hull's system was still the acme of psychological theorizing, John Cotton (unpub.) was able to show that there was no place in Hull's formulations at which one could go from the theory to actual behavior without making a host of subsidiary—otherwise wholly implicit—assumptions: none of these were considered part of Hull's system, although all were essential to its functioning; and all were more or less tautologous—and therefore implicit, unmentioned—in form. Similarly, Smedslund (1978) has recently shown that a theory of Bandura's consists of *nothing but* tautologies and implicit antiabsurdities.

Recognizing the quasi-tautological nature of theory does not, then, in any way discredit such theory; rather, it causes one to acknowledge that theories do the very best they can at any stage in the growth of human knowledge. Nor is this viewpoint one that is markedly novel in the literature of the philosophy of science, although it does fly in the face of such toilers as Bergmann or Popper. Holzkamp (1964, 1968), for example, has offered a largely similar conception of scientific theory, which he sees as focused on causal propositions which may paradoxically be described as "absolute conditionals" of the form: 'if p then q, *provided no other factors interfere.*' Since that proviso can never be wholly negated, such statements are inevitably bald tautologies, but no less useful for science for that reason. And again, the "evolutionary epistemology" of Campbell, geared to such notions as that of the plank-by-plank repair of the ship (theory) *while* it is navigating the sea of reality, seems to move fairly close to this conception also.

This view of theory, then, appears to encompass also most of the behavior of science as described by Kuhn (1962) but extends his view of the paradigm to include most of theorizing. If, that is, theories are quasi-tautologous in nature, scientific (intraparadigmatic) work then does become, as Kuhn suggests, the working out of soluble puzzles. Since the tautologies (or elements of a web which in its—local—entirety becomes quasi-tautologous) are largely implicit in the theory, and since scientists are largely empirical in their approach, such puzzle-solving activity typically articulates itself by extending the implicit proposition in a wholly implicit manner to reach empirical endpoints which appear novel (since the entire implicit web has never been addressed) but are in fact quite conservative (since they flow more or less directly from the initiating tautologies and therefore must be true—and vaguely banal—within the system). Real alternatives to the theory then appear as revolutions, since they implicitly challenge or supercede the implicit tautologies of the older formulations. And finally, the implicit nature of most of this procedure allows a lot of room for minor disagreement, countertheorizing, and extension.

What I am suggesting, simply, is that given, say, the Euclidean conception of space (within which the laws of plane geometry are tautologous), Newton's physics becomes quasi-tautologous, as any alternative conception leads to absurdities or (in other cases) simply appears absurd. Only a fundamental challenge to the tautologous base (by, e.g., shifting to non-Euclidean geometries) can successfully replace such a system (falsification, given Holzkamp's conditional, certainly cannot); and once this is enunciated, its physics become quasi-tautologous also, as illustrated by the present hunt, in physics, for ever more refined angels—quarks, beauties, charms—to dance on the point of electron microscopes. Similarly, given a state of knowledge of human behavior that can accept that not all behavior is rational, the Freudian model of the unconscious becomes quasi-tautologous, in that no serious alternative can appear anything but absurd; and this accounts very simply for the continued success of Freudian theory (and its variants, which include virtually all personality theories).

Let me try a somewhat more complex example and anecdote. Some years ago I found myself at a party with one of the more esteemed researchers and theorists practicing psychological science today. We had had a day of serious theoretical seminars and discussions and were now relaxing with enough alcohol to extend our usual empirical limits. Wouldn't it be wonderful, he suggested at one point, if we could find a pill that would raise everyone's IQ by 10 or 20 points? No, I allowed, it wouldn't: it was my impression, for example, that Japanese were averaging about 10 points higher on the Wechsler than the Americans, and all that happened was that the norms were revised for Japan. I saw no evidence that life in Japan, however, was any more desirable than in America, Canada, or Europe. But I didn't understand, he protested, everybody would be more equal and society would advance more rapidly. I was not convinced, the rank orderings would stay the same, and progress depended more on economic opportunity than the existence of new inventions. Well, he persisted, suppose the pill just raised the IQ of those below average in intelligence: we would get rid of stupidity, the need to care for large numbers of society, sloppy workmanship, and the like. No, I insisted, we would merely readjust the norms, or find some measure other than intelligence (say achievement motivation) to base our inequalities on.

My point is that my interlocutor here was working from an implicit theory of intelligence which viewed it, as most people do, as an attribute of personality, however defined (he leaned toward the physiochemical). Within this (quasi-tautologous) conception, the various deductions he made on the basis of a hypothetical IQ-raising pill were, similarly, quasi-tautologous and obvious. My somewhat more radical view of intelligence, on the other hand, as being simply a social ascription, or label, attached by some people (psychologists, educators) to others, undercut his quasi-tautologous paradigm and made it largely irrelevant; as Feyerabend would instantly recognize, our theories were wholly incompatible, although on the empirical end—the measurement and even the cause (heredity, learning) of IQ—we were in complete accord. And my theory, of course, was equally quasi-tautologous as his, with consequences that flowed from the implicit origins as naturally as his did, both of us being too well oiled at this point to really think hard.

I have, finally, seen little evidence to indicate that scientific theorizing in general, and psychological theorizing in particular, is much different from what we were doing, although neither of us garnered publication credits for our efforts.

How, then, can one best deal with theories, to compare, select, and above all use the best of them most fruitfully? Two approaches suggest themselves (quasi-tautologously). On the one hand, since much of what goes into theory is implicit in nature, the careful student could devote his energies to the attempt to make this implicit material explicit. This is, in essence, what Madsen has attempted here and in his other publications. Although his specific approach, apparently blind to the distinction between tautologous and nontautologous hypotheses (he again treats them as all equal and additive), may leave much to be desired, the hard work of translation that he has undertaken must be seen as wholly admirable, even heroic.

All of this, however, continues not to impress any empiricists, for whom this exercise amounts primarily to taking the fun (Kuhnian puzzle-solving) out of science: if all we are doing is elaborating tautologies, obviously this is no more fruitful than playing chess (which, too, represents merely the elaboration of tautologies).

An alternative approach to assessing theories has been suggested by Holzkamp, who noted the absolutely tautologous nature of the "interfering factors" conditional. In his view—here grossly oversimplified—successful confirmation of an empirical prediction counts as a plus point in favor of the theory. Although disconfirmation or falsification (or failure to realize the prediction) cannot count fundamentally against the theory (because of the "interfering factors" explanation as a way out), the theory can, in the face of poor results, be maintained only by invoking some specifiable "interfering factor." Invoking this counts as a minus point against the theory, but only until the existence of the interfering factor can be confirmed, at which time that minus point is erased. Thus, leaving aside the thornier issues of internal and external scope and domain of the theory, it appears possible to develop a metric, similar to Madsen's HQ, for the comparative assessment of theories, and one that appears, as I suggested earlier, both truer to the actual practice of scientists and freer of counterintuitive assumptions than do Madsen's formulations.

Unfortunately, the hard work of counting confirming, disconfirming, and eradicating instances for many theories, which the Holzkampian approach demands, has yet to be done. In this sense, Madsen is far ahead of the competition.

4. References

Agnew, N. M. On data: Does science have holes in her pockets? *Ontario Psychologist*, 1977, *9*(2), 33–62.

Campbell, D. T. Descriptive epistemology: Psychological, sociological, and evolutionary. The William James Lectures, Harvard University, 1977.

Feyerabend, P. K. Against method: Outline of an anarchistic theory of knowledge. In M. Radner & S. Winokur (Eds.), *Minnesota studies in the philosophy of science*. Minneapolis: University of Minnesota Press, 1970.

Freud, S. The interpretation of dreams. In J. Strachey (Ed. and Trans.), *The standard edition of the complete psychological works of Sigmund Freud* (Vol. 5). London: Hogarth Press, 1953. (Original work published 1900)

Holzkamp, K. *Theorie und Experiment in der Psychologie*. Berlin: W. de Gruyter, 1964.

Holzkamp, K. *Wissenschaft als Handlung*. Berlin: W. de Gruyter, 1968.

Kuhn, T. S. *The structure of scientific revolutions*. Chicago: University of Chicago Press, 1962.

Ross, L. The intuitive psychologist and his shortcomings. In L. Berkowitz (Ed.), *Advances in social psychology*. New York: Academic Press, 1977.

Smedslund, J. Bandura's theory of self-efficacy: A set of common sense theorems. *Scandinavian Journal of Psychology*, 1978, *19*, 1–14.

4

The Hypotheses Quotient

Reply to Commentators

K. B. Madsen

1. Introduction

After reading the critical comments by Dr. Brandt, Dr. Ettin, and Dr. Schaeffer, I deeply regret that I selected two of Freud's theories for the demonstration of the calculation of the Hypotheses Quotient (HQ), because I think that almost all the critical remarks are related to the example (Freud) and not to the general problems of the HQ. Therefore, I shall divide my reply into two parts: (1) the HQ in general and (2) the HQ of Freud's theories.

2. HQ in General

I think that there would have been a much more positive reception of my idea of the HQ if I had selected as an example the theory of C. L. Hull—or another formalized theory, such as those of Atkinson, Cattell, or Lewin. There would have been no problem of explication of the implicit hypotheses in the theory, because the authors themselves have already done the explicit formulation and systematic organization of the hypotheses. However, some of the critical comments are more general, and I shall try to answer them.

2.1. Classification

Even with a formalized theory like Hull's, we have to classify the hypotheses into two categories: theoretical hypotheses (*H–H*) and partly

K. B. Madsen • Royal Danish School of Educational Studies, Copenhagen N. V., Denmark.

empirical hypotheses (*S–H* and *H–R*). But in the case of a formalized theory there would not be any real problem because the terms of the theory would be clearly defined as hypothetical (transempirical) or descriptive (empirical). In the case of Hull's theory, this classification of hypotheses has been done by a computer program designed for the purpose (inspired by the program called the General Inquirer; see Stone, 1966).

In connection with this, I think that Brandt has misunderstood me when he states that I do not distinguish between unobserved and unobservable variables. In my definition of hypotheses I mention both unobserved and unobservable intervening variables, along with explanatory constructs. The reason for *not* making a lot out of the distinction is that unobservable variables may, after invention of new techniques, be observable.

Another misunderstanding of Brandt's concerns the relationship between variables. He states that I "disregard propositions in which the relationship is hypothetical or transempirical." This is not true, because many of the hypothetical variables are processes, and that means that a causal explanation has to introduce a hypothetical variable as intervening variable between the antecedent and the consequent variable. The antecedent and the consequent variables may be either empirical or hypothetical variables. Thus, most explanations require at least three variables, with at least one hypothetical variable.

Formulations of the relationship between two empirical variables (one independent variable and one dependent variable) are not hypotheses, but datatheses, according to my systematological metatheory. We shall return to the datatheses later.

Finally, I shall reply to comments by Ettin and Schaeffer, both of whom criticize the summation of hypotheses in the HQ formula. Ettin writes that the "formula gives an even weighting to all the statements chosen." This is not true; the formula distinguishes between theoretical hypotheses (*H–H*) and empirical hypotheses (*S–H* and *H–R*), and usually the theoretical hypotheses are the most important hypotheses—axioms or postulates. Thus the HQ formula does distinguish between hypotheses.

2.2. Reliability of the HQ

As a consequence of these remarks on classification, we may conclude that in the case of formalized theories like Hull's theory, in which the classification of hypotheses and the calculation of HQ can be done by a computer, *the reliability of the HQ is about the same as the reliability of most psychological tests.*

However, I admit, as already stated in my paper, that problems occur with nonformalized theories, those theories without explicit formulations. Yet even in these cases, I think, the HQ is "a quantitative *estimation* of the testability of a theory" (see the title of my paper). In connection with this, I can reply to Ettin's statement that "Madsen presents no studies or evidence that interrater reliability is possible" with the following information: There are no formal studies of the interrater reliability. However, I have taught the method to students of psychology for 19 years. First-year students are taught the method as a part of the introductory course in psychological metatheory (2 hours per week for a whole academic year). After the year's teaching and training in metatheoretical analysis of psychological texts, they have to pass a written examination. Usually between 75% and 90% of the students pass this examination. I think it is fair, therefore, to compare the interrater reliability of the calculation of the HQ in cases of nonformalized theories to the interrater reliability of scoring for the Thematic Apperception Test in the Atkinson and McClelland version (which also may be done by the same computer program, the General Inquirer) and to say that it is about the same.

In this connection, I may reply to Schaeffer's statement "that authors are perhaps the poorest judges of their own theories" with the following: I do not think that the author is the poorest judge of the main content of the theory. And I have only asked the authors of the analyzed theories to accept my systematic reconstruction of the main content of their theories. This systematic reconstruction is done as explicit formulations of implicit hypotheses and (in Madsen, 1974) as graphic models (in the cases wherein there are no graphic representations). The authors were not required to accept any evaluations (including the HQ) of their theories.

2.3. Validity of the HQ

Before answering the critical comments on the validity of the HQ, I may point to the following fact: the HQ is not constructed or invented as a measure of testability. It was found empirically as one of the results of my first comparative metatheoretical studies (Madsen, 1959). And I have since wondered what this relationship expressed or what sort of estimation it represented. As you can see from my paper, I have come to the conclusion that it is an estimation of the testability of a theory.

In this connection I should like to reply to Brandt's critical remarks about the exclusion of the datatheses. If we include the datatheses (the *S–R* laws), then the HQ would be an estimation of the explanatory power of the theory. This could also be an important characteristic of a theory, in addition to its testability (which is equal to the potential explanatory power, the HQ without datatheses). Some philosophers and scientists perhaps believe that the explanatory power is more important than the testability. However, I have chosen to interpret the HQ as an estimation of testability because the leading philosopher of science, Karl Popper, has claimed that testability is the criterion for a theory's being *scientific*. I admit that it could be worthwhile to estimate both the testability and the explanatory power of a theory and that it could be done by calculating the HQ with and without the datatheses.

However, even with the calculation of the explanatory power, we do not have an estimation of the amount of testing done for the theory nor of the degree of correspondence with observations (or the truth value) of the theory. That has to be done by empirical research.

Finally, I should like to reply, also to Brandt, that I have not "abandoned testability for heuristic value." I have always found heuristic value to be the most important characteristic of a theory. It is Popper who has made testability the most important.

Now we may turn to the case of Freud.

3. HQ of Freud's Theory

I have already stated in the paper four reasons for selecting Freud's theory as a demonstration case. After the critical comments, I would like to add one more reason: *Freud's theory is of the highest heuristic value among psychological theories,* and therefore I wanted to demonstrate that *Freud's theory is also a testable theory* (contrary to Popper's evaluation). The reader must remember that even an HQ of 2.0 demonstrates that a theory is testable (*if* the HQ is accepted as an estimation of testability). The reader must further remember that we selected the topographical theory because it is among the most abstract of Freud's many theories. In addition, it was easier to demonstrate the procedures of calculation of the HQ with the topographical theory than with the other, later, structural theory in *The Ego and the Id* (1923).

Freud's theories may be classified into three levels of abstraction:

1. The most abstract theories are the theories about the structure of the psychic system: the old but not completely abandoned topographical theory of 1900 and the new, so-called structural theory of 1923.
2. The next level of abstraction contains the so-called economical and dynamic theories about drives (in several revised versions),

about anxiety (of the new one, 1926), and about the primary and secondary processes (already included in the *Project* of 1895 and later in other versions).

3. On the lowest level of abstraction, we have the theories applied to explanation of *The Psychopathology of Everyday Life* (1901), dreams (several versions) and neuroses (several versions).

If this classification of Freud's theories is correct, we should expect a decreasing HQ as an expression of an increasing testability. And that is just what I have found. I presented in my paper the HQ of the topographical theory (=2.0) and the HQ of the theory of anxiety (=1.25). Since the writing of the paper, I have continued my analysis of Freud's theories and found the HQ of *The Psychopathology of Everyday Life* to be .33. (My metatheoretical analysis of Freud's theories will be published in a forthcoming book: *A History of Psychology in Metascientific Perspective*.)

I think that the three calculated HQs demonstrate that there are different levels of abstraction in Freud's theoretical production and that the least abstract theories have the same degree of testability as the majority of other psychological theories. Even the most abstract theories are, after all, testable, an HQ *can* be calculated, and an HQ of 2.0 is not so much higher than what is otherwise the highest found (Tolman's HQ = 1.43).

But, of course, all this depends upon the acceptance of the HQ in general as an estimation of the testability, and especially of my metatheoretical analysis of Freud's theory. Therefore, I shall now reply to the critical comments on the analysis of Freud's theory, answering the critics in alphabetical order.

Brandt asks me, among other things, about the distinction between hypothetical variables like perception and dependent *R*-variables like conscious experience. I conceive conscious experiences as empirical data. They are, however, private data, unless they are verbalized and by this process made public data. However, conscious experiences as both private and public data must be distinguished from the conscious—and unconscious—processes, which produce the conscious experiences. These processes, both the conscious and the unconscious, are hypothetical (transempirical) variables. You may directly experience percepts, images, thought, feelings and so on, but these conscious experiences are products of hypothetical (conscious and unconscious) processes. I think that I am in accord with Freud on this matter.

To Brandt's footnote I should like to reply that the quotation from Freud—selected by Brandt—is a very concentrated formulation of both Freud's philosophy of the world, and his philosophy of knowledge (influenced by Kant).

Ettin has made some general critical comments on my procedure which I think I have already answered. In addition, he presents a long exposition of his conception and evaluation of Freud's theories, which I find very interesting and with which I am mainly in agreement.

Schaeffer has made a very careful analysis of my systematic reformulation of the implicit hypotheses in Freud's text. However, I do not think that we can eliminate implicit hypotheses 4 and 5. My systematic reconstruction of an unformalized theory is based upon (1) the selected quotations and (2) the whole text and my background knowledge of Freud's production (this is the *Vorverstehen* in hermeneutic interpretations).

4. References

Freud, S. *The standard edition of the complete psychological works of Sigmund Freud* (J. Strachey, Ed. and Trans.). London: The Hogarth Press, 1953–1974.

Madsen, K. B. *Theories of motivation.* Copenhagen: Munksgaard, 1959 (4th ed., 1968).

Madsen, K. B. *Modern theories of motivation.* New York: Wiley, 1974.

Stone, P. *et al. The general inquirer: A computer approach to content analysis.* Cambridge, Mass: M.I.T. Press, 1966.

5

What Is Necessarily True in Psychology?

Jan Smedslund

Abstract. The position is taken that psychology is not an empirical science and that generally valid propositions in psychology are explications of common sense and hence necessarily true. A proposition in a given context belongs to common sense if and only if all competent users of the language involved agree that the proposition in the given context is true and that its negation is contradictory or senseless. Studies attempting to test necessarily true propositions are labelled pseudoempirical. The paper presents numerous examples of necessarily true propositions and pseudoempirical studies taken from various fields of psychology.

The position taken in this paper is that psychology is not an empirical science as ordinarily believed. All generally valid propositions in psychology are seen as successful explications of common sense and as being logically necessary. Empirical psychology propositions can, according to this point of view, at most have local validity.

The position will be described and some of its implications will be explored. Then, two detailed illustrations will be given of how explications of psychological common sense may be arrived at. Finally, in the bulk of the paper, examples are given of necessarily true propositions in psychological literature. Earlier versions of the position are described in Smedslund (1972, 1978a,b, 1979, 1980).

1. Description of the Position

The fundamental role of ordinary language in science is taken for granted. Observations must be described and hypotheses and theories

Jan Smedslund • Institute of Psychology, University of Oslo, Box 1094, Blindern, Oslo 3, Norway.

must be formulated in a language shared with those with whom one wishes to communicate. It is also taken for granted that all technical languages must be translatable into ordinary language in order to be taught and in order to sustain communication. For the psychological researcher, then, ordinary language is a *given* that structures all descriptions and all theorizing.

In every language there are limitations on what are meaningful ways of describing and explaining. These constraints form a highly organized system such that, given one set of propositions, others follow necessarily or are necessarily excluded. In other words, there is a *logic* of ordinary language. For extensive discussions, from a point of view similar to the present one, of the ontological and epistemological problems involved, see Israel (1979).

Another way of stating the preceding is that ordinary language is a conceptual system in which the concepts are related in various ways. Every proposition formulated by means of these concepts implies certain other propositions and implies the negation of certain others: Being a person implies having a body, having a body implies having a location at a given time, performing an act at a given time implies that one *can* perform that act at that time. Conversely, it makes no sense to state that a person was nowhere while performing an act or was unable to perform it while performing it. Competent users of English presumably agree about these implications.

Common sense may now be defined as *the system of implications shared by the competent users of a language.* Furthermore, *psychological common sense* is taken to refer to that part of the common sense of a language that pertains to psychological phenomena.

In an earlier publication (Smedslund, 1972, p. 78) I have described three important features of common sense as follows: "It is normally unreflected or unconscious, it is shared by all 'ordinary' persons, and, when made explicit, it is self-evident in a compelling way." This conception of common sense differs from another quite popular one. According to the latter, common sense involves testable propositions about factual matters and, hence, may be shown to be wrong. Introductions to textbooks sometimes emphasize how scientific psychology may correct ill-founded commonsense beliefs.

According to the definition proposed here, common sense does not have empirical content but involves necessarily true propositions. Let us use the technical terms *contingent* and *noncontingent.* A contingent proposition is one which may be true and may be false. A noncontingent proposition is one which is not contingent, that is, which is either necessarily true or necessarily false (Bradley & Swartz, 1979, pp. 13–24).

Accordingly, common sense in the new version consists exclusively of noncontingently true propositions.

The new concept of common sense cannot be relegated to the sphere of metatheory or philosophical psychology. On the contrary, it involves criteria which allow one, with the usual margin of uncertainty of empirical methods, to decide whether or not a given propositon in a given context belongs to the psychological common sense of a language. In summary, *a proposition in a given context belongs to common sense if and only if all competent users of the language involved agree that the proposition in the given context is true and that its negation is contradictory or senseless.*

Since the researcher is also a competent user of the language, he or she may take a shortcut by simply reporting that he or she regards a proposition as being true and its negation as being senseless or contradictory. If this conclusion is supported by a few spot checks on available individuals, it may be regarded as tentatively established. However, the ultimate test will always be the response distribution in a large sample of competent users of the language.

The preceding means that the explication of psychological common sense, which itself is nonempirical, can be controlled by ordinary empirical methods. Two such studies of, respectively, the propositions in Bandura's theory of self-efficacy and seven suggested rules of psychological treatment are now being conducted. The position that psychological theory must, ultimately, be based on consensus is also in agreement with Schutz's important methodological postulate of adequacy for the social sciences:

> Each term in a scientific model of human action must be constructed in such a way that a human act performed within the life-world by an individual actor in the way indicated by the typical construct would be understandable for the actor himself as well as for his fellow-men in terms of commonsense interpretation of everyday life. Compliance with this postulate warrants the consistency of the constructs of the social scientist with the constructs of commonsense experience of the social reality. (Schutz, 1967, p. 44)

The view that psychological theory consists of noncontingent propositions only raises several intriguing questions.

The first one is how to reconcile this interpretation with the fact that for a century psychology has had the outer appearance of an empirical science. The theoretical propositions that have been forthcoming have appeared in the context of experiments, clinical experience, or informal observations and have been the subject of apparent empirical testing. In order to understand this discrepancy, it is necessary to make several important distinctions. The first one has already been made, namely between contingent and noncontingent propositions. This distinction is a *modal* one (referring to what is possible and necessary) and is not an *epistemic* one (referring to source of knowledge, i.e., experience or reasoning). Furthermore, in analyzing the epistemic status of propositions it is necessary to distinguish between *empirical* and *experiential*. Following Bradley and Swartz (1979, p. 150), I shall define *'P is knowable empirically'* as "It is humanly possible to know *P only* experientially." Similarly, I shall define *'P is knowable a priori'* as "It is humanly possible to know *P other than* experientially." The preceding means that empirical knowledge is seen as knowledge that cannot be arrived at by reasoning, and *a priori* knowledge is seen as knowledge that can be arrived at by reasoning. If we now combine our modal and epistemic classifications we get the four possible combinations contingent and empirical, contingent and *a priori*, noncontingent and empirical, noncontingent and *a priori*. Without going into the philosophical subtleties and controversies involved, I shall here simply take the fairly generally accepted position that only two of the combinations can actually exist, namely contingent and empirical and noncontingent and *a priori*. See the discussion by Bradley and Swartz (1979, pp. 156–175). This means that while contingent truths can only be arrived at through experience, noncontingent truths can be arrived at through reasoning and/or through experience. In other words, there is a category of possible propositions which are experientially based, yet noncontingent. An example from the history of mathematics is the case of the seven bridges of Königsberg. It had long been known experientially that there is no route by which one can cross over all seven bridges without recrossing at least one bridge (Bradley & Swartz, 1979, pp. 151–152), but not until the early eighteenth century was it proved mathematically that no such route could possibly exist. Another example is that experientially based knowledge of the quantitative relationships expressed in the Pythagorean theorem is documented in cuneiform tables from a period more than a thousand years before the theorem was proved (van der Waerden, 1954, p. 76). Later in this paper, many examples from the history of psychology will be given.

The preceding analysis, showing the possibility of experientially based, yet noncontingent propositions, provides an explanation of how the image of psychology as an empirical science has been maintained over a century. Apparently, it has been taken for granted that propositions arrived at in the context of data gathering and being consistent with data must be contingent. Given this unreflective and unwarranted presupposition, there has been almost no systematic checking of the modal status of theoretical propositions in psychology. As we shall see later, the outcome has been disastrous as far as psychology's understanding of itself is concerned.

Another problem raised by the present view of psychology as explicated common sense is how to understand the interplay of noncontingent theoretical propositions, methods, and data. First, it should be clear that noncontingent propositions cannot be weakened or strengthened by data. Therefore, studies that attempt to test necessarily true propositions by empirical methods will here be labeled *pseudoempirical*. It is not implied that such studies are always entirely devoid of value. They may, sometimes, be instrumental in leading to a recognition of the noncontingent character of the propositions involved. However, the term *pseudoempirical* calls attention to their particular modal-epistemic status and to the fact that this is not recognized by the researcher. The latter is evident, because no one would undertake costly, time-consuming work in order to support a proposition known in advance to be necessarily true.

One consequence of the preceding is that in order to avoid the proliferation of pseudoempirical studies one should *always* and *routinely* check the logical status of the propositions in which one is interested. However, a decision about logical status can be made only by considering the definitions of the terms involved, as well as the definitions of other related terms and the content of propositions to be taken for granted. Hence, the requirement to check logical status entails a general improvement in the quality of theorizing.

However, even if data cannot strengthen or weaken noncontingent propositions, they can serve to test the validity of *methods* (procedures). A noncontingent proposition states under what conditions something must necessarily occur or be the case. This means that, if the expected outcome is obtained, one may surmise, but not know for certain, that the procedure was indeed efficient. On the other hand, if the expected outcome is *not* obtained, then one knows with certainty that the required conditions have not been established and, therefore, that the procedure was not efficient. In research with contingent propositions, negative outcomes are ambiguous, since there is the additional possibility that the propositions themselves are false. This means that noncontingent propositions offer unique advantages in the testing of psychological procedures. Indeed, it would seem that procedure-testing research should, whenever possible, rely on noncontingent propositions for maximal efficiency.

A final question about the possible advantages of noncontingent propositions concerns their predictive and heuristic function. Psychologists have, traditonally, been inclined to dismiss formal truths as useless. In my view, this is a severe mistake. Since noncontingent propositions state under what conditions something must occur or be the case, they permit the prediction that, given these conditions, a certain outcome must be found. They also function heuristically, by indicating the procedures for achieving a certain outcome, namely those that establish the required conditions. If the concepts involved in such propositions are logically related, they may be used to formulate other necessary propositions too. I believe that the common sense of ordinary language forms one single system in which everything is related (directly or indirectly) to everything else. By making this system explicit one may, therefore, be able to develop a sort of calculus. The advantage of a calculus is that it permits derivation of outcomes in a great number of situations, namely, all those described by the combinations allowed by the set of concepts involved.

The relationship of noncontingent propositions to data and methods and their potentialities for calculation will first be illustrated by a nonpsychological example which has the merit of being clear and relatively noncontroversial. The example is the Pythagorean theorem, which is taken to be noncontingent. Attempts to test the validity of the theorem empirically by measuring the sides of right-angled triangles or by measuring the largest angle in triangles with sides taken to be respectively 3, 4, and 5 units would be pseudoempirical and senseless. Observations can neither strengthen nor weaken the theorem. On the other hand, it may be used to estimate the precision of available measuring techniques. Thus, one may estimate the precision of a technique of measuring length from deviations from expected values, when measuring the sides in a triangle taken to be right-angled. Conversely, one may estimate the precision of a technique of measuring angles by the deviation from expected values in triangles with sides taken to be respectively 3, 4, and 5 units. Alternatively, the two preceding sets of findings may be taken to measure the joint precision of one's available measuring techniques. Finally, there exists a trigonometrical calculus of great practical value in surveying, engineering, and navigation. This calculus enables one to predict angles and distances from one's measurements. Many formally analogous psychological examples will be given below.

The position outlined here leads one to expect many noncontingent propositions and pseudoempirical studies in psychological literature. However, even if this turns out to be true, it does not mean that one cannot also find genuinely empirical principles and studies with some validity. I do not deny that this is the case. However, the present position entails that such truly empirical propositions as may be formulated in psychology cannot be expected to have more than local validity (in certain populations, classes of situations, cultures, time periods, etc.). The main reason for this is that psychological phenomena are historical and that historical processes always contain a random component and, hence, are irreversible (Gergen, 1973, 1976; Smedslund, 1972, 1979).

Each individual is the product of a life story which is punctuated by arbitrary events. "If you had not happened to go to that party, we would surely never have met and. . . ." As a consequence, each individual becomes to some considerable extent unique and hence unpredictable and unexplainable, except by reference to a series of unique historical events. The similarities between people in the same subculture or general culture also stem from history. Again, customs, technology, and language have developed through innumerable random events, the major and most visible ones being such things as invasions, wars, revolutions, plagues, famines, and technological innovations. The fact that the Norwegians in the 1980s sit on chairs rather than on floormats is a *given* that cannot be explained by any eternally and universally valid principles of behavior, but only historically. The remaining similarities between people across cultures can be referred to as characteristics of the species *homo sapiens.* These characteristics can only be very abstract capacity limits and functional tendencies, since human beings are so modifiable and innovative. The characteristics of this modifiability are also known to be modifiable as illustrated in the concepts of learning, learning to learn, learning to learn to learn, and so forth. The very function of memory is structured by the social fabric. As a consequence, it is asserted here, knowledge of possible universal and eternal capacity limits of *homo sapiens* is of very limited value in everyday life and in professional psychological practice. Psychology needs to develop strategies for dealing with the historically unique and with changing circumstances. This can be done by explicating the common sense that is implicit in ordinary language, which, again, constitutes the social reality in which people live. To what extent the principles and strategies arrived at through explication of the common sense of a language can be applied across cultures is a complicated question that will not be discussed here. However, it would appear that cross-cultural studies can only be conducted to the extent that there actually is a cross-cultural common sense.

2. Two Examples of the Explication of Common Sense

According to the research program presented above, the task of the theoretical psychologist is as follows: to try to formulate propositions that are regarded as true and their negations as senseless or contradictory, by all competent users of a language. This is not at all what psychologists have been doing up to now, and it is, therefore, necessary to give a number of different examples in order to clarify what is involved.

The first example is taken from my own research career (Smedslund, 1964) and shows what I now think is a typical case of pseudoempirical work.

In the context of a larger investigation of concrete operational reasoning in children, I also wanted to ascertain the relationship between the acquisition of conservation of length and the acquisition of transitivity of length. In the planning of this study very great care was taken to control irrelevant factors such as direct perception, guessing, forgetting, comprehension of instructions. The aim was to ensure as far as was practically possible that the presence of a given response in a subject meant the presence of a certain concrete operation and the absence of that response meant the absence of that same concrete operation. In planning the study I took for granted, without checking, that the relationship between conservation and transitivity was a contingent one and hence to be discovered by empirical methods.

The study yielded pass-fail scores of 160 children on tests of conservation of length and transitivity of length. Transitivity was more difficult than conservation, and only one child was scored as "pass" on transitivity and "fail" on conservation. My conclusion was that apparently children cannot acquire transitivity of length before they have acquired conservation of length.

Each test had two subitems. Transitivity of length was studied by the subject's being shown by juxtaposition that stick *A* was longer than stick *B* and that stick *B* was longer than stick *C*. Finally, sticks *A* and *C* were presented on a background which made *C* look longer than *A*, and the subject was asked to indicate which one was longer. The selection of stick *A*, accompanied by an adequate explanation, on at least one subitem, was taken as an indication of the presence of a transitive inference and yielded a "pass" score. Conservation was studied by the subject's first being shown by juxtaposition that stick *A* was longer than stick *B*. Then *A* and *B* were placed on a background which made *B* look longer than *A*, and the subject was asked to indicate which stick was longer. The selection of stick *A*, accompanied by an adequate explanation in at least one subitem, was taken as an indication of the presence of a conservation inference and yielded a "pass" score. The reader is referred to the original article for further details about the procedure and scoring.

After the study had been published, it gradually became clear to me that transitivity implies conservation and that, in the present terminology, the study was pseudoempirical. Let us call the comparison stick B_1 when juxtaposed to stick A and B_2 when juxtaposed to stick C. The inference from $A > B_1$ and $B_2 > C$ to $A > C$ is valid without qualifications only if $B_1 = B_2$, that is, if the length of the comparison stick is conserved over the displacement from the vicinity of stick A to the vicinity of stick C. If there were no conservation, the strict transitive inference would not be possible. This means that the proposition that conservation must be acquired before transitivity is noncontingent and *a priori* rather than contingent and empirical.

As seen in retrospect, the regret at having worked so hard to establish a noncontingent proposition is counterbalanced by the satisfaction of having constructed an apparently reliable procedure. This possibility of reinterpreting pseudoempirical studies as involving the testing of procedures is universally present. If the data do not conform closely with what is expected, one must conclude that the procedures were not successful in establishing the required conditions.

Although the technical concepts involved in this study may be unknown to most competent users of English, the outcome nevertheless corresponds to an explication of psychological common sense. This becomes clear if we consider the descriptions of the actual procedures. Given these descriptions and the necessary context, it should be possible to establish consensus about the necessary relationship between transitivity and conservation.

After having considered an example of an actual pseudoempirical study, let us now consider directly the question of how one can proceed to explicate psychological common sense. It should be noted that in asking this question the last remnants of a psychology inspired by nineteenth-century natural science disappear. We are proposing to study what the physicist Bohm (1980) has called an "implicate order," that is, an order in which everything is enfolded or implicated in everything else within a whole. This is radically different from realms constituted of entities which are *outside of each other,* in the sense that they exist independently of each other and are locatable in different regions of space and time. Psychological common sense consists of a whole system of implications and is not locatable in any separable spatiotemporal regions. We are approaching the study of psychology by investigating something of another order than individual actions and interactions. In fact, we will be following Shotter and Newson's (1982) dictum "ask not what goes on inside people, but what people go on inside of."

Let us imagine a simple experimental situation, not unlike many that have been used in psychological research. A chair is placed in front of a panel of drawers, each of which can be pulled out and shuts automatically when released. Our first subject, P_1, is seated in the chair. P_1 pulls out drawer 1, finds a donut and eats it. The problem we shall pose, quite naively, is this: Is it possible to formulate a general psychological principle that allows us to predict what the person will do next? We observe that P_1 pulls out drawer 1 once more, finds another donut, and eats it. This leads us to formulate the following tentative hypothesis.

Hypothesis 1. *If P does A with result R and likes R, then P will repeat A.*

However, this hypothesis is clearly insufficient, since it predicts that *P* will continue to repeat *A* indefinitely. Actually, we may observe that P_1 ceases to open the drawer after three or four repetitions, claiming that he or she is not hungry any more. In order to account for this, we may formulate a revised hypothesis.

Hypothesis 2. *If P does A with result R and likes R, and if P still wants to get R, then P will repeat A.*

The next subject, P_2, is now placed in the chair. P_2 pulls out drawer 1 and eats the donut. Then P_2 pulls out drawer 2. When asked about this, P_2 explains: "I did not pull out drawer 1 again, because I have already eaten the donut in it." It becomes apparent that, whereas P_1 believed that drawer 1 was a donut-automat, P_2 believes it to be an ordinary drawer. If you empty an ordinary drawer, there is no reason to pull it out once more. Apparently, hypothesis 2 is also insufficient and must be reformulated as follows.

Hypothesis 3. *If P does A with result R and likes R, and if P still wants to get R and believes that A will continue to lead to R, then P will repeat A.*

P_3 now sits in the chair, pulls out drawer 1, and eats the donut. She is just going to pull out drawer 1 again, when she suddenly looks at her watch, jumps up and leaves the room, explaining: "Oh, I have forgotten my appointment with the dentist, I must run." P_3 fulfills all the requirements of hypothesis 3, yet does not behave as expected. What makes the difference is that another want, stronger than the want to get another donut, interferes. This may be incorporated as follows:

Hypothesis 4. *If P does A with result R and likes R, and if P still wants to get R and believes that A will continue to lead to R and the want to get R is the strongest of P's wants, then P will repeat A.*

P_4 sits in the chair, pulls out drawer 1, eats the donut, and is beginning to pull out drawer 1 once more, when she suddenly lets her hand sink. We ask her to proceed, but she says, "I have dislocated my shoulder again, I just cannot lift my arm." P_4 fulfills all the requirements of hypothesis 4, yet does not repeat A, simply because she cannot do it, or, more precisely, because she believes (in this case with good reasons and conviction) that she cannot do it. In many other situations in everyday life the belief that one can or cannot do something is decisive with respect to whether or not one actually tries. Again, we must reformulate our hypothesis:

Hypothesis 5. *If P does A with result R and likes R, and if P still wants to get R and believes that A will continue to lead to R and the want to get R is the strongest of P's wants and P believes he or she can do A, then P will try to repeat A.*

P_5 gets seated, pulls out drawer 1, eats the donut, and prepares to pull out a drawer again, when the whole panel of drawers collapses and goes to pieces. At that moment, P_5 fulfills all the requirements of hypothesis 5, but she does not try to repeat *A* because there is no panel of drawers any more—the situation is entirely changed. This brings out the importance of specifying the situation (S).

Hypothesis 6. *If P does A in S with result R, and likes R, and if P still wants to get R, and believes that A in S will continue to lead to R, and the want to get R is the strongest of P's wants, and P believes he or she can do A in S, then P will try to repeat A in S.*

The preceding hypothesis is derived from a specific situation and still needs some obvious changes in order to become acceptable. First, it is not necessary for *P* to eat the donut, taste it, or even see it in order to predict *P*'s behavior. It is sufficient that *P*, for some reason, believes that there is a donut in drawer 1 and that it will be tasty. This can also be achieved by having a credible person tell *P* that there is a tasty donut in drawer 1, as well as in indefinitely many other ways. We can, therefore, simplify and generalize the hypothesis by merely assuming that *P* has somehow acquired a belief about the content of drawer 1. However, it is also clear that we must also know what *P* believes about the *other* drawers in the panel, if we are to predict her behavior. It must be assumed that *P* will always select the drawer that has the highest subjective likelihood of leading to the most wanted result.

Two other comments can be made. First, we can leave out the assumption that *P* likes *R*, because this is implied by the assumption that *P* wants *R*. Second, it may well be that *P* wants *R* for more than one reason. Wants add up, and instead of the strongest want we shall speak of the strongest combination of wants. Finally, it should be considered that *norms* are important determiners of behavior. However, they need not be included in our predictive formula because it is taken for granted that '*if N is a norm for P in S, then P wants to comply with N in S.*'

Taking the preceding comments into consideration, and adding a specification of time (*t*), we arrive at the following final formulation of the initial hypothesis:

Hypothesis 7. *If P's strongest combination of wants in S at* t *is to achieve R, and if P believes that A in S at* t *is the act with the highest likelihood of leading to R, and if P believes he or she can do A in S at* t, *then P will try to do A in S at* t *in order to achieve R.*

Hypothesis 7 is a special case of the principle of maximization of expected utility. It is silent about situations in which values and likelihoods indicate different actions, and it does not specify the subjective likelihood for *P* that he or she *can* do *A* in *S* at *t*.

Let us consider the modal status of this hypothesis. It has the form of a conjunction of three propositions implying a fourth one. The question is whether the negation of the proposition makes sense and is noncontradictory. The negation means that it could be the case that the three antecedent conditions were present and the consequent was absent. This means that a person does *not* try to perform an act, even though this act is believed to be the one with the highest likelihood of leading to the most wanted outcome and the person believes he can perform the act. It would seem as if any such explanation would have to invoke some other stronger combination of wants and/or some other stronger combination of beliefs. But this would contradict the assumptions of hypothesis 7. Therefore, it would appear that hypothesis 7 is a noncontingent proposition. If, in addition, all competent users of English, after having been introduced to the matter, agree that hypothesis 7 is true and that its negation is contradictory, then it is a part of the psychological common sense of English.

The preceding may be summarized as follows: Attempting to predict behavior, we first formulated a series of hypotheses (1–6) that are contingent and false. Finally, we arrived at a proposition that is noncontingent and true, and, we hope, a fair approximation to psychological common sense. I believe that is what psychologists in general have tended to do, albeit without recognizing it, because of the massive empiricist tradition.

Before turning to a survey of literature, let me briefly comment on a possible misinterpretation of the present position. This would be to equate the construction of propositions such as hypothesis 7 with what has traditionally been labeled *axiomatization* of psychological theory, as attempted, for example, by Hull. There is a superficial similarity since explication of common sense and axiomatization both result in the formulation of first principles from which others may be derived. However, there is also a very sharp difference. This is that the derivations from axiomatic psychological theory are, in principle, empirically testable, whereas this is not the case with commonsense propositions. Another way of phrasing this is that ordinary axioms may be negated, whereas negations of common sense principles are senseless or contradictory. Traditionally conceived axioms summarize what is believed to be the case in psychology, whereas commonsense principles summarize what must be said and thereby what cannot be said about psychological phenomena.

Axiomatization of the type attempted by Hull also does not take into account the reflexivity inherent in psychological knowledge. Since scientific psychologists are people, they are themselves part of the subject matter of their science (Smedslund, 1972, pp. 18–23). The analyses prompted by the present position can without difficulty be applied to the psychologist's professional activities too (Smedslund, 1981).

Finally, explication of common sense rejects as irrelevant all "externally imposed" explanations; that is, it presupposes that scientific psychological understanding is an elaboration of our naive making sense of other persons and ourselves. Hence, it is also possibly an antidote against basic category mistakes, such as trying to understand human beings as machines or physical systems.

3. Common Sense in Psychological Literature: A Selection of Instances

The preceding discussion leaves open the question of the actual incidence of unrecognized noncontingent propositions and pseudoempirical studies in research literature. The prevailing empiricist ideology would have it that there are very few such instances and that these must be regarded as unintended and uninteresting errors in the construction of hypotheses and theories. From the present point of view one would, on the contrary, expect that there are very numerous instances. Actually, I believe that whenever psychologists have regarded a general proposition as highly plausible, this is not because of data but because it approximates an explication of a part of psychological common sense. Conversely, when a suggested proposition becomes highly controversial, this is also not primarily because of data but because it is not a close enough approximation to any single part of common sense.

In this part of the paper, I will undertake to present a series of examples of allegedly noncontingent propositions and pseudoempirical studies. Admittedly, the prospects of convincing any believers in empirical psychology are none too good. If the examples given are few, it may be objected that they do not show that such propositions and studies are widespread. If the examples are very numerous, within the restricted space of an article, a likely objection is that they are not at all convincing or that they are too brief to be evaluated. If the selections are not from publications of high standing, they may be dismissed as being instances of ordinary bad research; and if they are not recent, they may be regarded as being of historical interest only. In order to meet such objections, I have decided to present a moderate number of examples relatively briefly, but with proofs that can at least be provisionally evaluated. The selections are taken from recognized publications and the majority concern topics of some current interest.

The first example goes far back in the history of psychology but is particularly clear and concerns a very important principle.

3.1. The Law of Effect

In Thorndike's well-known line drawing experiments (1931), it was found that no learning occurred when the subjects were blindfolded and got no feedback. Thorndike seemed to think that the experiments provided empirical support for the law of effect. However, if performance is to change in the direction of a criterion, then there *must* be something causing the change, and this something *must* involve information about the relation of the performance to the criterion. If a blindfolded subject improves his or her performance significantly, this means that he or she must utilize some information relating to the criterion (the blindfold is not tight, there are sounds, ESP, etc.). If a subject does not improve, then there is no utilization of information related to the criterion. In conclusion, data can neither confirm nor disconfirm the law but only show whether or not the conditions for utilization of feedback are actually established.

The possible circularity of the law has been a matter of dispute ever since the time of Thorndike (Postman, 1947). It has been argued that the law cannot be circular since (a) learning can allegedly occur in the absence of feedback (perceptual learning) and (b) learning can be absent in the presence of unambiguous feedback (training for absolute pitch, etc.). However, all such arguments rest on the assumption that a feedback can be defined *independently* of its effects on the subject. But this leads to unsurmountable difficulties. Auditory stimuli act as feedback for hearing subjects but not for deaf; written messages act as feedback for literate but not for illiterate subjects; a brick acts as feedback for a slum dweller outside Mexico City, but not for a Norwegian student. It is clear that the very same event sometimes acts as feedback and sometimes does not, the only difference residing in the psychological state of the person involved.

Meehl (1950) tried to rescue the empirical status of the law. I have tried to show that even within Meehl's causal, objectivistic frame of reference this effort failed and that the law is really devoid of empirical content (Smedslund, 1972, pp. 193–194).

On the other hand, within a frame of reference where actions are seen as intentional and as involving beliefs and wants, the law of effect merely expresses a commonsense principle that was formulated by Husserl and later by Schutz as follows:

> I trust that the world as it has been known to me up until now will continue further and that consequently the stock of knowledge obtained from my fellowmen and formed from my own experience will continue to preserve its fundamental validity. . . .From this assumption follows the further and fundamental one: that I can repeat my past successful acts. So long as the structure of the world can be taken to be constant, as long as my previous experience is valid, my ability to operate upon the world in this and that manner remains in principle preserved. (Schutz & Luckmann, 1974, p. 7)

From these and other related considerations by Husserl and by Schutz, the step is short to the formulation of the following common sense proposition: *The likelihood for P in S at t that R will occur is based on that part of P's retention of previous experience that he or she takes to be relevant.*

The consequences of negating this proposition are clearly absurd. If predictions of the future are not based on the past and/or if they are not based on judgments of relevance, they cannot be understood at all.

The proposition described above is, I believe, a central part of psychological common sense. It is used constantly in everyday life, both in inferring from a person's expectancies to his or her past, from his or her past to his or her expectancies, and in devising procedures for establishing expectancies through manipulating a person's experiences.

3.2. The Frustration–Aggression Hypothesis

This hypothesis was originally formulated as follows: "The occurrence of aggressive behavior always presupposes the existence of frustration and, contrariwise, . . . the existence of frustration always leads to some form of aggression" (Dollard, Doob, Miller, Mowrer, Sears, Ford, Hovland, & Sollenberger, 1939, p. 1). Frustration was defined as "that condition which exists when a goal-response suffers interference" (p. 11), and aggression was defined as "an act whose goal response is injury to an organism (or organism surrogate)" (p. 11). (Interestingly, the authors also provide definitions in which the concepts are logically dependent, but without making use of these.)

There is no apparent necessary linkage between the two concepts as defined above. They allow one to state, without contradiction, that a person is frustrated, yet behaving unaggressively, and that a person is not frustrated, yet behaving aggressively. Hence, the hypothesis, as it is formulated in behavioristic terminology, may be regarded as contingent. Subsequent research has shown empirically that there are, indeed, many ways of reacting to frustration other than by aggressive behavior. It has also been demonstrated that aggressive behavior may be the outcome of imitation or operant conditioning and therefore can be elicited without frustration. This means that both parts of the hypothesis as originally stated have been shown to be false as general propositions. See, for example, Schneider (1976, pp. 434–456).

Even so, there remains a core of the hypothesis which, I think, accounts for its continued intuitive appeal and which is clearly noncontingent. This is the conceptual link between the condition of frustration and the state of being angry, the latter being different from aggressive behavior. You can be angry without behaving aggressively, since this may be impossible or inadvisable. Also, you can behave aggressively without being angry, both as an automatic learned response and by deliberate pretense. However, normally one tends to think of an aggressively behaving person as being angry, and this, I believe, constitutes the intuitive appeal of the frustration–aggression hypothesis.

In ordinary language, a person is said to be angry *about* something. It makes no sense to say that *P* is angry but that there is nothing that *P* is angry about, or that nothing led *P* to become angry. Furthermore, it makes no sense to say that *P* is angry about something that involves only pleasant consequences, nor does it make sense to say that *P* is angry about something unpleasant as such, unrelated to *P*'s wants (e.g., about the alleged ugliness of a statue in a plaza in some town in South America, which fact has no connection whatsoever with any of *P*'s wants). What one is angry about must always involve frustration in order to make sense.

As expressed in this way, the frustration–aggression hypothesis becomes an explication of common sense, and instead of a contingent and false proposition we get one that is noncontingent and true.

3.3. Transfer as a Function of Similarity

The term *transfer* refers to the effects of learning in one situation on the performance and learning in another situation. It has been generally agreed that one basic factor in determining the amount of transfer is the degree of similarity between the two learning situations. The phenomena of similarity have been the subject of intensive analysis by psychologists (see Tversky, 1977). However, in traditional experiments on transfer, one need only make two assumptions, built into ordinary language. One of them is that amount of similarity and amount of difference in a particular respect are strictly inversely related. The other is that both similarity and difference vary monotonically from a maximum to a minimum value. It is not clear that any further assumptions have any bearing on the classical studies of transfer.

Half a century of research was summarized by Osgood (1949) in his famous transfer and retroaction surface. Later research has added several complications and distinctions (Martin, 1965), but without contradicting the main conclusions of Osgood as far as they went. In a previous publication, I have tried to show that the laws of Osgood are noncontingent, even within the objectivistic, causal frame of reference in which they were formulated (Smedslund, 1972, pp. 194–196).

Here, I will merely point out that the effects of transfer may be seen as consistent with the following common sense proposition: *Other things equal, the more similar in some respect two situations appear to be, to P, the more similarly will P tend to deal with them, in this respect.* The negation of this proposition makes no sense, since it would mean that *P* would not behave in accordance wtih his or her interpretations of the world. We cannot assume that a person regards two situations as highly similar in some particular respect, if the person deals with them as if they are very different in this respect, and if we can exclude deception. See also the comment on Seligman's theory below.

In conclusion, the principles relating amount of transfer and similarity are true because it could not possibly be otherwise. As a consequence, much of the research reported in this area is pseudoempirical.

3.4. Perceptual Recognition as a Function of Number of Alternatives

The concepts and mathematics of information theory have had considerable impact on psychological research. It remains to ascertain the modal status of the propositions involved.

One of the first experiments applying information theory to psychology was performed by Miller, Heise, and Lichten (1951) and dealt with auditory recognition of spoken words heard in different levels of masking noise. The results of this experiment showed that perceptual recognition performance is a function of the number of alternatives, in such a way that the fewer the alternatives, the more accurate the recognition.

Insofar as the findings are taken to support this conclusion, the study must be characterized as pseudoempirical. The authors refer to Shannon's mathematical theory of communication and it is very hard to see at what point there is any opening for empirical hypotheses. Their argument goes as follows:

> Imagine a many-dimensional space with a separate coordinate for each one of the different frequencies involved in human speech sounds. Along each coordinate plot the relative amplitude of the component at that frequency. In this hyperspace each unique speech sound is represented by a single point. Each point in the hyperspace represents a single acoustic spectrum. The group of similar sounds comprising a phoneme is represented by a cluster of points in the hyperspace. If a language utilized only two different phonemes the hyperspace would be split into two parts, one for each phoneme. The distinction between the two phonemes could be made as large as the vocal mechanism permits and discrimination would be relatively easy. But suppose the number of different phonemes in the language is increased from two to ten. With ten different phonemes the hyperspace must be divided into at least ten subspaces, and the average distance between phonemes must be smaller with ten phonemes than it is with two. The discriminations involved must be correspondingly more precise. . . . In other words, the ease with which a discrimination of speech sounds can be made is limited according to the number of different speech sounds that must be discriminated. From this line of reasoning it follows that the number of alternatives can be used to gauge the difficulty of discrimination. (Miller *et al.*, 1951, p. 332)

The preceding is no empirical psychological theory but simply deduces certain consequences that follow from the nature of the task. Given a constant total range, the number of subdivisions must necessarily influence the ease of discrimination. However, this noncontingent proposition is highly useful in two ways. First, it permits us to evaluate the reliability of a procedure: If the expected results do not obtain, something must be wrong with the procedure. Furthermore, the principle permits us to influence the accuracy of a person's discrimination by varying the number of equiprobable alternatives.

3.5. The Ames Demonstrations in Perception

One of the oldest research problems in experimental psychology has been how to explain the phenomenon that we perceive size, color, and form of objects as invariant over changes in respectively distance, illumination, and visual angle. One precondition of all this research has been the basic assumption that *one cannot perceive distant conditions by means of one's sense organs except through the mediation of some kind of cues.* This is self-evident and, hence, part of the psychological common sense, since by cues is meant *conditions that impinge on the sense organs.* Unless this assumption is made, no empirical work is possible in this area, and no controls and experimental manipulations make any sense. One interesting aspect of the famous Ames demonstrations in perception (Ittelson, 1952) is that some of them illustrate the truth of a corollary of this basic assumption: If the *cues for different distant conditions are made indistinguishable from each other, then a person cannot discriminate between these distant conditions.* An example of this is the chair demonstration:

> *Apparatus*
> The chair demonstration consists of a large wooden box containing three peepholes. Behind each peephole and visible through it is one of three different arrangements of white strings. These three groups of strings in different arrangements and at different distances have only one property in common; they all produce the same image on the retina when viewed from the peepholes.
> *Viewing conditions*
> The apparatus is viewed through each of the peepholes in turn. The small size of the holes insures monocular observation.
> *Typical observations*
> A similar object is perceived through each of the three peepholes. This object is generally described as a chair, seemingly constructed out of wire, three dimensional, rectangular, of definite size, and at a definite distance. (Ittelson, 1952, p. 26)

It should be clear from the preceding that the typical observations mentioned do not support any contingent statements about perception, as far as the discrimination between the three arrangements is concerned. However, the study illustrates once more the usefulness of noncontingent propositions in evaluating the reliability of procedures. If one fails to obtain the expected findings, one knows that the apparatus has not been well enough constructed and that the strings must be readjusted. The Ames demonstrations may also permit empirical studies. However, the chair demonstration is not one of them.

3.6. Questionnaires

The questionnaire is probably the most widely used research instrument in psychology. Also, it is an instrument where the danger of confounding contingent and noncontingent relationships is maximal. Therefore, it is important to analyze carefully the assumptions involved in this type of research.

The use of questionnaires presupposes that subjects understand the questions in the way intended by the researcher and that the researcher understands the answers in the way intended by the subjects. If a subject misunderstands a question and this is not discovered, then the researcher is bound to misunderstand the answer and hence draw an invalid conclusion. It is also possible for the researcher to misunderstand an answer, even though the subject has correctly understood the question. In this case, too, an invalid conclusion will be drawn. Hence, mutual understanding is a precondition for a valid use of questionnaires and should be controlled.

In order to decide whether or not questions and answers have been properly understood, one must have a serviceable definition of understanding and apply it to the findings. A definition of understanding may be formulated as follows: One decides whether or not one has correctly understood a behavior (verbal or nonverbal) of another person, by determining whether or not there is explicit or implicit agreement as to (1) what other behaviors or states of affairs are *equivalent* to the given one, (2) what is *implied by* the given behavior, (3) what is *contradicted by* the given behavior, and (4) what is *irrelevant to* the given behavior (Smedslund, 1970).

According to this definition, the meaning of a behavior is revealed by its logical relationship to other behaviors, that is, its place in a system of behaviors. This means that any given interpretation of a behavior entails a definite pattern of other behaviors or states of affairs. Therefore, given proper understanding, any given answer to a question logically entails certain answers to other questions. If the expected pattern of answers is obtained, this is no empirical finding of relationships between logically independent entities but merely shows that the questionnaire is a valid instrument in this respect.

It is quite easy to find examples of logically related items in actual questionnaires. Consider the following three items from Taylor's Manifest Anxiety Scale (1953, p. 286):

A. I am happy most of the time. True——False
B. I worry over money and business. True——False
C. I am inclined to take things hard. True——False

These three questions are somewhat vaguely formulated. However, it is not unreasonable to suppose that subjects tend to interpret them in the following way:

A′. In my everyday life, I am happy most of the time.
B′. In my everyday life, I worry over money and business much of the time.
C′. In my everyday life, I am taking things hard much of the time.

It seems doubtful that these three statements are completely logically independent. Apparently, A′ and B′ tend to mutual exclusion of each other, and also A′ and C′. Furthermore, an affirmation of B′ would seem to imply an affirmation of C′. The element of noncontingency that is involved means that measures of interitem relationships become at least partially spurious. One may also expect factor analyses to reveal more of the common meaning of items than of empirical relationships between independent entities.

The apparent absence of routine controls to separate logical and empirical dependence in the construction of questionnaires is a strong indication that psychologists are not generally aware of the problem involved. Apparently, the empiricist bias is so strong that no distinction is made between relationships that could have been otherwise and relationships that *must be what they are, given the assumption that the subjects have correctly understood the questions.* In conclusion, propositions about interitem relationships in questionnaires may often be noncontingent and the corresponding data may be pseudoempirical.

3.7. The Double-bind Hypothesis

The double-bind hypothesis about schizophrenia was originally formulated by Bateson, Jackson, Haley, and Weakland (1956) and has given rise to a large body of theoretical discussion and empirical research. For surveys see, for example, Schuman (1967) and Weakland (1974). Here I will try to show that the double-bind hypothesis is noncontingent and an explicaton of common sense. In order to do this, it is necessary to distinguish clearly between the hypothesis itself and its applications to complex family situations in which a host of other factors and contextual influences are involved. It is also necessary to distinguish between the abstract formulation of the hypothesis and its application to the problem of schizophrenia, the conceptual definition and clinical diagnosis of which are notoriously controversial and muddled, and also involve many other factors.

The double-bind hypothesis is formulated as follows by Watzlawick, Beavin, and Jackson:

1. Two or more persons are involved in an intense relationship that has a high degree of physical and/or psychological survival value for one, several, or all of them.
2. In such a context, a message is given which is so structured that (a) it asserts something, (b) it asserts something about its own assertion, and (c) these two assertions are mutually exclusive. Thus, if the message is an injunction, it must be disobeyed to be obeyed; if it is a definition of self or the other, the person thereby defined is this kind of person only if he is not, and is not if he is. The meaning of the message is, therefore, undecidable.
3. Finally, the recipient of the message is prevented from stepping

> outside the frame set by this message, either by metacommunicating (commenting) about it or by withdrawing. (1967, pp. 212–213)

The double-bind hypothesis states that, given a situation with characteristics (1), (2), and (3), a subject cannot give an adequate response. The necessity of this can easily be proved. If a direct response is adequate to component (a) of the message, then it must be inadequate to component (b) and vice versa. Therefore, no direct response *can* be adequate to the message as a whole. Responses which transcend the frame cannot, by definition, occur. Therefore, the situation must, by necessity, give rise to an inadequate response.

It may be concluded that the double-bind hypothesis is noncontingent and indicates a class of procedures for making it impossible for a person to respond adequately. This may account for its continued prominence, even though it has not been possible to demonstrate convincingly a connection between the incidence of double-bind situations in families and the incidence of diagnoses of schizophrenia in the family members.

3.8. Dissonance Theory

There is a whole group of theories which treat psychological change as a function of the consistency of different part processes, whether these be cognitions, attitudes, sentiments, or schemata. To this group belong, among others, Festinger's cognitive dissonance theory (1957), Osgood and Tannenbaum's theory of congruence (1955), Abelson and Rosenberg's balance model (Abelson & Rosenberg, 1958; Rosenberg, Hovland, McGuire, Abelson, & Brehm, 1960), and Piaget's theory of equilibration of assimilation and accomodation of schemata (1952). Here, I will discuss cognitive dissonance theory as a representative of the group.

Cognitive dissonance theory assumes that there are three types of relations between cognitive elements, namely *consonance, dissonance,* and *irrelevance.* Dissonance was originally defined as follows: "Two elements are in dissonant relation if, considering these two alone, the obverse of one element would follow from the other . . . x and y are dissonant if not-x follows from y" (Festinger, 1957, p. 13). The basic assumption of the theory was expressed as follows: "The presence of dissonance gives rise to pressures to reduce or eliminate the dissonance" (Festinger, 1957, p. 18).

I think that the dissonance hypothesis expresses an essential aspect of our commonsense conception of what is a human being. A proof of this must show that a negation of the hypothesis leads to absurd or to contradictory conclusions. There are many possible ways of proving this, but only a few of them can be sketched within the allotted space.

One proof presupposes that human behavior is based on beliefs and wants. But one cannot act on the basis of dissonant beliefs, that is, beliefs of the type 'I believe A and I believe not-A.' Hence, in order for goal-directed behavior to occur, dissonance reduction must take place. It is easy to show in concrete cases that an organism without a tendency to dissonance reduction could not survive.

Another proof presupposes that human beings are able to communicate and interact with each other efficiently most of the time. But if an actor has no tendency to dissonance reduction, he could not develop orderly views of anything, and, therefore, others could not predict his behavior, since no consistency could be found. Therefore, such a person could not participate in orderly and efficient social interaction.

A third proof presupposes that human beings can learn from experience. Suppose that a person believes that doing *A* leads to *B*, and then experiences that doing *A* actually leads to *C*. Absence of dissonance reduction would then mean that the person now simply believes both that *A* leads to *B* and that *A* leads to *C*. In other words, ordinary learning to revise one's opinions could not take place, and, as pointed out above, no orderly action would be forthcoming.

The empirical work on dissonance theory, mostly concerned with postdecision processes, has involved very complex and insufficiently analyzed situations. See, for instance, Schneider (1976, pp. 376–379). In my opinion, further progress can be made only after one has distinguished between what is necessarily true given the commonsense conceptual framework and what is open to empirical study. The dissonance reduction hypothesis itself is clearly an explication of common sense and indicates a class of useful procedures for changing a person's beliefs. If such a procedure succeeds in establishing dissonance, then changes must occur. If no changes are observed, then dissonance cannot have been produced.

3.9. Sequences in Cognitive Development

If the tasks involved in a developmental sequence are logically unrelated, then the data on sequence of acquisition are empirical. However, if the tasks are related by implication, then one knows that one task *must* be acquired before the other. Studies attempting to test hypotheses about implicational sequences are clearly pseudoempirical, and their data may be reinterpreted as evidence for the adequacy of the procedures.

Flavell and Wohlwill (1969, pp. 85–87) discuss developmental sequences in which successive stages are related by implication, and they note that these make up "a very substantial portion of our subject matter" (p. 86). They mention several such studies. My own study of conservation and transitivity of length, described earlier in this chapter is another clear example. Flavell and Wohlwill (1969, p. 87) conclude as follows: "Testing 'hypotheses' about sequential relations in such cases does seem to be an unprofitable exercise where the investigator is sure that the relation is of this type."

Finally, it should be mentioned that a recent discussion of the stage concept (Brainerd, 1978) contains evidence of an increasingly widespread awareness of the existence of noncontingent relationships. In the same vein, Brandtstädter (1980) has explicitly described the necessary character of certain sequences in Kohlberg's theory of moral development and the pseudoempirical status of the corresponding data.

3.10. Piaget's Theory of Centering and Decentering Applied to Social Interaction

Piaget's conception of centering and decentering (originally labeled "egocentrism") (Piaget & Inhelder, 1969) has often been interpreted as an empirically testable contribution to psychological theory. One example of this is a study of Feffer and Suchotliff (1966, p. 415–416) in which the following theoretical proposition is put to test: "Effective social interaction is a function of each participating individual's ability to consider his behavior simultaneously from different viewpoints." The authors conclude (p. 421) that the results may be interpreted as "providing support for the extension of the decentering concept to social interaction."

It appears that the hypothesis involved is noncontingent and that, consequently, the data are pseudoempirical. Since individuals do not have identical previous experiences and identical goals, they will often differ in their initial interpretation of a situation. Furthermore, since interpretations are not directly accessible to others, they must be communicated. But communications are also interpreted in the light of one's background. Therefore, communication can be efficient only when it is aimed at being intelligible to someone with a different background. This aim can be successful only to the extent that the other one's background is adequately understood. It follows that, insofar as there is a need to interact with others, and insofar as interpretations of a situation differ, one *must* take the point of view of the other one into account. It is equally clear that, if one is to pursue one's goals, one *must* also retain one's own perspective. Therefore, social interaction by necessity presupposes taking more than one perspective simultaneously into account.

The conclusion is that Feffer and Suchotliff's results must be reinterpreted as measuring the adequacy of their procedures, taking for granted a noncontingent proposition about the necessity of decentering. This proposition indicates a class of commonsense procedures for making social interaction efficient, namely ensuring that the actors take the different perspectives into account.

3.11. Bentler and Speckart's Model of Attitude–Behavior Relations

Bentler and Speckart (1979) have proposed a generalized model linking past behavior, attitude, subjective norm, and intention to future behavior. Their conclusions are based on questionnaire data and are critical of earlier work of Fishbein and Ajzen (1975). Nevertheless, they accept Fishbein and Ajzen's definitions of the main variables and present them as follows:

> A person's attitude toward a behavior can be estimated by multiplying his evaluation of each of the behavior's consequences by his subjective probability that performing the behavior will lead to that consequence and then summing the products for the total set of beliefs.
>
> Subjective norm . . . corresponds to the individual's *beliefs* regarding whether those referents who are important to him or her think that he or she should perform a given behavior. A person's subjective norm for a behavior may be determined by first obtaining such beliefs from an individual concerning each relevant referent, then multiplying each belief score by the motivation to comply with the given referent, and finally summing these products across all relevant referents. Hence, each normative belief is given a weighting (the motivation to comply) in much the same way as outcome beliefs are associated with subjective probabilities in formulating attitude.
>
> The behavioral intention is defined as one's own subjective probability that he or she will perform a given behavior. (Bentler & Speckart, 1979, pp. 452–453)

These definitions appear to violate the requirements of logical independence between variables to be related empirically.

First, the definition of attitude toward a behavior refers to the subjective probability and evaluation of all the consequences of a given behavior. But some of the consequences are specified by the definition of subjective norm, namely, what the important referents will think about the behavior. The motivation to comply with the referent's wishes will determine the evaluation of these consequences. The stronger the motivation to comply, the less favorably will an outcome contrary to the wish of the referent be evaluated. It follows from this that one cannot possibly have a maximally strong negative subjective norm about that same behavior. Similarly, one cannot, according to these definitions, have a maximally strong negative attitude toward a behavior and a maximally strong positive subjective norm about that same behavior. In other words, some proportion of any observed relationship between these two variables is not empirical, but follows from the definitions.

Second, the definition of intention (which appears to me extremely strange) links it to the subjective probability of performing the behavior, and the subjective probability is based on the information available to and utilized by the subject. But some of this available information is specified by the definitions of attitude and subjective norm, namely, the likelihood and evaluation of all the consequences of a given behavior, including the motivation to avoid or obtain some of these consequences (complying with the wishes of important referents). It is taken for granted here that the execution of a behavior is influenced by the subjectively available information about its consequences. It follows from this that the definitions of attitude and subjective norm specify information available to the subject that influences probability of execution of behavior. Therefore, one cannot have a maximally strong positive attitude toward a behavior and/or a maximally strong positive subjective norm about that behavior *and* a very low subjective probability of performing the behavior, if situational obstacles are eliminated. Conversely, one cannot have a maximally strong negative attitude toward a behavior and/or a maximally strong negative subjective norm about that behavior *and* a very high subjective probability of performing the behavior. In other words, a part of any observed relationship between, respectively, attitude and intention and subjective norm and intention is not empirical, but follows from the definitions. It may be concluded that if the variables of this formal model are applied in research in strict conformity with the definitions, then the forthcoming data will represent a confounding of logically necessary and empirical components. The model includes several commonsense components, masked as testable hypotheses. Therefore, the study reported by Bentler and Speckart and other studies relying on the same model may be classified as, at least partly, pseudoempirical.

3.12. Seligman's Revised Theory of Learned Helplessness

The revision was first proposed in Abramson, Seligman, and Teasdale (1978). Here, I will merely consider a basic assumption in the model as formulated in a recent statement: "Put simply: we assume that people react to events in accordance with their interpretations" (Peterson &

Seligman, 1981, p. 7). The negation of this proposition is "It is not the case that people react to events in accordance with their interpretations." If the negation is true, then people must act independently of or against their interpretations. But if this were the case, we could not determine people's interpretations, since we must depend on their reactions (verbal and nonverbal) to infer how they intepret things. For instance, we infer that they interpret a situation as aversive if they attempt to escape from it. Without belaboring the obvious, we simply conclude that the proposition that people do *not* act in accordance with their interpretations has contradictory and absurd implications. Therefore, the original statement that people react to events in accordance with their interpretations is noncontingent and is to be regarded as an explication of common sense. This means that studies attempting to test it are pseudoempirical and should be reinterpreted as tests of procedures. Positive outcomes cannot support Seligman's revised theory and negative outcomes cannot weaken it.

4. Is Man Rational?

Psychologists have generally viewed man's rationality as being an empirical question. This may be briefly illustrated by reference to two review articles.

A great many studies of decision making have centered around two principles which formulate the conditions of rational behavior, namely, the principle of maximizing expected utility and Bayes's Theorem. Experiments have tried to investigate the extent to which subjects actually maximize expected utility and hence behave rationally in this respect. Edwards (1968) summarizes the findings up to the time of his review as follows:

> All in all, the evidence favors rationality. . . . The main thrust of psychological theory in this area is likely to be a detailed spelling out of just how nearly rational men can be expected to be under given circumstances. (p. 41)

In a recent review, Evans (1980) also discusses the question of whether or not people reason logically. Much of the debate has centered on Henle's (1962) position that errors in reasoning tasks always are the results of faulty understanding of the premises and that people do not, strictly speaking, make logical errors. Evans concludes that the accumulated evidence does not support the Henle position; that is, the evidence suggests that people do not always reason logically.

Although their conclusions are somewhat different in emphasis,

Edwards and Evans both appear to interpret the problem of rationality as an empirical one, and both believe that the degree of approximation to rationality can be measured.

I think they are mistaken and that the reason for this can be traced to a failure to consider the premises for distinguishing between rational and irrational behavior. There is space here only for an outline of the argument.

Let me first, again, remind the reader of the difference between the behavior of an object involving causal processes and the behavior of a subject involving beliefs and wants. The discussion about rationality presupposes a view of behavior as involving a subject. It makes no sense to describe one set of objective causal processes as logical or rational and another set as illogical or irrational.

When activity is seen as behavior of a subject, wants and beliefs are necessarily involved. Consider the following example: "What is he doing out there? He is searching for the sunken city of Atlantis." The description of *what* is being done involves a specification of a *want* (to find Atlantis) and a *belief* (that it may be found in the given locality). This means that want, belief, and act are not separately locatable entities but are mutually constitutive aspects of the same whole. Knowledge of two of them permits inference about the third. If *P* wants to find Atlantis and is active in a certain vicinity, we infer that *P* believes that Atlantis may be located in that vicinity. If *P* is active in a certain vicinity and believes Atlantis is in that vicinity, we infer that he wants to find Atlantis. Finally, if *P* wants to find Atlantis and believes that Atlantis is in a certain vicinity, we expect to find *P* in that vicinity. The preceding may be derived from the following commonsense principle: *The behavior of P in S at t follows from (expresses) P's wants and beliefs in S at t.*

A negation of this proposition would mean that P does not act according to his wants and beliefs. This makes no sense and leads to absurd conclusions.

There are three types of apparently irrational behavior.

1. Cases of deviation from what the psychologist has defined as optimal (logical) performance. These may be explained along the lines suggested by Henle and others, namely, as resulting from misunderstanding. In the section on questionnaires, above, understanding was defined as explicit or implicit agreement as to what is equivalent to, implied by, contradicted by, and irrelevant to a given expression. It follows from this definition of understanding and a definition of logicality as ability to evaluate correctly relationships of equivalence, implication, contradiction, and irrelevance that understanding and logicality mutually presuppose each other (Smedslund, 1970). One cannot

understand a subject's behavior without presupposing that he or she is logical.

2. Cases that resist efforts to localize any misunderstanding at a conscious level. Apel (1965), with particular reference to psychoanalysis, has attempted to characterize analysis of such behavior as containing quasi-natural components. These involve attempts to understand the apparently irrational by reference to *unconscious* beliefs and wants. However, this still conforms with the stated position that one cannot understand the illogical, and merely shows how psychoanalysis has extended the commonsense model of action to cover even the apparently incomprehensible.

3. Finally, there are observations resistant to any psychological explanation in terms of (conscious or unconscious) beliefs and wants. Examples of this would be persistent tremors of the extremities, deficits in short-term memory and, in general, limitations in capacity. However, these limitations must be given causal explanations in terms of objective (bodily) processes and hence cannot be characterized as either logical or illogical. It is neither rational nor irrational of a tremor patient to spill coffee or of a senile person to forget to switch off the light.

In conclusion, it is doubtful that the question "Is man rational?" has any empirical content. Rationality is a built-in premise of common sense. The reader is also referred to a recent article by Cohen (1981), who raises the question of whether rationality is empirically testable and arrives at the conclusion that it is not, although from premises somewhat different from the present ones.

5. Concluding Remarks

The preceding completes my presentation of evidence for necessary true propositions in psychology. The reader is also referred to my translation of Bandura's theory of self-efficacy into 36 common sense propositions (Smedslund, 1978a,b; Bandura, 1977, 1978), an analysis of some experiments on the justice motive (Smedslund, 1979; Miller, 1977), and a formulation of seven commonsense rules of psychological treatment (Smedslund, 1981). Two empirical studies of common sense are under way, dealing with the amount of consensus about the propositions of Bandura's theory and about the logic of psychological treatment.

A few others have also been concerned with the nonempirical in psychology, albeit with very different starting points. Heider's work (1958) on commonsense psychology was a starting point for the present efforts and also contains a beginning insight into the necessary character of some commonsense principles (Heider, 1958, p. 297). Chein (1972) has emphasized the importance of what he calls "verities" in science, and Laucken (1973) has tried to show that naive psychology forms a system that is not empirically testable, yet is eminently useful. Ossorio and Davis (1968) and Ossorio (1973) have attempted to develop a kind of combinatorial calculus of components of psychological processes; and, finally, Shotter (1981) has tried to show that a recent study of legal and ethical attributions is pseudoempirical.

In this paper, I have tried to develop the beginnings of a new paradigm for psychology. There is no need for the discussion to remain at the level of programmatic generalities. For each allegedly general theoretical proposition in psychology two questions may be asked: (1) Is the proposition contingent or noncontingent? (2) Do all competent users of a language agree that the proposition is true and that its negation is absurd or contradictory? These questions can be answered by scientific methods.

ACKNOWLEDGMENTS

I am grateful to many colleagues at the University of Oslo for critical comments and stimulating discussions. In particular, I am indebted to Waldemar Rognes, whose continuous commentary and support has made it easier for me to pursue the present line of thought.

6. References

Abelson, R. P., & Rosenberg, M. J. Symbolic psychologic: A model of attitudinal cognition. *Behavioral Science*, 1958, 3, 1–13.

Abramson, L. Y., Seligman, M. E. P., & Teasdale, J. D. Learned helplessness in humans: Critique and reformulation. *Journal of Abnormal Psychology*, 1978, *87*, 49–74.

Apel, K. O. Die Entfaltung der 'sprachanalytischen' Philosophie und das Problem der 'Geisteswissenchaften'. *Philosophische Jahrbuch*, 1965, 72, 239–289.

Bandura, A. Self-efficacy: Toward a unifying theory of behavioral change. *Psychological Review*, 1977, *84*, 19–215.

Bandura, A. On distinguishing between logical and empirical verification. A comment on Smedslund. *Scandinavian Journal of Psychology*, 1978, *19*, 97–99.

Bateson, G. G., Jackson, D., Haley, J., & Weakland, J. Toward a theory of schizophrenia, *Behavioral Science*, 1956, *1*, 251–264.

Bentler, P. M., & Speckart, G. Models of attitude–behavior relations. *Psychological Review*, 1979, *86*, 452–464.

Bohm, D. *Wholeness and the implicate order*. London: Routledge & Kegan Paul, 1980.

Bradley, R., & Swartz, N. *Possible worlds. An introduction to logic and its philosophy*. Oxford: Basil Blackwell, 1979.

Brainerd, C. J. The stage question in cognitive-developmental theory. *The Behavioral and Brain Sciences*, 1978, *2*, 173–213.
Brandtstädter, J. Sprachanalysen und Verhaltenserklärungen in der Psychologie. *Trierer Psychologische Berichte*, 1980, *7*, 3.
Chein, I. *The science of behavior and the image of man.* New York: Basic Books, 1972.
Cohen, L. J. Can human irrationality be experimentally demonstrated? *The Behavioral and Brain Sciences*, 1981, *3*, 317–370.
Dollard, J., Doob, L. W., Miller, N. E., Mowrer, O. H., Sears, R. R., Ford, C. S., Hovland, C. I., & Sollenberger, R. T. *Frustration and aggression.* New Haven: Yale University Press, 1939.
Edwards, W. Decision making. Psychological aspects. In D. L. Sills (Ed.), *International Encyclopedia of the Social Sciences* (Vol. 4). New York: Macmillan, 1968, pp. 34–42.
Evans, J. St. B. T. Current issues in the psychology of reasoning. *British Journal of Psychology*, 1980, *71*, 227–239.
Feffer, M., & Suchotliff, L. Decentering implications of social interactions. *Journal of Personality and Social Psychology*, 1966, *4*, 415–422.
Festinger, L. *A theory of cognitive dissonance.* Evanston, Ill.: Row, Peterson, 1957.
Fishbein, M., & Ajzen, I. *Belief, attitude, intention and behavior: An introduction to theory and research.* Reading, Mass.: Addison-Wesley, 1975.
Flavell, J. H., & Wohlwill, J. F. Formal and functional aspects of cognitive development. In D. Elkind & J. H. Flavell (Eds.), *Studies in cognitive development.* Oxford: Oxford University Press, 1969.
Gergen, K. J. Social psychology as history. *Journal of Personality and Social Psychology*, 1973, *36*, 309–320.
Gergen, K. J. Social psychology, science and history. *Personality and Social Psychology Bulletin*, 1976, *2*, 373–383.
Heider, F. *The psychology of interpersonal relations.* New York: Wiley, 1958.
Henle, M. On the relation between logic and thinking. *Psychological Review*, 1962, *69*, 366–378.
Israel, J. *The language of dialectics and the dialectics of language.* Copenhagen: Munksgaard, 1979.
Ittelson, W. H. *The Ames demonstrations in perception.* Princeton: Princeton University Press, 1952.
Laucken, U. *Naive Verhaltenstheorie.* Stuttgart: Klett, 1973.
Martin, E. Transfer of verbal paired associates. *Psychological Review*, 1965, *72*, 327–343.
Meehl, P. E. On the circularity of the law of effect. *Psychological Bulletin*, 1950, *47*, 52–75.
Miller, D. T. Personal deserving versus justice for others: An exploration of the justice motive. *Journal of Experimental Social Psychology*, 1977, *13*, 1–13.
Miller, G. A., Heise, G. A., & Lichten, W. The intelligibility of speech as a function of the context of the test materials. *Journal of Experimental Psychology*, 1951, *41*, 329–335.
Osgood, C. E. The similarity paradox in human learning. A resolution. *Psychological Review*, 1949, *56*, 132–143.
Osgood, C. E., & Tannenbaum, P. H. The principle of congruity in the prediction of attitude change. *Psychological Review*, 1955, *62*, 42–55.
Ossorio, P. G. Never smile at a crocodile. *Journal for the Theory of Social Behaviour*, 1973, *3*, 121–140.
Ossorio, P. G., & Davis, K. E. The self, intentionality, and reactions to evaluations of the self. In C. Gordon & K. J. Gergen (Eds.), *The self in social interaction.* New York: Wiley, 1968.

Peterson, C., & Seligman, M. E. P. Helplessness and attributional style in depression. *Journal of the Norwegian Psychological Association*, 1981, *18*, 1–18, 53–59.
Piaget, J. *The origins of intelligence in children*. London: Routledge & Kegan Paul, 1952.
Piaget, J., & Inhelder, B. *The psychology of the child*. London: Routledge & Kegan Paul, 1969.
Postman, L. The history and present status of the Law of Effect. *Psychological Bulletin*, 1947, *44*, 489–563.
Rosenberg, M. J., Hovland, C. I., McGuire, W. J., Abelson, R. P., & Brehm, J. W. *Attitude organization and change*. New Haven: Yale University Press, 1962.
Schneider, D. J. *Social psychology*. London: Addison-Wesley, 1976.
Schuman, A. I. The double bind hypothesis a decade later. *Psychological Bulletin*, 1967, *68*, 409–416.
Schutz, A. *Collected papers. I. The problem of social reality*. The Hague: Martinus Nijhoff, 1967.
Schutz, A., & Luckmann, T. *The structure of the life-world*. London: Heinemann Educational Books, 1973.
Shotter, J. Critical notice: Are Fincham & Shultz's findings empirical findings? *British Journal of Social Psychology*, 1981, *20*, 143–145.
Shotter, J., & Newson, J. An ecological approach to cognitive development. In G. Butterworth & P. Light (Eds.), *Social cognition: Studies of the development of understanding*. Sussex: Harvester, 1982.
Smedslund, J. Concrete reasoning: A study of intellectual development. *Monographs of the Society for Research in Child Development*, 1964, *29*, No. 2.
Smedslund, J. Circular relation between understanding and logic. *Scandinavian Journal of Psychology*, 1970, *11*, 217–219.
Smedslund, J. *Becoming a psychologist. Theoretical foundations for a humanistic psychology*. New York: Halsted Press and Oslo: Universitetsforlaget, 1972.
Smedslund, J. Bandura's theory of self-efficacy: A set of common sense theorems. *Scandinavian Journal of Psychology*, 1978, *19*, 1–14. (a)
Smedslund, J. Some psychological theories are not empirical: Reply to Bandura. *Scandinavian Journal of Psychology*, 1978, *19*, 101–102. (b)
Smedslund, J. Between the analytic and the arbitrary: A case study of psychological research. *Scandinavian Journal of Psychology*, 1979, *20*, 129–140.
Smedslund, J. Analyzing the primary code: From empiricism to apriorism. In D. Olson (Ed.), *The social foundations of language and thought. Essays in honor of J. S. Bruner*. New York: Norton, 1980.
Smedslund, J. The logic of psychological treatment. *Scandinavian Journal of Psychology*, 1981, *22*, 65–77.
Taylor, J. A. A personality scale of manifest anxiety. *Journal of Abnormal and Social Psychology*, 1953, *48*, 285–290.
Thorndike, E. L. *Human learning*. New York: Appleton-Century-Crofts, 1931.
Tversky, A. Features of similarity. *Psychological Review*, 1977, *84*, 327–352.
van der Waerden, B. L. *Science awakening*. Groningen, Holland: Noordhoff, 1954.
Watzlawick, P., Beavin, J. H., & Jacson, D. D. *Pragmatics of human communication*. New York: Norton, 1967.
Weakland, H. H. The double bind hypothesis by self-reflexive hindsight. *Family Process*, 1974, *13*, 269–277.

5

What Is Remarkable in Psychology?

Herman Tennessen

> When an exploded philosophy dies it goes to psychology where it is resurrected and presented as the latest of insights.[1]
>
> With apologies to C. D. Broad

> The universe is not in accord with common sense ideas.[2]
>
> Carl Sagan

In psychology—as well as in any other scientific or nonscientific context—the remarkability of a sentence, *S*, is a direct function of the degree to which members of the language community in question, *L*, would be inclined to interpret *S* in the direction of a proposition, *P*, with which a maximal amount of *L*-members would most assuredly disagree; whereas the *P*-proponents, to everyone else's stupefaction, were actually able to demonstrate convincingly that *P* (as formulated by *S*) should indeed be considered tenable in view of (now) available but hitherto astoundingly

[1] The exploded philosophy to which we refer is, needless to say, the so-called ordinary-language philosophy which in the late 1950s was perpetrated by members of quite a few respectable universities. "Its heyday was short," though, according to P. F. Strawson's commemorative flashback in *The Times Literary Supplement*, September 9, 1960, p. LX. What has prompted its prodigious second coming, however, is totally incomprehensible. But there it is, I suppose, as Smedslund candidly discloses: "I believe that the common sense of ordinary language forms one single system in which everything is related (directly or indirectly) to eveything else."

[2] Here quoted after Ferris (1979), p. XII.

Herman Tennessen • Department of Philosophy and Center for Advanced Study in Theoretical Psychology, University of Alberta, Edmonton, Alberta, Canada T6G 2E9.

unexpected evidence, or through some scintillating piece of profound, ingenious ratiocination. The prototypal example of a remarkable proposition is Einstein's conjecture that a ray of light coming from a star and passing near the sun would be deflected through an angle of about 1.75 seconds of arc, which would in fact render such stars visible as are actually (seen from the Earth) located "behind" the sun. The conjecture was tested during the solar eclipse of May, 1919, and the results appeared in good accord with Einstein's prediction.

By the same token, a sentence is *un*remarkable to the extent to which it either states a plain and uncontestable truth or falsity or merely reflects some trivial, common sensical proposition in some of the more common (dictionary) senses of *common sense* as well as in the somewhat exotic sense suggested by Smedslund: "*A proposition in a given context belongs to common sense if and only if all competent*[3] *users of the language involved agree that the proposition in a given context is true and its negation is contradictory or senseless.*"[4]

There is a general guiding maxim for plausible interpretations of all such linguistic locutions or sentences which have the slightest degree of communicative presumptions. It is the central point of the so-called Tennessen–Searl hypothesis[5]: *We all intend to remark what is remarkable.*

One integral constituent of 'remarkability' is, as we have seen, audacity, boldness. Were we to envisage a continuum of all possible propositions from extreme audacity at one ultimate point, to extreme nugacity at the other pole, we should most likely find that outrageously audacious propositions would evoke reactions like "shocking," "repugnant," "incredible," "amphigoric," "senseless," or "absurd," whereas the response

[3] In the heyday of so-called ordinary language philosophy the only speaker worthy of attention was the *native* speaker. That made things simple: A copy of your own birth certificate was all you needed to be accepted. On the other hand, this procedure barred the linguistic voting rights of Joseph Conrad (and Jan Smedslund). So *native*, which seemed clear but was perhaps misleading, has been replaced by *competent*, which although possibly a trifle less inappropriate, is devastatingly flummoxing and forbiddingly elusive and obscure.

[4] Even if Smedslund has neglected to drop any hint as to what he conceivably could have meant to intimate by "competent," one can scarcely avoid the impression that he thinks there should be quite a few "competent [language] users" within a language community. Thus the mere task of establishing the consensus of a total population (of "competent users") may seem like a prohibitively tall order. It comes as no surprise, therefore, that Smedslund is perfectly willing to settle for a test consisting of "the response distribution in a large sample."

[5] Gullvåg, 1967, pp. 449–452.

to an excessively nugacious proposition would rather prompt responses like: "big deal!" or "so what!" or such characterizations as "trifling," "piddling," "immaterial," "valueless," "insignificant," "goes without saying," or "just plain common sense." Both truth and falsity may *in principle* be assigned to audacious as well as to nugacious propositions. Unfortunately, however, as a general rule, veracity seems to correlate positively with nugacity, as will falsity with audacity. And an audacious proposition is, as we have seen, remarkable only to the extent to which it is also tenable; by the same token, a nugacious-sounding sentence can acquire a degree of remarkability only if, plausibly interpreted, it states such a proposition as can—at some time or other—be shown to be false.[6]

By extending our continuum beyond the ultimates of audacity and nugacity, we shall find ourselves confronted with propositions which are not only unremarkable but *totally devoid of any interest whatsoever:* so-called necessary, analytic propositions which, if formalized, would show either exclusively *T*s or exclusively *F*s in their truth tables. Transcending the *nugacity* end of the continuum, we would enter into the area of necessarily, analytically, logically, notationally, demonstratively, absolutely-*true* propositions, or *tautologies.* By transcending the other end, the ultimate *audacity, false* propositions of the necessary, analytical, logical—kind would be encountered—so-called (logical) *contradictions.* Neither kind of propositions, needless to say, is ever intended to be explicitly stated by anyone. No sentence, therefore, should plausibly—let alone charitably—be interpreted to state any proposition in these directions.

Quite another thing is that a speaker can easily be beguiled into believing that he or she is producing propositions which are both (absolutely) reliable and extremely audacious; not the least if the discourse is carried out within the quotidian vernacular (also called "everyday

[6] Consider, for example, the formulated proposition *P*: "The night must be dark" (at least between the two polar circles). *P* is bound to be considered nugacious in the extreme: Just plain common sense! And common sense it was in 1683 as well, when Edmund Halley advanced *anti-P,* the so-called Olbers' paradox: "Given the universally accepted cosmology, the night sky must be as bright as the surface of an average star." No doubt *anti-P* should be wholly, unanimously, and emphatically rejected as senseless and absurd by all common sense philosophers, contemporaries like Samuel Johnson and George Berkeley ("the man of common sense is eminently the man who trusts his senses"), as well as by the rich variety of later representatives (Reid, Hamilton, Sidgwick, Stout, Moore). Halley had to wait nearly 250 years before he was shown to be right, and Heinrich Olbers' paradox was finally dissolved by the Hubble-Humason interpretation of the red shift (*vide,* e.g., Ferris, 1979, p. 30; Sagan, 1980, pp. 254–256; and Sandage, 1979, pp. 5–11).

speech" or "ordinary language"), which is notorious for its extravagant amorphousness and total lack of semantic aspirations.[7] The ambiguous formulation borrows, so to speak, audacity from one possible direction of precising, and tenability from another. One familiar instance is the hedonistic apothegm: Everybody seeks pleasure. It resorbs audacity from a precising direction ultimately leading to the (implausible) interpretation: Everybody is wallowing in delicious, concupiscent, carnal desires. But, pressured with regard to the tenability of such or cognitively similar propositions, the hedonists will be inclined to regress into something safer, a more nugacious proposition, say: Everybody seeks what he really likes. Not to mention (the implausible): Everybody seeks whatever he (everybody) seeks: '$p \supset p$')! Similar examples are legion; a favorite of mine is "Human nature is always the same," as exploited by, *inter alia,* opponents of peace negotiations or arms reduction.

What, then, about technical language? Surely by predominantly employing technical terms one would escape the embarrassing pitfalls provided in such abundancy by the quotidian vernacular, the everyday speech, the ordinary language. Although there can be no doubt as to the enormous importance of well-defined technical (theoretical) terms for the growth of so-called hard sciences (microastrophysics, for example),[8] the apparent advantages for the softer sciences are not all that well founded. More often than not, the technical verbiage serves to disguise the cognitive paltriness of a formulated hypothesis. If it would permit, without essential distortion, a translation into the quotidian vernacular, the chances are that its trivial, common sensical, everyday-life reasonableness would become so embarrassingly unmistakable that any

[7] The general rule here is that the more conspicuously *the external form* of a sentence exhibits the features commonly recognized as, say, analytic or senseless, the less plausible is an interpretation in that direction. Thus, the remark: "The coffee in the machine is not coffee" is never interpreted as stating anything comparable to '$p \& \sim p$'. Similarly, "A dollar is a dollar" bears no cognitive similarity to:'$p \supset p$'. And—to mention only one example—to interpret, as Smedslund does, a remark about a person "that he was nowhere while performing an act" as stating a senseless proposition reveals not only a peculiar, linguistic rigidity and semantic naiveté but demonstrates such a lack of communicative imagination and general communicative competence as would scarcely warrant a passing grade in the propaedeutic courses in so-called logic at the University of Oslo.

[8] Although anyone concerned with "common sense" or "ordinary language" is bound to be baffled by such notions as *tachyons,* defined as "physical particles whose masses can only be represented by imaginary numbers and whose speed can never decelerate below the velocity of light". Or consider *positrons,* conceived as "electrons traveling backwards in time" (or the idea of *reversed causation,* etc.; Tennessen, 1976, pp. 121 & 122).

suggestion for conducting an empirical test of the hypothesis would often rule itself out as fatuously futile and thoroughly ludicrous.

But Smedslund apparently wants to go much further. He seems inclined to view absolutely every psychological hypothesis (or "valid proposition") in any language as a mere "explication of common sense," in point of fact: as so trite a truism that as a matter of complete indifference it would be surpassed only by a hypothesis (or a proposition) exposed as necessarily (logically, notationally, etc.) true (or false).[9]

As "evidence," ostensibly designed to support this heinous accusation against psychology, Smedslund cites a dozen examples of theory formulations which he chooses, quite arbitrarily, to interpret in such a direction that they assume the guise of what Smedslund takes to be "common sensical," "necessary true propositions in psychology." The interpretations are implausible, as judged according to the Tennessen–Searl hypothesis: We remark what is remarkable. And even if they could have been accepted as plausible, Smedslund makes the elementary error of confusing evidence and illustration (of his point of view).[10]

Smedslund's position as to what is necessarily true in psychology is eminently audacious. It would have been such a gloriously exciting thesis had there only been *some symptoms* of corroborative encouragement. Much to be deplored: No discernible ones are offered.

1. References

Dubinsky, D. The impossibility of formal knowledge. *Methodology and Science,* 1980, *13*(1), 28–38.

Ferris, T. *The red shift.* New York: Bantam, 1979.

Gullvåg, I. *Referanse, mening og eksistens.* Oslo: Oslo University Press, 1967.

Sagan, C. *Cosmos.* New York: Random House, 1980.

Sandage, A. R. The red shift, *Cosmology + 1* (Readings from *Scientific American.*) San Francisco: Freeman, 1977, pp. 4–12.

Tennessen, H. Note on the confusion of evidence and illustration. *The Journal of Philosophy,* 1959, *56*(18), 733–736.

Tennessen, H. Ordinary language *in memoriam. Inquiry,* 1965, *8,* 225–249.

Tennessen, H. On knowing what one knows not. In J. R. Royce & W. W. Rozeboom (Eds.), *The psychology of knowing.* New York, Paris, London: Gordon & Breach, 1972, pp. 111–175.

---

[9] That (obviously) necessary, demonstrative, explicative, etc. truths are not even remarkable within the so-called formal sciences like mathematics is argued in Tennessen, 1972, pp. 116–118, Tennessen, 1980, pp. 16–24; and Dubinsky, 1980, pp. 28–38.

[10] Tennessen, 1959, pp. 733–736.

Tennessen, H. Scientists in vain wants of world views. *Methodology of Science*, 1976, *9*, 120–128.
Tennessen, H. Quandaries of quotidianism. *Methodology and Science*, 1978, *11*(2), 114–123.
Tennessen, H. *Problems of knowledge*. Assen, Holland: van Gorcum, 1980.

5

On the Limitations of Commonsense Psychology

Fred Vollmer

1. Introduction

The psychological concepts that form part of our ordinary language are, according to Smedslund, logically related to one another. To be a *person*, for instance, implies having a *body*. And the proposition *'p is a person'* implies *'p has a body'*. Every psychological concept is in this way logically related to some other psychological concepts, and consequently there are many conceptual truths to be discovered by analyzing and explicating ordinary language. Psychological common sense, according to Smedslund, consists of those psychological conceptual truths that everyone knows and agrees upon. Which propositions are of this kind must be ascertained by empirical methods.

Many conceptual truths, according to Smedslund, have an empirical basis and have been discovered by experience. Only subsequent conceptual analyses have shown them to be noncontingent. Examples, according to Smedslund, are the seven bridges of Königsberg and the Pythagorean theorem. It is, furthermore, Smedslund's viewpoint that within psychology this is a very widespread phenomenon that has been going on since the nineteenth century and is still taking place in current research. Psychologists, that is, have by empirical methods been arriving at propositions that are necessarily true. As psychologists rarely do conceptual analysis, however, they have not known this but have mistakenly assumed their propositions to be contingent.

Fred Vollmer • Department of Psychology, University of Bergen, Bergen, Norway.

It would be outrageous, however, to assume that *all* principles, propositions, assertions, statements, predictions, descriptions, hypotheses that have been and can be formulated in psychology are of a noncontingent kind. And Smedslund, of course, does not assume this. What he does seem to believe, however, is that genuine empirical propositions in psychology, regarding conceptually independent events or characteristics and their interrelationships, can never have the same high level of validity as commonsense propositions. While conceptual truths are assumed to have general validity, presumably because the meanings and logical relationships of ordinary language concepts stay the same across time and space, indeed are "not locatable in any separable spatiotemporal regions" and are "of another order than individual actions and interactions," empirical truths about human life can, according to Smedslund, only be valid for certain people at certain times, not for all people forever. The reason why empirical psychological propositions can have only local validity, Smedslund believes, is that psychological phenomena are historical. Every individual, that is, is the outcome of a complex series of historical events and can only be understood and explained by reference to his or her history. As no two life stories are alike, it follows that each person has his or her own explanation and cannot be explained by any generally valid principles of behavior. Psychology, according to Smedslund, "needs to develop strategies for dealing with the historically unique and with changing circumstances." Smedslund, furthermore, seems to believe that "dealing with" historically unique events can be achieved by knowledge of commonsense propositions and that the only kind of theory psychology needs consists of commonsense propositions:

> Psychological theory consists of noncontingent propositions only.
>
> The task of the theoretical psychologist is as follows: to try to formulate propositions that are regarded as true and their negations as senseless or contradictory, by all competent users of a language.
>
> Psychological theory should consist of explications of common sense.

2. Comments

The first aspect of Smedslund's metapsychology to require comment is the assumption that empirical psychological propositions can have only local validity. Although persons as concrete wholes may require complex and unique historical explanations, there are still a number of basic psychological processes like perception, learning, thinking, emotion, and motivation that at least in some respects may be very much the same for many different people. And there are probably many empirical questions about such phenomena, or aspects of them, that can be answered without reference to complex and unique life stories. For instance, how strong must a sound be in order to be heard, how small a difference in stimulation can people perceive, how long must a picture be exposed in order to be seen, do certain stimuli, colors for instance, capture attention more readily than others? When do children normally acquire various perceptual abilities, how should traffic signs be constructed in order to be perceived with maximal clarity and speed? Smedslund asserts rather categorically that knowledge of such abstract processes "is of very limited value in everyday life and in professional psychological practice." No arguments for this assertion are put forth, however.

The second and major topic to be discussed is Smedslund's assertion that historically unique events can be "dealt with" by knowledge of commonsense propositions and that such propositions *only* can and/or should make up psychological theory. What is puzzling here is that Smedslund on the one hand assumes each individual to be the outcome of a unique history. He furthermore seems to conceive of such complex and unique human lives as consisting, at least in part, in a number of logically independent (extrinsically related) events where one arbitrary happening can be a causal condition of some other one. Smedslund must conceive of human lives in this way, for, according to him, it is about such complex series of unique historical events that genuine empirical propositions (contingently true ones) *can* be made in psychology. On the other hand, it is asserted that the *only* kinds of theories psychologists need bother with formulating should consist in general conceptual truths (every person has a body). Smedslund either must think that commonsense propositions will be sufficient to explain individual human beings or else must hold that psychology should not attempt to explain and understand individual human lives. The former alternative, however, must be false. If a person p is "unexplainable, except by reference to a series of unique historical events," then commonsense theory must be inadequate to explain p. For general conceptual truths surely do not contain reference to the series of unique historical events making up p's life story. And I take it to be self-evident that psychology should be interested in understanding and explaining individual human beings. I can only conclude that something is missing in Smedslund's picture of what psychology should be and what psychologists must do and know.

If human life consists in part in a large number of different and logically independent psychological events, that is, events not belonging to the same or logically related concepts, events such that one can be a cause of the other, why should there not also be causal theories in psychology, even if they are true for one person or group only? Why should, and how can, psychological theory consist of noncontingent propositions only? And even in cases in which we are not interested in causes, but in the meaning of something a person did, it is hard to see how necessary, true, commonsense propositions are going to provide us with concrete answers. Though common sense may give us a general answer to the effect that human actions are related to beliefs and intentions, it will not by itself provide us with a specific hypothesis as to *which* beliefs or intentions were being expressed in this or that particular action. To entertain a hypothesis about this, we must leave common sense and look at other parts of the life-world.

Whereas I agree, then, with Smedslund that ordinary language is the language for psychology to use, I fail to see why psychologists should use it to express conceptual truths *only*. And although I agree with Smedslund that it is necessary and important to analyze concepts, to reflect on what they *mean*, in order to achieve a proper understanding of psychological events, and avoid interpreting events that belong together and form internally related parts of the same whole as causes and effects of one another, surely such conceptual analysis cannot be *enough* and is not going to tell us what a person's life story is. Surely historical sciences (like history and psychoanalysis) are not going to find out about histories by way of conceptual analysis alone. In order to understand an event taking place in a historical-cultural-social context, one must know and understand the language, concepts, and ways of life of that culture. But then one must still look at the concrete phenomena. History, after all, is still a painstaking study of events.

But in asserting here that general propositions that are necessarily true are not going to do the whole job of explaining complex, individual human lives, and that one also has to study (empirically) concrete psychological phenomena in order to know why this or that has happened, am I not making the bad mistake of regarding psychological *phenomena* (events, reality) as one thing, and language as something totally different and only externally related to psychological reality (cf. Smedslund, 1982), and assuming that the one can be studied independent of the other? I hope not. At least I totally agree with Smedslund that ordinary language and psychological reality are internally related. Ordinary language is the language that mentions psychological events, the language through which psychological events are defined, understood, and constituted. Ordinary language must consequently have a privileged position in psychology as the only language through which psychological reality can properly be grasped and meaningfully spoken of. It is also through concepts of ordinary language (like belief and intention) that people understand and explain their own actions. So to the extent that psychology should be a hermeneutical science, interested in what the actor himself meant by doing so and so, psychology again must use ordinary language. And ordinary language is, of course, used by people in everyday life not only to explain but to do many other things like promise, ask, order, beg, warn, threaten, and apologize. Speaking is doing many things, and the use of ordinary language together with other behavior in situations provides the numerous language games of which life in part can be said to consist. Again, I agree that all this indicates that psychologists, or any others who are interested in understanding and explaining human life, must also be interested in ordinary language as a part of that life. What I still find problematic, however, is that psychologists' only interest in ordinary language should be to unearth propositions that are necessarily true. Ordinary language, after all, is much more than common sense (as defined by Smedslund). However one looks at the matter, it would seem self-evident that a major task for psychology as a science must be to explain psychological phenomena. Such explanation can be intentional or causal. And conceptual analysis has an important function here in helping us to decide whether we are dealing with cause and effect or with internally related phenomena. But how explicating necessarily true propositions alone is going to do the job of explanation—what made that happen, what did that mean—is difficult to see.

As a concrete example of the insufficiency of commonsense propositions, let us look at the phenomenon of learning. How do people learn? Smedslund shows us that we can know something about this by analyzing the concept of learning. Learning consists in a person's changing his own behavior in the direction of some criterion or goal. And a person cannot be doing such a thing, according to Smedslund, if he does not know about his own behavior in relation to the criterion. Learning, then, implies and cannot occur without feedback. It is, according to Smedslund, not necessary to do experiments (as Thorndike did) to know this.

But there is more to know about learning than that it consists in *p* changing his behavior in relation to a criterion. When someone knows how his performance relates to some standard of excellence and wants to be able to perform better, he does not then just start improving. Thus one does not tell someone kindly to start improving and then expect him to do so directly. Improving is not a basic action that a person does *simpliciter*. It is more like a result that a person can achieve if, and only if, certain other conditions have been fullfilled; *p* often does not know what will lead to improvement or when improvement will take place. He may have to wait patiently for it to occur: "I hope there will be some improvement in the next couple of days." And he may be surprised and glad when it does happen: "Suddenly I could do it much better." Learning, in short, has causal determinants. And if it has causal determinants, we are not going to find out what these causal determinants are and how they work by analyzing concepts and formulating commonsense propositions. What is needed is empirical research and formulation of empirical propositions.

But here it may be objected that we already know that "Übung macht Meister," that practice is what leads to improvement and skilled performance. The question is: Is "Übung macht Meister" a conceptual truth? I do not know what Smedslund would answer to this. But if we look at the world, the fact that a person cannot gain skill unless he practices is surely a matter of *natural,* not logical, necessity. We can imagine a world wherein people become good at doing A by taking a certain pill. And it is not a *logical* absurdity to say that someone is playing beautifully but has never practiced. It *is* absurd to suppose such things, I agree, but not absurd in the same way as it is to suppose that somewhere circles could be square. The former ideas are absurd because the world we live in is in fact quite different; the latter idea is *logically* impossible and could not be true in *any* world. I cannot help thinking, then, that although we already know that "Übung macht Meister," the truth of this proposition depends on facts being the way they are and not otherwise. And if, in everyday life, 'practice' and 'skill' *do* imply one another logically, may it not be that we have come to use and define the concepts in this way because we know, through thousands of years of experience, that no one can become skillful who does not practice? May it not also be the case that some other of the noncontingent truths Smedslund asserts psychologists have been discovering by empirical methods are really factual, causal truths, that subsequently have been formulated in commonsense propositions of a noncontingent nature? And if this is the case, certainly one cannot rely on common sense to contain all the factual truths there are in the world.

Finally, it should be pointed out that though "Übung macht Meister" tells us, contingently or noncontingently, that practice is the thing that will lead to improvememt and skill, it does not tell us what kind of practice will be most effective, or how much and often a person must practice, in order to reach a certain level of skill. Again, empirical research and formulation of empirical propositions are needed to deal with such questions.

My final comment concerns Smedslund's conception of necessarily true propositions. Part of what Smedslund writes about commonsense propositions, that they are "of another order than individual actions and interactions" and are "not locatable in any separable spatiotemporal regions," conveys the impression that such truths exist in some ideal Platonic reality of their own, detached from and independent of the shadowy, flickering, and ever-changing life-world. Though they are said to have an empirical basis, their truth does not seem to depend in any way on the world. Thus what is necessarily true, according to Smedslund, cannot be falsified by experience. But is this sharp distinction between conceptual and factual truths and the absolute independence of the former from the latter really tenable? Necessarily true propositions, it is held, follow from definitions of concepts, are in a way repetitions of definitions. Such definitions, in turn, express conceptions of what the nature or essence of a certain thing or event consists in. Something cannot be a person unless it has a human body. But do not such conceptions of essence, in turn, rest on how we experience the world to be? Does our notion of persons as beings having human bodies not rest on the empirical truth that all the living persons we know in fact *do* have such bodies? If shoes or dogs or trees could speak to us and understand what we said, or if we could speak to and be spoken to by persons no longer physically existing, as some claim can be done, then perhaps our concept of a person would change. Once it was thought to be an essential truth about hysteria that only women could be hysterics, and about the mental that nothing could be mental that was not conscious. Such conceptions as to the nature of hysteria and mind have subsequently been challenged, on the basis of facts.

Conceptions as to what things essentially consist in, I conclude, are not gained by intuiting some separate Platonic universe of unchanging ideas, but in reflecting on the world of everyday life and on how things in that world in fact are. Essential definitions express our understanding of the world we live in. If this is so, however, the notion of necessarily true propositions as assertions which cannot be falsified empirically, with truth values independent of the world, is somewhat dubious.

3. Conclusion

Commonsense psychology understood as an enterprise analyzing the meaning of ordinary language concepts and formulating necessarily true commonsense propositions can only be a part of psychology. It is a highly important and necessary part that psychologists should take much more seriously in order to achieve a better understanding of events they are studying. Wittgenstein's dictum that "in psychology there are experimental methods and conceptual confusion" (1953, p. 232) is still valid, and one of the chief and important merits of Smedslund's work is to point this out anew. But conceptual analysis still cannot be *all* there is to psychology. In human life there are infinitely many empirical questions that cannot be answered or dealt with by commonsense propositions alone, but for which research and empirical propositions are also needed. Part of what Smedslund writes does not appear to be in conflict with this conclusion, for example, the notion of an individual as the outcome of a series of unique historical events and the concession that truly empirical propositions may be found in psychology. Other passages, however, where it is stated, for instance, that psychology is not an empirical science, that psychological theory can consist of commonsense propositions only, and reference to "believers in empirical psychology" who are hard to convince of their mistaken views, and an "empirical ideology" that is all wrong, seem to indicate that empirical research in psychology is meaningless and a waste of time and that Smedslund does not believe in it. This position I find extreme and untenable. The notion that everything in psychology can be dealt with by commonsense propositions is a new kind of absolutism that is just as mistaken as the positivistic doctrine that everything can be dealt with by the Covering Law Model. The lesson to learn is that all doctrines asserting there to be one and only one valid approach that will solve all problems in some area of research are necessarily false.

4. References

Smedslund, J. Common sense as psychosocial reality: A reply to Sjöberg. *Scandinavian Journal of Psychology*, 1982, 23, 79–82.

Wittgenstein, L. *Philosophical investigations*. Oxford: Blackwell, 1953.

5

It Ain't Necessarily So[1]

K. V. Wilkes

Psychology often surprises us; moreover, it is a discipline which attracts many people to its banner. It is difficult to see how it could do either if the burden of Smedslund's central thesis were correct. I believe that he is wrong, although the issues he raises are of great interest and importance; I shall work my way toward my main argument by discussing a couple of points with which I am in at least partial agreement.

The first point of partial agreement is this: A dominant and attractive view of the scientific enterprise[2] now holds that the aim of the game is, *inter alia*, to uncover necessary truths. (Smedslund defines his *noncontingent* in terms of necessity, so my preference for talk in terms of necessity should not be misleading.) However, the view mentioned emphatically does not restrict the claim to psychology—indeed, and unfortunately, psychology is all too little discussed by theoreticians of science—it extends also to such sciences as physics, chemistry, zoology, and botany; thus agreement in this respect alone would not single out psychology from other empirical sciences. Defense of this view requires only a realist (rather than a positivist, or instrumentalist) construal of science, plus a commitment to some version of essentialism (which need not be a strong, metaphysical essentialism). One way—and there are several—of arguing that most sciences seek *inter alia* to discover necessary (noncontingent) truths runs as follows: The primary *analysanda*

[1] Title with apologies to Putnam (who got there first).

[2] Most realists would accept some version of this view, although many would want to abjure talk of necessity. It is Kripke, of course, who is primarily responsible for the revival of essentialist ideas; see his 'Naming and Necessity,' in D. Davidson and G. Harman (Eds.), *Semantics of Natural Language* (Dordrecht, Holland: D. Reidel, 1972, pp. 253–355).

K. V. Wilkes • St. Hilda's College, Oxford University, Oxford OX4 1DY, England.

for science are 'natural kinds' (for example, species, the elements, molecules, atoms, electrons). To be a natural kind is, precisely, to have an essential nature which serves to define the kind and which is such that it can be systematically unfolded by means of general laws. These general laws explain why the kind-members have the properties that they do; often the laws are microstructural, setting out the properties constitutive of that kind. Analysis of a natural kind thus identifies not only the contingent, but also the necessary, properties of all kind-members (contingent: zebras are striped; necessary: zebras are mammals). Thus, for example, anyone who has any sympathy at all with talk of 'necessity' would probably agree that all the following count as necessary propositions: that whales are mammals, that gold has atomic number 79, that light is a stream of molecules, that water is H_2O. All these statements are genuine and valuable products of scientific inquiry; therefore, if psychology were indeed seeking *(inter alia)* to discover necessary truths, that fact would not make it at all exceptional—most empirical sciences do so too. (My own reservation is rather that psychology has greater difficulty than the physical sciences in identifying 'natural kinds', and so it will have greater difficulty in formulating necessarily true propositions; but that needs a separate, and lengthy, argument.[3])

One can, then, agree with Smedslund and concede that part of the job of psychology, insofar as it resembles other natural sciences, is to uncover necessarily true propositions about the subject matter in its domain. On the other hand, the concession serves to highlight the substantial differences between this picture of the scientific enterprise and the picture of psychology which he offers. The most striking difference is, of course, that the necessary propositions illustrated above are all the product of *empirical* investigation and have to be; they are emphatically not discovered, or discoverable, *a priori*, and nobody could call the research 'pseudoempirical'. We can get closer to the central disagreement by examining Smedslund's account of 'noncontingent'. He defines a noncontingent proposition as a proposition of common sense which is such that "all competent users of the language involved agree that the proposition in the given context is true and that its negation is contradictory or senseless." The necessarily true propositions cited above will fit this definition reasonably well *if* after the phrase "of the language" we add "and the theory." For perhaps not all laymen would regard it as "contradictory or senseless" to deny that whales are mammals or that gold has atomic number 79; but the zoologist or physicist is likely to do

[3] I provide such an argument in *The Autonomy of Psychology* (forthcoming).

so. The crucial point is that unaided common sense informed us of none of these noncontingent truths—they had to be discovered empirically—and that (theoretically based) observation and experiment were essential.

The crux of the disagreement, then, shifts to the privileged status, where psychology *alone* is concerned, of common sense. Why is it just psychology that bans theory from joining common sense in the identification of noncontingent propositions? Is it indeed true that psychology, unlike most other sciences, must be confined within the straitjacket of *unaided* common sense? Put another way, why should we regard psychology as special in that it should abjure empirical methods and theoretical backing in its identification of noncontingent propositions? To explore this issue properly, we must turn to the second point in which Smedslund and I are in at least partial agreement: namely, that many (although surely not all?[4]) ordinary-language psychological statements are more or less noncontingent. The reasons given for agreeing with Smedslund on this point should also show why I think he is wrong to suggest that scientific psychology should be dominated by its nonscientific (commonsense) counterpart.

Let us agree that many generalizations in commonsense psychology come close, often, to being noncontingent. Indeed, philosophers working on the problem of the explanation of action are prone to talk in terms of the "*a priori* principles" of belief–desire explanation; this seems to me fairly close to some of Smedslund's arguments. However, before we swallow the noncontingent, *a priori,* status of commonsensical psychological principles we need to enter a few caveats. First: remarkably few such commonsense generalizations are ever explicitly formulated; we tend to assume that implicit general principles must be at work but are very feeble at stating them. There is a clear reason for this: the chief glory and triumph of commonsense psychology lies not in its ability to exploit or unearth generalities, but rather in its penetrating power (at its best) to explain why just this agent did precisely that action,in these particular circumstances, and at that specific time. Such explanations rarely use or need general laws (read Dostoievsky or Henry James!)—they concern the unique, the particular, the specific. On the rare occasions when common sense seeks to provide generalizations ("Why do people join neo-Nazi parties?" "What sort of person makes it to the top in Moscow's Politburo?") the resulting answers are almost invariably

[4] I find it hard to believe that Smedslund means to include all propositions of commonsense psychology in the category "noncontingent," as his italicized definition suggests; what, for instance, of "He cold-shouldered John because of his jealousy"? I assume he means general propositions.

boring, superficial, or unconvincing. A second caveat: the generalizations that laymen do indeed possess are immune from charges of falsehood *not* because they are substantial necessary truths but because they have such a protective shroud of all-embracing, wide-ranging, and indeterminate *ceteris paribus* clauses that they are true because their claims are so trivial and minimal. Armed with the *ceteris paribus* shield, one can readily agree that both "many hands make light work" *and* "too many cooks spoil the broth," both "out of sight, out of mind" *and* "absence makes the heart grow fonder" are true—yet the members in the two pairs are surely contradictory: it is the "other things equal" clause which reconciles them and which could reconcile almost anything one cared to produce. Thus it is near-vacuity, not insight, which makes commonsense generalities so uncontestable (cf. "If you heat water *enough*, it boils"), whereas everyday explanations of particular actions can, by contrast, be penetrating and enlightening. Third: commonsense generalizations are protected not only by these "other things equal" clauses, but also by the nature of the terms employed. Few ordinary language terms can be defined precisely, and we should not expect it to be otherwise. After all, we need them for hosts of tasks other than the (scientific) tasks of systematic description and explanation: we use them to praise, blame, warn, deter, order, encourage, assess, evaluate, judge, hint, joke—and they succeed in these diverse roles just *because* of their flexibility, range of nuance, capacity for metaphorical extension. ('Happiness', unwisely used in one example by Smedslund, is of course a notorious instance of a term which has defied centuries of attempts at definition.) In everyday action explanation, the uniqueness and particularity of circumstances, agent, action, and audience endow each term with an economically precise meaning (one can exploit context-dependent shades of nuance, ambiguity, etc.). Scientific statements, on the other hand, achieve clarity through the precision of the terms employed—experiments must be conducted in repeatable, not unique, contexts—and, since these terms are required only for the job of systematic description and explanation, sharp definition is possible; they do not need the flexibility of ordinary language terms. Of course, science borrows terms from ordinary language; but, having adopted them, it must adapt them to its purposes—consider the everyday and the scientific senses of *force, energy, mass.*[5]

Given these three points about ordinary language psychological

[5] I have expanded on this point in a number of places, most recently in "Functionalism, Psychology and the Philosophy of Mind," *Philosophical Topics*, 1981, *12*, 147–167. See also the excellent work by G. Mandler and W. Kessen, *The Language of Psychology* (New York: Wiley, 1959).

statements, we should reject Smedslund's contention that there is indeed any fixed, tight, or highly organized "logic of ordinary language"; only rarely will one proposition formally entail or exclude another. The position is surely that we operate with numerous (implicit) beliefs such as: 'Usually, whenever *p*, then *q*'; or 'Probably, if *p* then not-*r*'; or, of course, 'Other things equal, if *q* then almost certainly not-*s*'. Such propositions may indeed be noncontingent (it may well be senseless to deny such bland and minimal claims), but once precise meaning has been given to the terms used in formulating '*p*', '*q*', '*r*' and '*s*', and once the truth-preserving vagueness of qualifiers like "usually" and "probably" has been cashed out, then it becomes most improbable that one who merely has a competent grasp of the language would regard them as noncontingent, *even if empirical research established that they were*—even if a competent user of the *theory* came to accept that it was senseless to deny them. Hence, commonsense psychology cannot dominate its scientific counterpart; it provides at best a jumping-off point.

It might be helpful to illustrate the point with one of Smedslund's own examples. He claims that research into transference of learning as a function of similarity rests upon the following noncontingent commonsensical proposition: "Other things equal, the more similar in some respect two situations appear to be, the more similarly will *P* tend to deal with them, in this respect." He concludes, "The principles relating amount of transfer and similarity are true because it could not possibly be otherwise." For the sake of the argument I shall agree that the commonsensical proposition is noncontingent. But consider what the scientist has to do with it. First, the major task of spelling out the "other things equal"—finding the factors which are relevant for the facilitation or suppression of transference, plotting the extent to which they interact or cancel each other out. All this will vary according to the material to be learned, the nature of the subject, the experimental apparatus. Second, he must decide or discover how to interpret the term *similar*. Similarity is not a *given* in nature; different people (and different species) may vary widely in their assessments of similarity concerning both similarity of situations and similarity of response. Third, he must cash out the "appear to be." What determines the judgment of how something has appeared to an organism? How is "behaviorally silent" learning to be accommodated? To what extent are appearances age-, sex-, culture-, race-, or species-specific? Fourth, he must give precise meaning to the notion of *tending;* dispositional states are elusive and hard to measure. When all this has been done, he *may* have something that counts as a noncontingent truth, but it will emphatically not appear so to the layman; only someone familiar with the backing theory may realize that "it could not possibly be otherwise," just as only the physicist understands why gold *must* have atomic number 79. He may, however, have a contingently true statement (cf., zebras are striped).

Common sense, then (I claim), is where some scientific research begins. The crucial next stage of scientific activity, however, abandons common sense altogether. For in all sciences (not just psychology) the generalizations, laws, and regularities provided by common sense provide some of the initial *explananda;* then we see the first step away from common sense, as these regularities have to be tidied up and made precise and testable. But the next—the key—stage is what really matters: one seeks to provide an explanation for these generalizations, showing why they hold, and why they hold to the extent that they do. Water boils at 100°C at sea-level—why? X% of octogenarians show such-and-such deficits in short-term memory—why? Common sense alone does not and could not answer these questions, because very typically the explanations require the introduction of theoretical (postulated or inferred) entities and processes, and the hypothesizing of laws relating these postulates to each other and to observables.[6] In a word, they require *theory.* As far as I can see, Smedslund ignores completely the theoretical (exciting) aspects of psychology, and here (for me) his thesis is least plausible.

Let me offer an example of the sort of theoretical research about which Smedslund is silent, namely cognitive theories of learning. Here, in order to account for the (tidied-up!) observed behavioral regularities, theoreticians postulate a variety of internal psychological states, the internal relationships of which purport to explain why the organism behaves as it does. The postulated or inferred theoretical states may be as wildly remote from common sense as are quarks and positrons (consider, for example, those postulated in Gray's two-process model of learning.[7] It is with the development of the theoretical superstructure that we can see how it is that science not only describes and classifies but, more importantly, gets behind the observed regularities and explains them.

I turn now to a few less central points of disagreement. First, it is noteworthy that Smedslund leaves out all mention of animal psychology. If he had discussed that, I suspect that his thesis would seem far less tenable, because we have a rather feeble and anthropomorphic com-

[6] It will be clear that I am assuming that a tenable distinction can be drawn between theoretical and observational statements; this needs argument, but I think it can be done.

[7] J. A. Gray, *Elements of a Two-Process Theory of Learning* (London: Academic Press, 1975), p. 347.

monsense psychology when it comes to rats, monkeys, or goats, so there will be few interesting propositions about their behavior which "all competent users of the language" would agree to be noncontingent. Yet it is surely impossible to deny that the comparative assumption is, often, a helpful and indeed indispensable tool for the study of humans: we need this work.

Second, Smedslund, I think, underestimates the extent of the findings that could result from the study of cross-cultural psychological capacities, the "characteristics of the species *homo sapiens*." This is an important point, because it is just here that really startling and counterintuitive discoveries are being made about our mental organization. The data come from neuropsychology, frequently from studying brain-damaged patients; but the conclusions concern psychology directly, revealing facts about our cognitive structure that have wide-ranging implications. For instance, the phenomenon of "pure" alexia shows that reading and writing competences must be substantially separate; people can be alexic for their own but not for foreign languages, for words but not numbers, for letters but not words (or vice versa), thus showing how complexly structured language mastery must be; Gerstmann's syndrome reveals that apparently unrelated competences are in fact tightly knit; the differences between Broca's and Wernicke's aphasias are of enormous importance to psycholinguists. Examples could be multiplied indefinitely; the point is, however, that the species *homo sapiens* has an intricate, unpredictable, and fascinating psychological organization about which common sense is, and must be, silent.

Third: a methodological point. A developed science can afford to have a few core principles which are held to be immune to revision; they are held immune in the sense that rejection of these is, in effect, rejection of the entire theory (Newton's theory was supplanted along with the modification of the definition of energy as $\frac{1}{2}mv^2$). But psychology is not a developed science and has no overarching theory; there are, rather, hosts of partial minitheories all operating (and competing) in restricted domains. For such a science, multiplication of noncontingent propositions is at best restrictive and at worst disastrous; whether some proposition *is* indeed noncontingent should almost always be open to question (what if we had agreed with the ancient Greeks that atoms were by definition indivisible?). Smedslund's first, and autobiographical, example reveals that one child in his experiment paradoxically scored a "pass" on transitivity and a "fail" on conservation; surely no student of his should be discouraged, even though nothing may come of it, from exploring this apparently senseless result—it is by exploring anomalies that many of the most striking scientific advances have been made.

Furthermore, a point that cannot be emphasized sufficiently: intuitions about what is or is not "senseless" vary enormously and are substantially modified by new findings; personally, I find approximately half the propositions Smedslund calls senseless perfectly sensible—intuition is a highly fallible guide. Thus we should (a) be wary of holding any sentence immune to revision and (b) demand more than linguistic intuition before we (cautiously) do so.

Fourth, a minor point: Smedslund's citation of the Husserl/Schutz formulation of the Law of Effect is admittedly impressionistic; so impressionistic is it, in fact, that it could serve equally well as a vague formulation of the Principle of Induction (this is not, of course, surprising). Notoriously, dozens have found it far from senseless to deny the principle; although (*pace* Popper) one has to assume its truth, this methodological requirement does not establish it as noncontingent.

Were there space for more, I would have liked to examine Smedslund's view of rationality; it seems close kin to the equally implausible view held by Socrates, which gives it added interest, and my own belief is that cognitive psychology can cope rather well with less-than-optimally-rational behavior[8]. But that would require a lengthy discussion, so I shall instead conclude on a far more general note. The psychology Smedslund seems to be advocating—which relies on linguistic intuition and commonsense beliefs and which seeks out ordinary language entailments and noncontingent truths—is indistinguishable from the philosophy of mind as it was conducted not so long ago (roughly, until the later writing of Wittgenstein). Now there is nothing wrong with doing the philosophy of mind. However, psychology fought free of the dominion of philosophy about a century ago, to the benefit of both disciplines; they can now contribute to each other as they could not when it was philosophy that was defined by J. S. Mill as "the scientific study of man." To return the science to the domain of philosophy is to foreclose prematurely upon the future of the discipline; and just now it would be a somewhat paradoxical thing to do since philosophers of mind, at last reconciled to the independent existence of psychology, are starting to take a close interest in the theories and the (contingent!) propositions advanced by psychologists.

[8] See also D. Dennett, "Intentional Systems," *Journal of Philosophy*, 1971, *68*, 87–106.

5

Psychology Cannot Take Leave of Common Sense

Reply to Commentators

Jan Smedslund

My three critics have raised a number of metatheoretical questions, and I will try to comment on them at the metatheoretical level too. However, I also hope to make it clear that the issues can only begin to be clarified when the discussion moves to the level to actual psychological research.

1. Tennessen

Tennessen argues that my interpretations of theoretical formulations in psychology are implausible, as judged according to the hypothesis that "we all intend to remark what is remarkable." Remarkability is said to be a direct function of the degree to which members of the language community are inclined to disagree with the proposition involved, while it is, nevertheless, shown to be tenable. Some degree of remarkability may also come to very nugacious propositions if they can be shown to be false. Tennessen also states that, in accordance with his hypothesis, neither necessarily true nor necessarily false propositions were "ever intended to be explicitly stated by anyone. No sentence, therefore, should plausibly—let alone charitably—be interpreted to state any proposition in these directions." In accordance with this, Tennessen rejects my interpretations of psychological theories.

Jan Smedslund • Institute of Psychology, University of Oslo, Box 1094, Blindern, Oslo 3, Norway.

In reply, let me first point out that Tennessen does not provide any alternative interpretations of my examples. He appears to imply that, in view of his hypothesis, there must exist such plausible interpretations. His argument, therefore, is left to depend entirely on the tenability of his main hypothesis.

Philosophers, logicians, and mathematicians have long been preoccupied with necessarily true propositions. In view of this, I find it unconvincing that no sentence should ever be interpreted to state a necessarily true proposition. It is easy to find venerable propositions from the history of science that were surely intended to be necessarily true, yet could hardly have evoked massive disagreement among the contemporaries. One example would be the statement by Thales that "a circle is divided into two equal parts by its diameter" (van de Waerden, 1954, p. 87). In considering examples of this type, it would appear, therefore, that necessarily true propositions may be remarkable, at least in some contexts. In view of Tennessen's avoidance of specifics, and in view of the doubtful status of his hypothesis, I, consequently, cannot give much weight to this part of his criticism.

Tennessen criticizes me for not defining the term *competent* (language user) in my definition of common sense. My reply is that competence of this sort must be determined by consensus among other language users. A person is a competent user of a language if he or she is regarded as such by other competent users of the language, and so forth.

I agree with Tennessen that, in my chapter, I have not provided any evidence supporting my position, but only arguments and illustrations. What is needed are data indicating whether or not specific theoretical formulations in psychology actually conform with the suggested criteria of common sense. Two such studies have just been published: In one of them (Smedslund, 1982c), 36 alleged commonsense formulations relating to Bandura's theory of self-efficacy were used. Subjects were asked (1) to give a prediction of behavior involving each theorem, (2) to judge whether or not an alternative prediction is conceivable, (3) to judge whether or not an explanation based on the theorem is acceptable, and (4) to judge whether or not an explanation based on the negation of the theorem is acceptable. The average consensus on these four types of judgments was respectively 93%, 80%, 92%, and 96%. These findings appear to indicate that the given theoretical formulations, originally published in a highly prestigious psychological journal, come close to meeting the criteria of common sense, as I have defined it. Another, less elaborate, study involving clinical psychologists, yielded an average of 95% agreement (when obvious misunderstandings were eliminated) about the validity of seven rules of psychological treatment (Smedslund, 1982b). Apparently, these rules are fairly successful explications of common sense in the area of interpersonal relations. This is the evidence available at the moment. It is open to methodological criticism and therefore allows the discussion to be moved away from its present programmatic level.

2. Vollmer

Vollmer agrees that analysis of ordinary language concepts is important in psychology and also that ordinary language and psychological reality are internally related (see Smedslund, 1982a). However, he thinks that analysis of common sense is insufficient and that we must also have empirical psychological research. Vollmer's critique is threefold.

First, he argues that there are basic psychological processes that can be studied empirically and with generally valid results. By way of exemplification he mentions psychophysical and perceptual studies. My answer to this is that measurements of even a simple psychophysical threshold are highly sensitive to variations in the outcome matrix, that is, the perceived and evaluated consequences of answering one way or the other. Only be eliminating such psychological variables by elaborate averaging out procedures under rigidly controlled conditions can one approach stable measurements. But this averaging and control means that the findings are no longer representative of any performance of particular persons in particular contexts and with particular outcome matrices. Similarly, responses to colors, pictures, traffic signs, and the like are sensitive to contexts, to personal histories, and to cultural background (common history). At best, findings are locally valid and locally useful at the group level. After a century of experimental research, the harvest of generally valid and genuinely empirical psychological principles has been meager indeed. See Smedslund (1972, pp. 179–224) for a detailed treatment of this theme.

Vollmer's second argument deals with what he regards as a puzzle or contradiction in my position: On the one hand, psychological phenomena are historical and irreversible, and on the other hand, psychological theory should contain general conceptual truths only. Vollmer regards this as inconsistent since commonsense theory cannot explain historical processes and yet psychology should understand and explain human lives.

Consider the following statement about a part of person *A*'s life: "No one ever loved her so she came to believe that she was not lovable."

The fact that no one ever loved *A* cannot be accounted for by any psychological theory, empirical or nonempirical. (Perhaps *A*'s parents were killed in an accident, when she was a few days old, and the foster family happened to be entirely cold and uncaring.) However, *given* these facts, a nonempirical principle (the Law of Effect) explains the outcome.

A next step in explaining *A*'s life could then be: "Because *A* believed she was not lovable, she avoided all intimate contact with persons in order not to get hurt." This behavior could be explained by hypothesis 7. A third phase in *A*'s life occurred when she met *B*, who fell in love with her, despite her extreme reticence. Again, this meeting with *B* cannot be explained by any kind of psychological theory (*B* was transferred to the same office as *A* as a result of an administrative error). However, the effect on *A* of *B*'s falling in love with her, namely initial confusion and anxiety, can be explained by noncontingent principles (or could if more writing space were available). In conclusion, no psychological theory, empirical or nonempirical, can fully explain a human life. What psychological theory can do is explain why people behave and change in certain ways, given a set of antecedent conditions. Hence, I can see no contradiction in my position at this point.

Vollmer also argues that we must study concrete psychological phenomena and not only develop conceptual truths. I agree with this. The schematic life story presented above could not have been described and explained except on the basis of interviewing and interacting with *A* (observing *A*). However, in the present context, it is confusing to refer to psychological observation as "empirical." The opposition here is between empirical and nonempirical *theory*. Both of these can be applied to real situations and hence be used for prediction and control only through observation and intervention. In other words, there is no difference between the two types of theories in the amount of observation and intervention necessary for practical application.

Vollmer argues strongly for the necessity of establishing empirical principles of learning. He maintains that common sense is silent when it comes to questions such as, What kind of practice will be most effective? How much and often must a person practice in order to reach a certain level of skill? I think it remains an open question how silent common sense is in such matters. Among the 36 commonsense theorems in Smedslund (1982c), at least 21 yield definite predictions about learning, in a variety of conditions.

Vollmer's third major point is to question the conception of necessarily true propositions as assertions whose truth values are totally independent of the world. He also speculates that noncontingent truths really originate in causal truths summarizing thousands of years of human experience. I think these considerations, while interesting, have no immediate bearing on the issue at hand. As I see it, the central problem here is to evaluate the relative advantages of two types of theory in psychology: one containing empirically testable principles, allegedly disregarding common sense, and the other containing necessarily true principles, being explications of common sense, and to be applied in the testing of concrete psychological procedures. These types of theories remain different, irrespective of one's choice of broader evolutionary perspectives.

3. Wilkes

Wilkes agrees with me in that we may be looking for necessary truths. However, she takes this position not because she believes that psychology differs in any fundamental way from other empirical sciences (except by having greater difficulty in identifying "natural kinds") but because of general metatheoretical considerations.

She points out three alleged shortcomings of commonsense propositions. Firstly, very few commonsense principles have been made explicit. This is true. However, Wilkes explains this by arguing that common sense has little power to provide generalizations but is strongest in explaining the particular case. My alternative explanation is that common sense is mainly implicit, because there is no motivation to explicate it in ordinary life. I believe that the power of ordinary language in analyzing particular cases rests precisely on the implicit use of general common sense propositions. The matter can only be decided by the outcome of systematic efforts to explicate common sense.

Second, Wilkes argues that generalizations made by laymen are immune from charges of falsehood, not because they are necessarily true but because they are protected by all-embracing *ceteris paribus* clauses. Here, Wilkes inadvertently moves from my definition of common sense to the more usual one, as can be seen from her chosen examples. I have defined common sense as agreement about what *follows* from what (implications), whereas one usually talks about common sense as involving generalizations about what *leads* to what (causes). Proverbs in ordinary language, such as those quoted by Wilkes, are almost never examples of what I have defined as common sense. Hence, Wilkes's comments here are irrelevant for our discussion.

Incidentally, necessarily true commonsense theorems are also, sometimes, protected by *ceteris paribus* clauses. However, these are not all-embracing but contain a very limited number of specifiable factors that are to be excluded or kept constant (see Smedslund 1978, p. 102).

Wilkes's third objection to common sense propositions is that ordinary language terms cannot generally be precisely defined because they are needed for so many purposes, and that they are successful precisely because of their flexibility. It is indeed true that the meanings of terms in ordinary language are highly sensitive to context. However, it is never the case that the meaning of such terms derives exclusively from the context in which they occur. In other words, the flexibility is not unlimited. If that were the case, one could have no dictionaries.

If ordinary language terms, then, include components of relatively context-free meaning, it should be possible to give definitions in which some such context-free components are made explicit. Propositions expressing necessary relationships between these defined terms should then be possible. The outcome of such a project cannot be decided in advance but depends on studies of the kind referred to in my reply to Tennessen above. Wilkes has not provided any direct evidence for her conclusion that there is no "fixed, tight, or highly organized 'logic of ordinary language'." On the other hand, the data referred to above show some promise of supporting my project.

What I have said above does not imply agreement with Wilkes's allegation that psychology, as I see it, must be "confined within the straitjacket of *unaided* common sense." Unaided common sense is implicit and unformulated. The formulation of a system of explicit commonsense propositions, therefore, goes beyond unaided common sense and may become a genuine help for psychologists and laypeople. Furthermore, the systematization of common sense may lead to unexpected conclusions and developments. Even if the point of departure may seem trivial and self-evident, the ensuing deductions need not be uninteresting.

Wilkes discusses in some detail my example of similarity and transfer of learning. Her discussion reveals some degree of misunderstanding. I have not denied that one can launch empirical studies in this area or elsewhere, provided that one avoids conceptually related variables (pseudoempiricism) and provided one takes into account the historical nature of one's findings. These themes have been discussed before. Also, Wilkes apparently has not recognized that most of the tasks encountered by researchers would be common to the empirical and the commonsense position. In both cases one would need to assess what exactly is seen as similar in what respect and to what degree by the particular persons involved, before any actual predictions could be ventured.

Wilkes makes some statements which show that she has indeed not understood why common sense is unavoidable in psychology. She writes: "Common sense... is where some scientific research *begins*. The crucial next state of scientific activity, however, abandons common sense altogether." Compare also Tennessen's quotation from Sagan: "The universe is not in accord with common sense ideas."

Psychologists are persons who behave toward and have notions about persons, who themselves behave toward and have notions about persons (including the psychologist) and so on. The psychologists and their fellow human beings have become the persons they are by becoming socialized into a culture and a linguistic community. People can communicate and deal with each other, and psychologists can experiment with and treat people, precisely by virtue of this common background (common sense). People's (including psychologists') behavior is *channelized* by common sense, which therefore is an integral part of psychology's subject matter. However, common sense is also a condition for participating in a culture and hence must be presupposed by the psychologist. Starting from a perspective similar to the present one, Israel (1979, 1982) has characterized the task confronting the social scientist, who is always working *within* a culture, as that of "inflating a balloon from the inside."

One brief comment on two of Wilkes' less central points: I have no quarrel with research on animals and on brain damage and neurophysiology. These branches of research involve matters partly independent of culture and language and offer fascinating prospects. They may provide knowledge of boundary conditions of psychological events, particularly about what an individual *can* do.

Finally, Wilkes asserts that my approach is indistinguishable from doing the philosophy of mind. This is a grave misunderstanding. I am far from advocating a return to armchair psychology. On the contrary, systematic explication of common sense means the formulation of principles to be relied on in the development of better procedures in everyday practice of psychology. Necessarily true propositions are very useful in dealing with problems of prediction and control.

4. Concluding Remarks

In a recent article, I have tried to summarize the difference between the two views of psychology as follows:

Empirical hypotheses may be true or false. This means that both a hypothesis and its negation provide meaningful descriptions of a possible outcome. Empirical research, in principle, represents a venture into unknown territory.

Common sense is defined as the system of implications shared by all members of a culture. Therefore, explications of common sense involve matters which are already tacitly known to everyone. Common-sense research is a venture into known territory. The uncertainty involved concerns the degree of fit between the explication and common sense itself (Smedslund, 1982b, pp. 447–448).

The difference, then, is between a view of human beings as aliens with unknown characteristics, yet to be discovered, and a view of human beings as intimate acquaintances, but yet to be described. Defending the latter position, I will conclude as follows: The universe may not be in accord with common sense but human beings must be.

5. References

Israel, J. *The language of dialectics and the dialectics of language*. Copenhagen: Munksgaard, 1979.

Israel, J. *Om konsten att blåsa upp en ballong innifrån*. Göteborg: Korpen, 1982.

Smedslund, J. *Becoming a psychologist. Theoretical foundations for a humanistic psychology*. New York: Halsted Press and Oslo: Universitetsforlaget, 1972.

Smedslund, J. Some psychological theories are not empirical: Reply to Bandura. *Scandinavian Journal of Psychology*, 1978, *19*, 101–102.

Smedslund, J. Common sense as psychosocial reality: A reply to Sjöberg. *Scandinavian Journal of Psychology*, 1982, *23*, 79–82. (a)

Smedslund, J. Seven common sense rules of psychological treatment. *Journal of the Norwegian Psychological Association*, 1982, *19*, 441–449. (b)

Smedslund, J. Revising explications of common sense through dialogue: Thirty-six psychological theorems. *Scandinavian Journal of Psychology*, 1982, *23*. (c)

van der Waerden, B. L. *Science awakening*. Groningen, Holland: Noordhoff, 1954.

6

Interactionism and the Person × Situation Debate

A Theoretical Perspective

Michael E. Hyland

Abstract. Although largely a methodological debate, the person × situation issue is predicated on assumed theoretical differences. A review of the literature provides no evidence that personologism exists as a theoretical position. The label of situationism is often applied to radical behaviorism; yet a correct interpretation of radical behaviorism (Skinner, 1938, 1963, 1972) is quite different from the assumed theoretical position of situationism. Personologism and situationism are methodological, not theoretical positions.

The word *interaction* is used in different ways. Unidirectional interaction, person–situation interaction, and behavior–situation interaction are all analyzed. They are shown to be aspects of a single, more general interactive system with emergent properties. As such they should not be designated as separate interactions. It is argued that an interaction is a problem, not a solution.

In the 1960s and 1970s a number of articles appeared in psychological journals relating to an issue given the label "the person × situation debate" ($p \times s$ debate). The $p \times s$ debate was in the main a debate about methodological issues, but at the same time the methodology was usually assumed to be predicated on certain theoretical differences. In this paper I shall (a) briefly summarize methodological aspects of the debate, (b) examine the different theoretical positions, and (c) examine the different meanings of the word *interactionism.*

What is the person × situation debate? Ekehammar (1974), in a

Michael E. Hyland • Department of Psychology, Plymouth Polytechnic, Drake Circus, Plymouth, Devon PL4 8AA, England.

major review article, suggests that it can be understood within the context of three *theoretical* positions:

> Personologism is here used as a label for those views advocating stable intraorganismic constructs, such as "traits," "psychic structures," or "internal dispositions.". . . This position may generally be expressed as $B = f(P)$, where B stands for behavior and P for person. Situationism can be regarded as the antithesis of personologism and labels those views emphasizing environmental (situational) factors as main sources of behavioral variation . . . This position may generally be expressed as $B = f(E)$, where E stands for environment or some part thereof (e.g., situation). Interactionism can be regarded as the synthesis of personologism and situationism, which implies that neither the person *per se* nor the situation *per se* is emphasized, but the interaction of the two factors is regarded as the main source of behavioral variation. . . . This position may generally be expressed as $B = f(P,E)$. (Ekehammar, 1974, p. 1026)

Many contributors to the debate—particularly in its early stages—would agree with Ekehammar's interpretation, though the debate is usually recognized in terms of topics in psychology. For instance, Mischel, Jeffrey, and Patterson (1974) describe the issue this way:

> Advocates of trait theory seek to discover underlying, generalized dispositions that characterize persons relatively stably over time and across many situations, and search for behaviors that may serve as "signs" of such dispositions. Behaviorally oriented psychologists, on the other hand, focus on behavior directly, treating it as a sample from a wider repertoire rather than as a sign of generalized inner attributes. Unlike trait psychologists, behavioral psychologists see behavior as highly dependent of the situation in which it occurs and therefore do not assume broad generalization across diverse situations. (p. 231)

The $p \times s$ debate, then, is usually seen as a debate which is based on the different theoretical assumptions held by different groups of psychologists.

1. Empirical Basis for the Debate

The person × situation debate arose originally over a very specific issue in personality theory. It is found, empirically, that test scores—irrespective of the particular sort of personality test employed—are sometimes rather poor at predicting other behaviors. That is, they are rather poor at predicting scores on other tests which should be measuring the same sort of construct; and they are rather poor at predicting other nontest behavior. In the late 1960s personality theorists began to comment on the fact that tests correlated at best with a correlation coefficient of 0.3 or 0.4 (Mischel, 1968, 1973; Peterson, 1968; Vernon, 1964).

These low correlations had in fact been recognized for many years (Hartshorne & May, 1928, 1929; Hartshorne, May & Shuttleworth, 1930; Lehman & Witty, 1934) but had not attracted too much comment. The low correlations were often held to be the result of something wrong with the tests themselves or of contaminating variables such as mood, rather than the result of anything intrinsically wrong with the theory on which those tests were based. The person × situation debate arose when some psychologists (e.g., Mischel, 1968) suggested that the low predictability was not the result of the tests but something intrinsic to the study of personality. They argued that the reason for low correlations is that behavior is specific to situations—or at least more specific to situations than personality theorists assume. According to traditional personality theory, it was argued, behavior should reflect the personality construct, the "person variable" irrespective of the situation in which the individual is placed, the "situation variable." However, it was now suggested that behavior is also specific to the situation. The question therefore arose, to what extent is behavior determined by the situation and to what extent by the person.

This question formed the basis for a number of studies carried out in the 1960s and 1970s. The majority of work was carried out using the analysis of variance paradigm in which persons and situations are entered as factors in a two-way analysis of variance. The size of each of the main effects and the interaction between them was supposed to indicate to what extent behavior was caused by the person, the situation, or the interaction between the person and situation. The analysis of variance paradigm has been criticized as a method for resolving the contribution of the person and situation and person to behavior (Alker, 1977; Golding, 1975). However, even if the method were valid, the results are inconclusive. In brief, the results depend on the subject population employed, the particular situations employed, and the behavior or dependant variable measured. Depending on these parameters, it is possible to obtain almost any size of person effect, situation effect, or interaction effect (Mischel, 1973; Olweus, 1974).

It was soon recognized that results from a particular ANOVA experiment could not be generalized to other studies (Mischel, 1973; Wallach & Legget, 1972), though the conditions under which one factor might be expected to be greater than another could be discussed (e.g., Bem & Allen 1974; Lord, 1982; Sarason, Smith, & Diener, 1975). The $p \times s$ ANOVA experiments decreased in number in the later 1970s with gathering criticism that the issue was largely a pseudoproblem. The question, How much do the person and situation contribute to behavior? was meaningless, it was argued, given that any answer can be obtained. The question should be rephrased, How do the person and situation contribute to behavior? (Endler, 1973; Olweus, 1977)—the latter question, incidentally, was asked long before the emergence of the person × situation debate (e.g., Lewin, 1935, 1938).

Olweus (1977) sums up the changing orientation to the ANOVA studies thus:

> Even if it is clear . . . that estimates of variance components may provide useful information for particular purposes, it should also be stressed that such estimates may have very little help in finding clues to the mechanisms involved, that is, in getting answers to the fundamental question "how?" (p. 231)

By the late 1970s, attempts to find general estimates of the proportional contribution of persons and situations had largely been abandoned. Instead, there was an increased interest in theoretical issues relating to the debate and recognition that the debate had been largely atheoretical (Olweus, 1977)—a condition which historians of science (Kuhn, 1970; Lakatos, 1971) suggest is less likely to lead to scientific advance.

2. Personologism, Situationism, and Interactionism

Although much of the person × situation debate was over methodological issues, clearly the point of the debate was not simply to try to find out which of two factors contributed most to the variance of a particular dependent variable. Ekehammar (1974), in the excerpt quoted above distinguishes personologism situationism and interactionism as three theoretical positions. Other authors (Bowers, 1973; Mischel, 1973) make a similar distinction on theoretical grounds. In this section I will show that personologism and situationism have never been theoretical positions of the form outlined above. Personologism and situationism can only be identified as methodological positions.

2.1. Personologism

There are many different sorts of personality theory in the psychological literature (psychoanalytic, motivational, trait, physiological theories, to name a few), but in the $p \times s$ debate attention has focused on trait theories—almost to the exclusion of everything else. As an early advocacy of trait theories, Allport's work is often referred to as an example of personologism. In 1937 Allport introduced the words *idiographic* and *nomothetic* into psychological terminology. They were not Allport's invention, having first been suggested in 1904 by Windelbad.

> The philosopher Windelbad . . . proposed to separate the nomothetic from the idiographic disciplines. The former, he held, seek only general laws and employ only those procedures admitted by the exact sciences. . . . The idiographic sciences, such as history, biography, and literature, on the other hand, endeavor to understand some particular event in nature or in society. (Allport, 1937, p. 22)

Windelbad's original distinction of nomothetic and idiographic has to with the sort of prediction a scientist is trying to achieve. Thus, in nomothetic psychology the psychologist aims to construct general laws which are applicable to all people. Nomothetic psychology seeks laws which explain the behavior of any individual. Idiographic psychology, on the other hand, seeks laws which apply only to a single individual. Allport thought it quite feasible to construct laws for individual people: "Each person by himself is actually a special law of nature, so too is any structural occurrence within the pattern of his life" (Allport, 1937, p. 21).

Allport applied Windelbad's idiographic-nomothetic distinction to the theoretical concept of a trait. Traits may be either nomothetic (applicable to all individuals) or idiographic (applicable to just one individual). Allport constantly stressed the latter view and so his trait theory is also referred to as idiographic. Properly speaking, though, Windelbad's distinction is a distinction in terms of what is being explained (the *explanandum*), not what is doing the explaining (the *explanans*). Allport's distinction has until recently had little impact on psychological thinking. Allport (1966) writes, "Sanford (1966) has written that by and large psychologists are 'unimpressed' by my insisting on this distinction" (p. 9). However, some recent trait theorists (Bem & Allen, 1974; Bem & Funder, 1978; Kenrick & Stringfield, 1980; Kenrick & Braver, 1982) have returned to a modified form of idiographic trait theory as a way of demonstrating higher levels of personal consistency.

Allport's distinction of nomothetic and idiographic should not be confused with the distinction between personality and experimental psychology or between individual differences and situational differences—as is sometimes done (e.g., Underwood, 1975). Allport's two psychologies do not refer to personologism and situationism. Indeed, Allport suggests that the experimental method can be used in the study of personality, though he suggests that "some problems of individuality completely elude the experimental method" (Allport, 1937, p. 21).

Later authors have interpreted Allport's distinction as though he were prescribing a definite dichotomy between two different sorts of psychology. In fact, Allport argues strongly that the ideographic and nomothetic in psychology should be combined (Kenrick & Braver, 1982, advocate a similar position):

> It is more helpful to regard the two methods as overlapping and as contributing to one another. . . . One should now add that this "intermediate position" will fall properly within the scope of a broadened psychology. (Allport, 1937, p. 22–23)

In none of his writings does Allport suggest that behavior is unaffected by the situation. In fact, his version of the "behavior is a function of" formula is rendered explicitly as "Personality = *f*(Heredity) × (Environment). The two causal factors are not added together, but are interrelated as multiplier and multiplicand. If either were zero there could be no personality" (Allport, 1937, p. 106). Note the interactive implication in this statement. Strictly speaking, Allport cannot be accused, as Ekehammar (1974, fn 3, p. 1026) accuses trait theorists, of only adopting an additive combination of situation and person. It is, of course, perfectly true that Allport placed little emphasis on the situation in any of his more detailed theoretical discussions; so much so that he says in a later paper, "I have learned that my earlier views seemed to neglect the variability induced by ecological, social, and situational factors" (Allport, 1966, p. 9).

Later trait theorists also introduce the idea of the situation into their theories even if they may not do so in their empirical studies. Cattell (1965) offers the formula $R = f(S.P)$, referring to response, stimulus situation, and personality respectively. He points out, "Lack of allowance for the situation is one of the main causes of misjudging personality" (p. 27). Variation of behavior due to situational variation is in fact built into Cattell's theory.

Motivationally based theories of personality also introduce the idea of the situation. Murray introduces the idea of press as a description of the environment in "molar" and "psychologically relevant" terms: "we have selected the term *press* . . . to designate a directional tendency in an object or situation" (Murray, 1938, p. 118). McClelland's (1951) account of personality examines in some detail the situational factors which arouse motives. Psychoanalytically oriented authors of *The Authoritarian Personality* (Adorno, Frenkel-Brunswik, Levinson, & Sandford, 1950) wrote:

> Overt action, like open verbal expression, depends very largely upon the situation of the moment—something that is best described in economic and political terms—but individuals differ widely with respect to their readiness to be provoked into action. (p. 4)

Psychoanalysis, does, of course, make a definite genotype–phenotype distinction between underlying personality constructs and behavior.

These examples demonstrate that the simple formula of $B = f(P)$ was never seriously entertained by many personality theorists who were writing before the advent of the $p \times s$ debate—a similar point is made by subsequent defenders of the trait position (Block, 1977; Epstein, 1977; Krauskopf, 1978). Given that personality theorists accept the role of the situation in determining behavior, why is this idea so notably absent from (at least the traditional) empirical studies? There are a number of reasons, the simplest of which is methodological. Correlational techniques used in the study of personality focus attention on individual differences rather than on situational differences. The neglect of situational effects in early studies of personality may reflect, in part, the methodology used. One of the consequences of the $p \times s$ debate is that modern personality theorists show much greater awareness of the situation in their empirical studies (e.g., Brownell, 1982; Erkut, Jaquette, & Staub, 1981; Nygard, 1981; Schuster, Murrell, & Cook, 1980). The $p \times s$ debate has at least had this methodological consequence.

A second reason for not introducing the situation lies in what personality theorists are actually trying to explain. Epstein (1979a) suggests that the trait position, situationism, and interactionism are not different solutions to the same problem but different problems. That is, they are trying to explain different phenomena. If personality theorists are only trying to explain transituationally consistent behavior, then, naturally, the situation will not attract much attention in their theories. It is unfair to accuse personality theorists of trying to explain all behavior. Kelly (1955), for instance, prefaces his theory with the statement, "The system or theory which we are about to expound and explore has a limited range of convenience, its range being restricted, as far as we can see at the moment, to human personality and, more particularly to problems of interpersonal relationships" (p. 11). Murray (1938, pp. 3–5) provides a similar though more detailed account of the limits of his theory.

In the late 1960s the $p \times s$ debate arose in part (Bowers, 1977) over the assertion that transituational consistencies are very low. More recent work shows that some measures of behavior exhibit transituational consistency and some measures do not. Indeed the original Hartshorne and May studies, which have been used as a critique of conventional personality theory (Mischel, 1968), are often misinterpreted (Epstein, 1979a; Rushton, Jackson, & Pannonen, 1981). Hartshorne and May found little evidence of personal consistency on some measures of behavior—but more on others.

Transituational or personal consistency is not an all-or-nothing phenomenon: it varies in degree. When referring to stability in personality, Epstein (1979a) asks the question, "How stable is stable?" (p. 1123). There is no absolute answer to this question as it depends on assumptions about the sort of explanations held to be worthwhile. In particular, it depends on the degree of predictive power (Popper, 1963) thought necessary for a useful theory.

Apart from the degree of stability, there is, however, a more fundamental sort of question, namely, What sort of stability are we looking for? Allport (1937) introduces his theory of personality with a discussion of the scientific dictum "scientia non est individuorum." According to Allport, a science is concerned with general laws about classes of events rather than unique events (see Bateson, 1980, for a more recent discussion outside the context of psychology). Epstein (1979a) draws a similar conclusion—but from the $p \times s$ debate. He says:

> The observation that it is not possible to predict single instances of behavior, but that it is possible to predict behavior averaged over a sample of situations and/or occasions has important implications not only for the study of personality but for psychological research in general. (p. 1097)

When estimating the stability of behavior our estimate will naturally be affected by the size and sort of class of behavior which forms the unit of analysis. Larger units will tend to be more stable than smaller units—as error variance is compensated for. This point is used (or rediscovered) by recent authors favoring a traditional nomothetic trait theory. Krauskopf (1978) and Epstein (1979a) point out that personal consistency is obtained when using nomothetic traits—but only when an adequate sample of behavior (over time) is obtained. In other words, the classes of behaviors used as units of analysis in the early ANOVA studies were too small (over time) to demonstrate cross-situational consistency. The idea that personality should involve the study of behavior over time was suggested well before the emergence of the $p \times s$ debate. Murray (1938) concluded that "without some notion of the whole there can be no assurance that the processes selected for intensive study are significant constituents" (p. 5; see also Epstein, 1979b).

If the $p \times s$ debate has any consequence to the theory of personality, it is in the realization that the size of class of behaviors used as the unit of prediction is an important feature of theory construction. No theory predicts all behavior. Any theory predicts a certain unit of behavior, a unit which is made up of a class of behaviors. Scientific theories do not predict unique events. A theory should make quite explicit the size and sort of class of behavior which it employs as the unit of prediction. Failure to do so in the past has led to at least partly unnecessary criticism of personality theories.

2.2. Situationism

Situationism is the name given—usually by its opponents (Allport, 1966; Bowers, 1973)—to a group of rather varied theories. Bowers (1973) suggests that although there are many different sorts of situationist theory, they can be discussed under one heading. One of the extreme situationist theories, according to Bowers, is that advanced by Skinner (1963, 1972) in his radical behaviorism. According to common interpretation, radical behaviorists insist that behavior is a function of the situation, hence, $B = f(S)$.

Radical behaviorism—at least the form so ably argued by Skinner—is often misinterpreted by its critics. Skinner never suggests that individual differences do not occur, nor does he suggest that the mind does not exist. Skinner states quite explicitly (Skinner, 1963) that his argument is based on a philosophy of science.

Hempel (1958) suggests that there are two levels of scientific systematization. At the level of empirical generalization only observational terms are admitted into an explanation; at the level of theory formation, theoretical terms are allowed as well as observational terms. There have been advocates of the former atheoretical approach for many years. Operational definition (Bridgman, 1928), Craig's theorem (Craig, 1956), and Ramsey sentences (Ramsey, 1931) are all ideas which have been developed outside the field of psychology. They all show how theoretical terms may be eliminated in favor of observational terms—and they are all considerably less popular now than they were at the time of publication. Skinner's contribution is to suggest that psychology should adopt an atheoretical approach to explanation. As theoretical terms in psychology are person variables, in effect this means that person variables should be reduced to situational terms—that is, observational terms. The radical behaviorism versus theoretical psychology debate—quite separate from the $p \times s$ debate—is over whether it is practical or a good thing to eliminate theoretical terms (Hempel, 1958; Hyland, 1981). There are a number of disadvantages in the elimination of theoretical terms, one of which relates to the prediction of person variance (Hyland, 1981). In brief, radical behaviorists argue that person variation can be adequately predicted from a knowledge of those situational influences which have occurred during the individual's life time and knowledge of the situational effects which have influenced the individual's genetic constitution. Radical behaviorists must provide a historical interpretation of personality. Trait theorists, on the other hand, can provide an ahistorical account of personality. Radical behaviorism does not deny the existence of the person or person variables. What it does is to advocate a way of representing those person variables in situational terms. Skinner does not argue that the situation is more important than the person; he argues that it is a mistake to introduce person variables in a theory—because they are theoretical terms.

Although from a theoretical point of view radical behaviorism does not deny the existence of individual differences, from a methodological and practical point of view it usually does. Behaviorists employ the experimental method—there is, indeed, often a logical confusion between the two (Bowers, 1973). The experimental method focuses attention on situational differences rather than on individual differences. The criticism that situationist theories ignore individual differences is certainly true from a methodological point of view, although there is no theoretical position which states that individual differences do not occur. A similar point is made by Zlotowicz (1977), who in defending Skinner's position concludes:

> In fact Bower's criticism is mostly aimed at the diffuse but massive trend of experiments in which individual differences are drowned in group means and where the basic procedure is to compare control and experimental groups. (p. 387)

Situationism is a methodological, not a theoretical position.

Interestingly, the shortcoming of the experimental method in focusing attention away from individual differences has been discussed—but outside the context of the person × situation debate. Hyland and Foot (1974) discuss the different consequences which can occur when there are individual differences in treatment effects in an experiment. Underwood (1975) discusses the theoretical advantages of introducing a consideration of individual differences into theories normally investigated only in terms of situational differences. To be fair to experimental psychology, though, there are authors who employ the experimental method with a consideration of individual differences—and find it sufficiently normal not to provide any methodological justification or comment. More general criticisms of the experiment and laboratory studies are found (Claxton, 1980; Harré & Secord, 1972), but they derive little inspiration from the $p \times s$ debate. To the extent that the $p \times s$ debate has had a methodological effect in psychology, it is limited to personality studies.

2.3. Interactionism

A few authors do not accept Ekehammar's (1974) assertion that personologism and situationism exist as theoretical positions. Goldberg (1972) writes:

> In the name of science, an enormous amount of poppycock has recently been expressed to the effect that (a) all behavior is "situational" in character, and/or (b) that psychometricians and/or trait theorists have never considered situational influences on human behavior. (p. 550)

However, irrespective of whether personologism and situationism are believed to have existed as theoretical positions, the alternative, interactionism, is now seen as the only acceptable solution.

A number of authors note that the word *interaction* is used in different senses. Pervin (1968), Overton and Reese (1973), and Magnusson and Endler (1977) distinguish two uses, not necessarily the same. Krauskopf (1978) distinguishes three uses and Olweus (1977) four. Pervin and Lewis (1978) distinguish five different meanings, including the common language social meaning of the term. Altogether, six different interpretations of the word *interaction* have appeared in relation to the $p \times s$ debate.

The first sense in which the word *interaction* is used is purely methodological or statistical. We talk of an interaction when two independent variables, each of which affects a dependent variable, do so in a way which is nonadditive (i.e., the two-dimensional behavior surface is curved). One way in which the word is used is simply as a statistical effect obtained in a two-way ANOVA paradigm. The remaining five meanings of the word are all theoretical.

The second meaning of the word *interaction* is often referred to as "unidirectional interactionism" (Olweus, 1977). Unidirectional interactionism is the sort of interaction which might be inferred from Lewin's statement that behavior is a function of the person and his environment, $B = f(P,E)$. According to this view, behavior is caused by two different sorts of entity, person variables and situation or environment variables (Figure 1). There is an obvious parallel between unidirectional interactionism and the first statistical sense of the word described above. The statistical interaction can be used to test the theoretical idea of unidirectional interaction. These two sorts of interactionism are, however, quite different ideas and should not be confused.

A third meaning of the word (used, for instance, by Bowers, 1973, 1977) is in the sense of an interdependency between the person and the situation: the situation is interpreted by the person as well as affecting

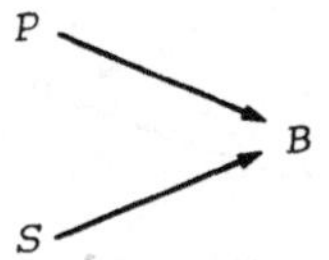

Figure 1

Figure 2

Figure 3

the person (Figure 2). Magnusson and Endler say, "Actual behavior is a function of a continuous process of multidirectional interaction or feedback between the individual and the situations he or she encounters" (p. 4).

Fourth, the word *interaction* is used in the sense of interdependence of the situation and behavior. Not only does the situation affect behavior, but also an individual's behavior has the effect of altering the situation. This can be expressed as an interaction between behavior and the situation (Figure 3). Interaction between behavior and the situation is sometimes called reciprocal causation (Overton & Reese, 1973; Endler & Magnusson, 1976), transaction (Pervin, 1968; Pervin & Lewis, 1978), and, by Magnusson and Endler (1977), within-situation interaction.

The remaining two uses of the word are suggested by Krauskopf (1978). These are within-situation interaction (Figure 4) and within-person interaction (Figure 5). It is easy to agree with Krauskopf's assertion that we need to invent new adjectives for these different sorts of interaction, but at the same time we do need to use these adjectives consistently!

The most important theoretical uses of the term are the first three theoretical uses described above, and these are in fact the three theoretical uses identified by Olweus (1977). These three different meanings of *interaction* are not, however, peculiar to the $p \times s$ debate; they have all been used before. Not only have they been discussed separately, but they are also combined together by Bandura (1977)—who writes outside the context of the $p \times s$ debate—in his account of reciprocal determinism. In Figure 6, Bandura illustrates what he means by reciprocal determinism.

What are the consequences of these different sorts of interaction to theory construction in psychology? Are there really three separate sorts of theoretical interaction?

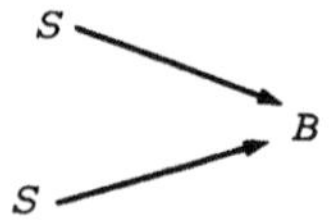

Figure 4

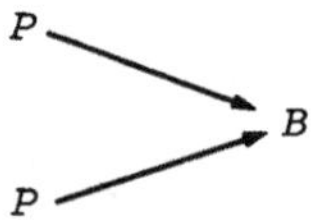

Figure 5

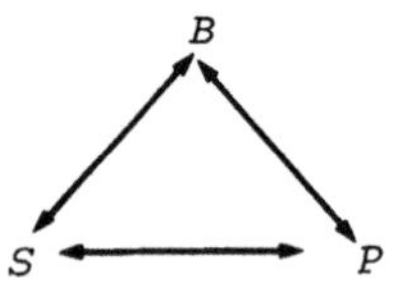

Figure 6

3. Three Sorts of Interaction

3.1. Unidirectional Interactionism

There are, in fact, two forms of unidirectional interactionism, which I will call *naive* and *theoretical*. According to the naive viewpoint, there are two factors, the person and the situation, which both contribute to behavior. Naive unidirectional interactionism is theoretically misleading, based as it is on an error of logical types (Bateson, 1980). The person and the situation are not two logically equivalent factors which affect behavior; the situation affects the person to a greater or lesser extent and it is then the person who behaves. The scheme in the upper part of Figure 7 would be better represented (for the present purpose only) as that in the lower part, where *S* and *B* are observables and *P* the theoretical function. The naive viewpoint is the result of superimposing the ANOVA paradigm—a methodological paradigm—onto a theoretical perspective.

Theoretical unidirectional interactionism is an idea found in the work of, for instance, Mischel (1973), who distinguishes five different person variables—we could call them hypothetical constructs. The situation affects behavior only to the extent that it affects these person variables:

> Situations thus affect behavior insofar as they influence such person variables as the individual's encoding, his expectancies, the subjective value of stimuli, or the ability to generate response patterns. (p. 276)

According to the theoretical viewpoint, hypothetical constructs are the immediate determinants of behavior. To the extent that those hypothetical constructs are altered by the situation, the situation also affects behavior. In the case of theoretical unidirectional interactionism there is no error of logical types (as there can be in the naive sort); the person

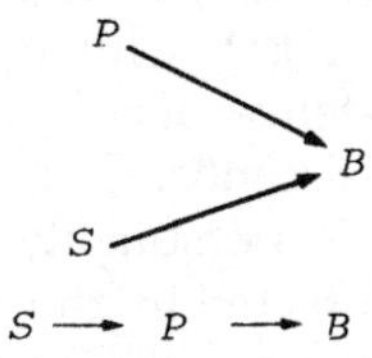

Figure 7

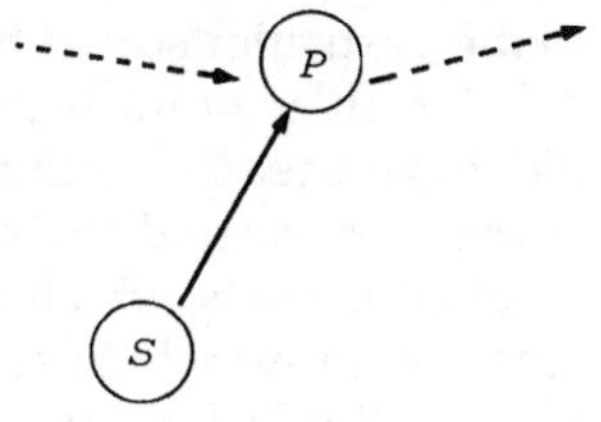

Figure 8

is a theoretical function connecting two observables, behavior and the situation.

Hyland (1981) expresses the unidirectional viewpoint in terms of an analogy. Imagine two billiard balls, *P* and *S*, where *S* hits *P*. The eventual position of *P* determines behavior (Figure 8). If *P* is stationary, then the eventual position of *P* will be determined by the direction and velocity of *S*. If, on the other hand, *P* is moving and *S* is only a very light ball, then *S* may hardly deflect the course of *P* at all. Evidently the contribution of *S* and *P* to behavior can vary from one extreme to the other, depending on the momentum of *P* and *S*. If we extend the analogy to person and situation, it is evident that the extent to which situations affect hypothetical constructs will depend on the nature of the situation and on the nature of the hypothetical constructs. We might expect that in some circumstances the situation would have a very marked effect on the eventual state of the hypothetical constructs and in other circumstances very little. To calculate an average contribution of the person and situation to behavior is not only meaningless but also positively misleading as it hides the fact that the situation can affect person variables in many different ways.

Hypothetical constructs are inferred theoretical entities. A theorist may infer the existence of a hypothetical construct for a number of different reasons. For instance, a personality theorist may infer the existence of hypothetical constructs which have the property of transituational consistency—because it is precisely that sort of hypothetical construct that will achieve the sort of prediction he wishes to achieve. An experimental psychologist, on the other hand may, but need not, infer hypothetical constructs which are consistent across people but vary in different situations. Thus the objectives of the theorist will determine, in part, the sort of hypothetical constructs he hypothesizes. Underwood (1975) makes the point that hypothetical constructs which are inferred to explain intersituational differences may, at the same time, exhibit interpersonal differences (and presumably vice versa). In other words, if hypothetical constructs have existential status—which they do by definition (MacCorquodale & Meehl, 1948)—then their effects need not be limited to the particular sort of behavioral variation they were postulated to explain. A similar point is made by Hyland (1981), who argues that we should have greater confidence in the existence of a hypothetical construct the more varied the evidence for its existence. Of course, the hypothetical constructs which are of most use to personality theorists will be precisely those which are not too much affected by the situation. However, hypothetical constructs inferred by personality theorists can never be said to be in principle different from those inferred by experimental psychologists. I would certainly not agree with Heilizer's (1980) contention that personologism and situationism are different Kuhnian paradigms. Although personality theorists and experimental psychologists may have different objectives, the theoretical assumptions remain basically the same. As stated earlier, personologism and situationism can only be described as methodological positions; they are not based on different theoretical assumptions.

3.2. The Person and the Situation

Philosophers of science commonly distinguish between observables, that is, "things and events which are ascertainable by direct observations," and theoretical entities, which are "presumptive objects, events, and attributes which cannot be perceived or otherwise directly observed by us" (Hempel, 1958, p. 41). Nowadays most people accept that the observable-theoretical distinction is one of degree rather than a dichotomy (Maxwell, 1962). All theoretical entities have an observational component, what in psychological terminology is often referred to as the operations of the entity. At the same time, all observables are theory-laden, in that the description of the observable involves interpretation and the interpretation is influenced by the sort of theories the describer has about the observable he is describing (Chalmers, 1978; Hanson, 1958).

In psychological theories person variables are theoretical entities and nowadays are understood in the sense of hypothetical constructs (MacCorquodale & Meehl, 1948). The situation, on the other hand, is treated as an observable. Many theories operate on the basis that the situation is something which happens outside an individual's skin, whereas hypothetical constructs are somehow located "inside" (e.g., Pervin, 1968, p. 56). We must appreciate, nonetheless, that the situation is theory-laden just as hypothetical constructs have operations.

Some psychologists distinguish the external situation, which is assumed to be objective and is located outside the individual's skin, from the psychological or subjective situation, which is some sort of internal representation of the external situation (e.g., Koffka, 1935; Murray, 1938). Lewin, for instance, argues that it is the psychological situation rather than the physical situation which determines behavior, a position followed by authors writing both within (Endler & Magnusson, 1976) and outside (Ittelson, Proshansky, & Rivlin, 1974) the $p \times s$ debate. The difference between the external and internal situation has a very important consequence for the debate. The external situation is the situation in the $p \times s$ formula, whereas the internal situation, by virtue of its being a theoretical entity belonging to the individual, is a person variable. This idea forms the basis of Bowers' (1973, 1977) argument against situationism. Bowers does not refer to the literature on the theory-ladenness of observables but makes a very similar point and therefore concludes that behavior cannot be the sole consequence of an objective situation. A related point is made by Rausch (1977), who suggests that the difference between person variables and situation variables may be more vague than is often appreciated.

Although the difference between the external and internal situation may be relevant to any theory explaining behavior, it is nevertheless an idea which is studied specifically under the heading of perception. When studying perception, we try to find out how the external situation comes to be represented internally in consciousness. In the traditional terminology, we try to find out how sensation becomes perception. The internal representaton of the situation is an interpretation of the external situaton. However, we have just said that all observables are theory-laden, in other words, that all observables are based on interpretation. What, then, is the difference between the theory-laden external situation and the theory-laden internal situation? Not all theories introduce the idea of an internal representation of the situation. When must we introduce the idea of a difference between internal and external situation into a theory?

The difference between the external and internal situation easily leads to misunderstandings. Correctly interpreted, the difference has nothing to do with the degree of theory-ladenness of the situation, nor has it to do with the position of a person's skin. Nor has the difference anything to do with the degree of organization of the units of the situation. The criterion for deciding what is inside and what outside the individual is based on an assumption the researcher makes about the generality of descriptions of the situation. If an individual's interpretation of the situation is assumed (note: *assumed*) to be the same as the researcher's, then the researcher is using the concept of the situation as a concept which, though theory-laden, has general application to all individuals. Because a situation has this assumed universality (or near universality) of interpretation, the situation is treated, by convention, as something which is external or independent of the individual. This is not to say that we pretend that there is an objective reality outside the individual; only that the individual does not provide any unique contribution to the interpretation of the situation. We say that the situation is external to the individual not because we believe that there really are an inside and outside but because of the sort of assumption we hold about the generality of interpretation which is built into the theory.

If, on the other hand, the researcher assumes that an individual's interpretation of the situation may differ from that of his own, or that there are different possible interpretations of the situation, then it follows that an individual's own interpretation must be taken into account. Because different individuals may interpret the same situation differently, it is useful to have a theoretical device which provides an internal representation of the situation. The theoretical device becomes necessary when we assume that there may be differences in the way individuals interpret situations.

In sum, the old distinction between the psychological and physical situation has, if correctly interpreted, nothing to do with the exact position of an individual's skin. There is no inside and outside. The distinction has to do with the assumed generality of theory-laden situational descriptions. Internal representations of the situation are a theoretical device which must be used whenever we assume that a given interpretation of a situation is not universal and that the individual's own interpretation is important to the interpretation of the situation.

The internal representation of the situation is a theoretical concept; it is a person variable. However, in the early stages of the person × situation debate the idea of an internal representation of the situation, as well as the idea of perception has with a few exceptions been completely absent. Certainly from a methodological point of view there has never been any suggestion that the situation which appears in an ANOVA paradigm is anything other than an objective reality.

In the psychological literature some authors introduce fairly complex units of the situation into their theories but tend to use the situation in the sense of the external situation. For instance, Murray (1938) introduces the idea of environmental press as a molar unit of the situation. Murray's environmental press is similar to Tolman's *manipulanda* expectations (Tolman, 1932) and more recently to Gibson's (1979) affordances. Gibson argues that affordances (what an object can do or be used for) are "out there" in the environment and can be perceived directly. We do not have to build up an internal affordance from atomistic units of the situation. Gibson suggests that events are "primary realities."

There is a very important criticism of locating complex, theory-laden concepts in the situation. As Neisser (1976) points out in a critique of Gibson, such theories are incapable of accounting for individual differences. By defining a particular interpretation of the situation as universal, a theory naturally cannot explain differences of interpretation.

Complex, theory-laden units of the external situation have a very obvious implication to the person × situaton debate. What according to one author might be an internal representation of the situation and hence a person variable is to another the external situation and hence a situation variable. Hence person variables and situation variables can become hopelessly confused.

A good example of the confusion between person and situational variables is found in the book *Social Situations* by Argyle, Furnham, and Graham (1981). These authors are committed to a situationist-oriented interpretation of social psychology and the person × situation debate. In their introductory chapter they note that there have been different definitions of the concept of situation and they provide their own definition, which is "by situation we shall mean a type of social encounter with which members of a culture or subculture are familiar" (p. 4). One should merely note that this sort of definition is almost completely meaningless. They then go on to list nine features of the situation:

1. Goals and goal structure. "People enter situations because they anticipate being able to attain certain goals. These goals can be regarded as a feature of the situation, and the main goals attainable in a situation can be assessed."
2. Rules. "Rules are generated in social situations in order to regulate behavior so that the goals can be attained."
3. Roles. "Nearly every situation has a number of specified roles."
4. Repertoire of elements. "Games all have a limited repertoire of acts that are permitted and count as meaningful moves. Some situations, such as auction sales, have very restricted repertoires."
5. Sequences of behavior. "Rituals and formal situations have strictly ordered sequences of events."
6. Concepts. "Characteristic shared concepts are developed for handling many situations. . . . Situations of intergroup conflict produce constructs derogatory of the out-group."
7. Environmental setting.
8. Language and speech. "There are certain features of language that are situation-specific while others are applicable to all or many situations."
9. Difficulties and skills. "Social situations, like jobs, often require certain skills or talent in order to be successfully executed" (p. 6).

Goals and concepts are almost always interpreted elsewhere as person variables. The same usually applies to repertoire of elements, sequences of behavior, and skill. For instance, there is a very clear overlap between Argyle *et al.'s* (1981) situations and Mischel's (1973) list of five person variables. What Argyle *et al.* do is quite simply to define all person variables in terms of the situation (cf. Ramsey sentences, Ramsey, 1931), thereby demonstrating their assertion that behavior is caused by the situation! Of course, by locating these fairly complex units in the situation Argyle *et al.* must assume that the particular interpretations of the situation are common to all people. Their theory cannot accommodate individual differences in interpretations of the situation—but then social psychologists are commonly uninterested in individual differences. Not surprisingly, theories often reflect what an author is trying to explain.

The relationship between the person and situation is referred to, in the context of the $p \times s$ debate, as an interaction (Bowers, 1973, 1977; Endler & Magnusson, 1976). Endler and Magnusson (1976) say, "Actual behavior is a function of a continuous process or multidirectional interaction (feedback) between the individual and the situation that he or she encounters" (p. 968). Is there a multidirectional interaction in the sense used above? To use the word *interaction* for the relation between person and situation can, in the present context, be positively misleading. It is perfectly true to say that individuals determine the meaning ascribed to a situation, but we mean by this that the internal situation is specific to the individual, not the external situation. Evidently we cannot say that there is a peculiar person contribution to the external situation, as the theory-ladenness of the external situation is assumed common. Perception can be represented (though need not be; see Neisser, 1976) in terms of a lineal form (see Figure 9).

To talk of an interaction between the person and the situation involves a confusion of two different sorts of situation, external and internal. The internal representation of the situation is a person variable!

The reason for this confusion lies in the word *person* or *person variable*. *Person variable* is a methodological term which has been introduced as a consequence of the use of the ANOVA paradigm in the $p \times s$ debate. However, the term *person variable* is also given a theoretical interpretation in the sense of a theoretical property of an individual, that is, in the sense in which the term *hypothetical construct* (Hyland, 1981) is used. Theoretical terms in psychology are properties of individuals. If we use the term *hypothetical construct,* then no confusion arises: the hypothetical construct is a property of the person; behavior is a property of the person; the person is located in the situation. To use the words *person, situation,*

External situation → Internal situation → Some other hypothetical construct → Behavior

Figure 9

and *behavior* implies that behavior and the situation are not properties of the person—which is, of course, nonsense. The person is placed in an external, theory-laden situation: this leads to an internal representation of the situation. The internal representation of the situation has various theoretical consequences which lead eventually to behavior. The behavior then leads to a change in the situation, a point we cover in the next section, where I will discuss a single general interaction between hypothetical constructs, situations, and behavior. It is incorrect to use the word *interaction* as an additional special case of reciprocal causality between the person and situation. A person only causes changes in the external situation by virtue of his behavior—not by virtue of just thinking.

3.3. The Situation and Behavior

Many authors in the $p \times s$ debate use the term *interaction* in the sense of reciprocal causation between behavior and the situation (Olweus, 1977). Mischel (1973) says, "The person continuously selects, changes, and generates conditions just as much as he is affected by them" (p. 278). This particular sort of interaction is also called reciprocal causation (Endler & Magnusson, 1976; Overton & Reese, 1973), transaction (Pervin, 1968; Pervin & Lewis, 1978), and within-situation interaction (Endler & Magnusson (1977).

As long ago as the mid-1930s, Skinner (1938) made the important observation that behavior is controlled by its consequences. The relationship between behavior and its effects on the environment has been emphasized by learning theorists ever since. Bandura (1977) says:

> Though the potential environment is identical for all animals, the actual environment depends upon their behavior. Is the animal controlling the environment or is the environment controlling the animal? What we have here is a two-way regulatory system in which the organism appears either as an object or an agent of control, depending upon which side of the reciprocal process one chooses to examine. (1977, p. 196)

The situation and behavior are not independent entities which cause each other in some way or other. Rather, they are descriptions given to events, and depending on the particular sort of interpretation the event can be described either as situation or behavior. The situation and behavior are part of the same system. It is for this reason that the word *transaction*, instead of *interaction*, is sometimes used (Pervin, 1968; Riegel & Meacham, 1978). Pervin suggests that:

> Transactionalism has three properties.
>
> (a) Each part of the system has no independence outside of the other parts of the system or the system as a whole.

> (b) One part of the system is not acted upon by another part, but instead there is a constant reciprocal relationship. They are not cause–effect relationships but transactions.
> (c) Action in any part of the system has consequences for other parts of the system. (p. 64)

Most accounts of behavior–situation interactions, both inside and outside the person × situaton debate, are merely assertions that an interaction or transaction of some form occurs. There is seldom any attempt to go beyond this simple assertion to discover the properties of the transactional system. It is as though by uttering the word *interaction* we can conveniently forget what the problem was all about. A notable exception is Powers's (1978) cybernetic analysis of purposive systems. Powers distinguishes theories having a lineal form (e.g., the situation causes hypothetical construct X which causes behavior) from theories which have a cyclical form (e.g., the situation causes hypothetical construct X which causes behavior which causes a change in the situation which causes hypothetical construct X and so on). The cyclical form of theory is that embodied in transactionalism. Powers's paper entails an analysis of theories having a cyclical form. One of the points he makes is that it is not possible to understand theories having a cyclical form by employing a sequential analysis (see also Bateson, 1980). That is, if *A* causes *B* and *B* causes *A*, it will not be possible to understand the behavior of the *A*–*B* system by a sequential analysis of the form *A* causes *B* causes *A* causes *B* causes A and so on. The reason why this is not possible is that cyclical or feedback systems have emergent properties (Bateson, 1980) and in particular emergent properties which depend on time. Powers (1979) points out that "a sequential-state analysis . . . introduces time without taking into account phenomena that depend on time" (1979, p. 427). It is interesting to note that time as a variable is almost always absent from psychological theories—apart from, say, theories of reaction time (see also Fiske, 1977). Psychological theories show how a variable changes, not how fast it changes. Rate of change is an essential concept in cybernetics.

Powers suggests that a correct application of cybernetics to psychological theories could constitute a revolution in the Kuhnian (Kuhn, 1970) sense of the word. Powers's claim may not be overly grandiose. There are relatively few theories in the psychological literature which incorporate behavior situation feedback, but where they do occur (e.g., Miller, Galanter, and Pribram's TOTE, Neiser's 1976 theory of perception) they are almost invariably understood within the context of a sequential state analysis. If situation–behavior interactions do constitute an emergent whole, then we need to move away from a sequential state analysis to some alternative such as Powers's quasi-static analysis.

Some behavioral variation results in very little change in the situation. Other sorts of behavioral variation produce considerable change. The behavior of turning one's head leads to very marked change in information available. From the point of view of a longer time perspective, when a student enrolls for a psychology degree, his action leads to his being placed in a very different situation than if he enrolled for an archeology degree. These different situations—or external situations—lead to different perceptions of the situation—or internal situation. For instance, the action of attending a psychology degree course leads to the individual's having a very different sort of information than he would have if he went on the archeology course. We could say, therefore, that behavior determines the person. Of course, person variables are also seen as the immediate cause of behavior, so we can say that the person causes behavior which causes the situation which causes the person and so on. The point is, that the assertion of situation behavior interaction necessarily implies an atheoretical rather than a theoretical explanation. According to a theoretical pychology (Hyland, 1981), behavior is the consequence of theoretical person variables. From the point of view of theoretical pychology, it is misleading to pick out the behavior and situation as an independent unit of interaction. Such an interaction must always be seen as a person–situation–behavior interaction.

Earlier I suggest that the term *person variable* is rather misleading and that it would be better to return to the more traditional term of *hypothetical construct*. Hence, instead of person–situation–behavior interaction, it would be better to refer to it as an interaction between hypothetical constructs, situation, and behavior. Of course it is very easy to say that there is a situation–hypothetical construct–behavior interaction and then feel one has somehow solved the problem. However, the very assertion that feedback is involved implies, according to Powers (1978), that a lineal analysis is not possible. The challenge to theoretical psychology is how to move away from the lineal form of situation causes hypothetical construct causes behavior causes situation to some nonlineal form of theory.

4. Conclusion

A debate such as the $p \times s$ debate can contribute to psychology in two ways: methodology and theory. Its main methodological contribution has been limited to personality theory, where it has drawn attention to situational influences. There is now greater awareness of the level of cross-situational consistency which may be expected from particular personality scales.

The theoretical contribution is relatively small, since many of the points raised are also covered outside the debate. However, one important contribution has been to introduce the idea of the size of the class of behaviors to be explained as an important aspect of theory construction.

One way in which the $p \times s$ debate is quite misleading is in the use of the word *interaction*. I have argued that there is only one sort of theoretical interaction: a situation–hypothetical construct–behavior interaction, which should be treated as a single system having emergent properties. Unidirectional interactions, person–situation interactions, and behavior–situation interactions are all aspects of this more general interactive system. Unidirectional interaction is just a noncyclical version of a theoretical explanation in psychology; person–situation and behavior–situation interactions just focus on different parts of a single cycle. It is, nonetheless, quite wrong to divide a single system into these different aspects. Situation–hypothetical construct–behavior interaction is a single system; no part or parts function independently.

Krauskopf (1978) suggests that the word *interaction* should be avoided because it is used in different senses. There is another reason for avoiding the word and that is that it is a palliative for ignorance. The $p \times s$ debate has treated the word as though it were an answer. It is as though having said that there is an interaction we no longer need to find out anything more about it. Quite the reverse is true. To say that there is an interaction is an admission of ignorance. Too often the word *interaction* has been used as a kind of latter-day explanatory fiction (Skinner, 1972). The conclusion that an interaction exists should never be looked on as a solution, but as the starting point of a problem. Alker (1977) concludes his analysis of ANOVA studies: "These ANOVA studies are . . . solutions searching for a problem. That is the problem." An alternative interpretation is that the solution is the problem.

5. References

Adorno, T. W., Frenkel-Brunswik, E., Levinson, D. J., & Sanford, R. N. *The authoritarian personality*. New York: Harper & Row, 1950.

Alker, H. A. Beyond ANOVA psychology in the study of person–situation interactions. In D. Magnusson & N. S. Endler (Eds.), *Personality at the crossroads*. New York: Wiley, 1977.

Allport, G. W. *Personality: A psychological interpretation*. New York: Holt, 1937.

Allport, G. W. Traits revisited. *American Psychologist*, 1966, *21*, 1–16.

Argyle, M., Furnham, A., & Graham, J. A. *Social situations*. London: Cambridge University Press, 1981.

Bandura, A. *Social learning theory*. Englewood Cliffs, N. J.: Prentice-Hall, 1977.

Bateson, G. *Mind and nature*. London: Fontana, 1980.

Bem, D. J., & Allen, A. On predicting some of the people some of the time: The search for cross-situational consistencies in behavior. *Psychological Review*, 1974, *81*, 506–520.

Bem, D. J., & Funder, D. C. Predicting more of the people more of the time: Assessing the personality of situations. *Psychological Review*, 1978, *85*, 485–501.

Block, J. Advancing the psychology of personality: Paradigmatic shift or improving the quality of research. In D. Magnusson & N. S. Endler (Eds.), *Personality at the crossroads*. Hillsdale, N. J.: Lawrence Erlbaum, 1977.

Bowers, K. S. Situationism in psychology: An analysis and a critique. *Psychological Review*, 1973, *80*, 307–336.

Bowers, K. S. There's more to Iago than meets the eye: A clinical account of personal consistency. In D. Magnusson & N. S. Endler (Eds.), *Personality at the crossroads*. Hillsdale, N. J.: Lawrence Erlbaum, 1977.

Bridgman, P. W. *The logic of modern physics*. London: Macmillan, 1928.

Brownell, P. The effects of personality–situation congruence in a managerial context: Locus of control and budgetary participation. *Journal of Personality and Social Psychology*, 1982, *42*, 753–763.

Cattell, R. B. *The scientific analysis of personality*. Harmondsworth, England: Penguin Books, 1965.

Chalmers, A. F. *What is this thing called science?* Milton Keynes, U. K.: Open University Press, 1978.

Claxton, G. Cognitive psychology: A suitable case for what sort of treatment? In G. Claxton (Ed.), *Cognitive psychology*. London: Routledge & Kegan Paul, 1980.

Craig, W. Replacement of auxilary expressions. *Philosophical Review*, 1956, *65*, 38–55.

Ekehammar, B. Interactionism in personality from a historical perspective. *Psychological Bulletin*, 1974, *81*, 1026–1048.

Endler, N. S. The person versus the situation—A pseudo issue? A response to Alker. *Journal of Personality*, 1973, *41*, 287–303.

Endler, N. S., & Magnusson, D. Toward an interactional psychology of personality. *Psychological Bulletin*, 1976, *83*, 957–974.

Epstein, S. Traits are alive and well. In D. Magnusson & N. S. Endler (Eds.), *Personality at the crossroads*. Hillsdale, N. J.: Lawrence Erlbaum, 1977.

Epstein, S. The stability of behavior: 1. On predicting most of the people much of the time. *Journal of Personality and Social Psychology*, 1979, *37*, 1097–1126. (a)

Epstein, S. Explorations in personality today and tomorrow: A tribute to H. A. Murray. *American Psychologist*, 1979, *34*, 649–653. (b)

Erkut, S., Jaquette, D. S., & Staub, E. Moral judgment–situation interaction as a basis for predicting pro-social behavior. *Journal of Personality*, 1981, *49*, 1–14.

Fiske, D. W. Personologies, abstractions, and interactions. In D. Magnusson, & N. S. Endler (Eds.), *Personality at the crossroads*. Hillsdale, N. J.: Lawrence Erlbaum, 1977.

Gibson, J. J. *The ecological approach to visual perception*. Boston: Houghton Mifflin, 1979.

Goldberg, L. R. Some recent trends in personality assessment. *Journal of Personality Assessment*, 1972, *36*, 547–560.

Golding, S. L. Flies in the ointment: Some methodological problems in the analysis of the percentage variance due to persons and situations. *Psychological Review*, 1975, *82*, 278–288.

Hanson, N. R. *Patterns of discovery: An inquiry into the conceptual foundations of science*. London: Cambridge University Press, 1959.

Harré, R., & Secord, P. F. *The explanation of social behaviour*. Oxford: Blackwell, 1972.
Hartshorne, H., & May, M. A. *Studies in the nature of character*. Vol. 1. *Studies in deceit*. New York: Macmillan, 1928.
Hartshorne, H., & May, M. A. *Studies in the nature of character*. Vol. 2. *Studies in service and self-control*. New York: Macmillan, 1929.
Hartshorne, H., May, M. A., & Shuttleworth, F. K. *Studies in the nature of character*. Vol. 3. *Studies in the organization of character*. New York: Macmillan, 1930.
Heilizer, F. Psychodigms of theory in personality and social psychology. *Psychological Reports*, 1980, *46*, 63–85.
Hempel, C. G. The theoretician's dilemma. A study of the logic of theory construction. In H. Feigl, M. Scriven, & G. Maxwell (Eds.), *Minnesota studies in the philosophy of science* (Vol. 2). Minneapolis: University of Minnesota Press, 1958.
Hyland, M. E. *Introduction to theoretical psychology*. London: Macmillan, 1981.
Hyland, M. E., & Foot, H. C. Group data, individual behavior: A methodological note. *British Journal of Social and Clinical Psychology*, 1974, *13*, 93–95.
Ittelson, W. H., Proshansky, H. M., Rivlin, L. G., & Winkel, G. H. *An introduction to environmental psychology*. New York: Holt, Rinehart & Winston, 1974.
Kelly, G. A. *A theory of personality*. New York: Norton, 1955 Kenrick, D. T., & Braver, S. L. Personality: Idiographic and nomothetic! A rejoinder. *Psychological Review*, 1982, *89*, 182–186.
Kenrick, D. T., & Stringfield, D. O. Personality traits and the eye of the beholder: Crossing some traditional philosophical boundaries in the search for consistency in all people. *Psychological Review*, 1980, *87*, 88–104.
Koffka, K. *Principles of gestalt psychology*. New York: Harcourt Brace, 1935.
Krauskopf, C. J. Comment on Endler and Magnusson's attempt to redefine personality. *Psychological Bulletin*, 1978, *85*, 280–285.
Kuhn, T. *The structure of scientific revolutions*. Chicago: University of Chicago Press, 1970.
Lakatos, I. History of science and its rational reconstruction. In R. C. Buck & R. S. Cohen (Eds.), *Boston studies in the philosophy of science* (Vol. 8). Dordecht, Holland: Reidel, 1971.
Lehman, H. C., & Witty, P. A. Faculty psychology and personality traits. *American Journal of Psychology*, 1934, *44*, 490.
Lewin, K. *A dynamic theory of personality*. New York: McGraw-Hill, 1935.
Lewin, K. *The conceptual representation and the measurement of psychological forces*. Durham, N. C.: Duke University Press, 1938. Lord, C. G. Predicting behavioural consistency from an individual's perception of situational similarities. *Journal of Personality and Social Psychology*, 1982, *42*, 1076–1088.
MacCorquodale, K., & Meehl, P. E. On a distinction between hypothetical constructs and intervening variables. *Psychological Review*, 1948, *55*, 95–107.
Magnusson, D., & Endler, N. S. Interactional psychology: Present status and future prospects. In D. Magnusson & N. S. Endler (Eds.), *Personality at the crossroads*. Hillsdale, N. J.: Lawrence Erlbaum, 1977.
Maxwell, G. The ontological status of theoretical entities. In H. Feigl & G. Maxwell (Eds.), *Minnesota studies in the philosophy of science* (Vol. 3). Minneapolis: University of Minnesota Press, 1962.
Mischel, W. *Personality and assessment*. New York: Wiley, 1968.
Mischel, W. Toward a cognitive social learning conceptualization of personality. *Psychological Review*, 1973, *80*, 252–283.
Mischel, W., Jeffery, K. M., & Patterson, C. J. The layman's use of trait and behavioural information to predict behaviour. *Journal of Research in Personality*, 1974, *8*, 231–242.
McClelland, D. C. *Personality*. New York: Sloane, 1951.

Murray, H. A. *Explorations in personality.* New York: Oxford University Press, 1938.

Neisser, U. *Cognition and reality.* San Francisco: Freeman, 1976.

Nygard, R. Toward an interactional psychology: Models from achievement motivation research. *Journal of Personality,* 1981, *49,* 263–387.

Olweus, D. A critical analysis of the "modern" interactionist position. In D. Magnusson & N. S. Endler (Eds.), *Personality at the crossroads.* Hillsdale, N. J.: Lawrence Erlbaum, 1977.

Overton, W. F., & Reese, H. W. Models of development: Methodological implications. In J. R. Nesselroade & H. W. Reese (Eds.), *Life span and developmental psychology: Methodological issues.* New York: Academic Press, 1973.

Pervin, L. A. Performance and satisfaction as a function of individual–environment fit. *Psychological Bulletin,* 1968, *69,* 56–68.

Pervin, L. A., & Lewis, M. Overview of the internal-external issue. In L. A. Pervin & M. Lewis (Eds.), *Perspectives in interactional psychology.* New York: Plenum Press, 1978.

Peterson, D. R. *The clinical study of social behavior.* New York: Appleton-Century-Crofts, 1968.

Popper, K. R. *Conjectures and refutations.* London: Routledge & Kegan Paul, 1963.

Powers, W. T. Quantitative analysis of purposive systems: Some spadework at the foundations of scientific psychology. *Psychological Review,* 1978, *85,* 417–435.

Ramsey, F. P. *The foundations of mathematics.* New York: Humanities, 1931.

Rausch, H. L. Paradox levels, and junctures in person–situation systems. In D. Magnusson & N. S. Endler (Eds.), *Personality at the crossroads.* Hillsdale, N. J.: Lawrence Erlbaum, 1977.

Riegel, K. F. & Meacham, J. A. Dialectics, transaction, and Piaget's theory. In L. A. Pervin & M. Lewis (Eds.), *Perspectives in interactional psychology.* New York: Plenum Press, 1978.

Rushton J. P., Jackson, D. N., & Pannonen, S. V. Personality: Nomothetic or idiographic? A response to Kenrick & Stringfield. *Psychological Review,* 1981, *88,* 582–589.

Sarason, I. G., Smith, R. E., & Diener, E. Personality research: Components of variance attributable to the person and the situation. *Journal of Personality and Social Psychology,* 1975, *32,* 199–204.

Schuster, S., Murrel, S. A., & Cook, W. A. Person, setting, and interaction contributions to nursery school social behavior patterns. *Journal of Personality,* 1980, *48,* 24–37.

Skinner, B. F. *The behavior of organisms.* New York: Appleton-Century-Crofts, 1938.

Skinner, B. F. *Behaviorism at fifty.* Science, 1963, *140,* 951–958.

Skinner, B. F. *Beyond freedom and dignity.* London: Jonathon Cape, 1972.

Tolman, E. C. *Purposive behavior in animals and men.* New York: Appleton-Century-Crofts, 1932.

Underwood, B. J. Individual differences and theory construction. *American Psychologist,* 1975, *30,* 128–134.

Vernon, P. E. *Personality assessment: A critical survey.* New York: Wiley, 1964.

Wallach, M. A., & Leggett, M. I. Testing the hypothesis that a person will be consistent: Stylistic consistency vs. situational specificity in size of children's drawing. *Journal of Personality,* 1972, *40,* 309–330.

Zlotowicz, M. Situations, interactions, and comparative psychology. In D. Magnusson & N. S. Endler (Eds.), *Personality at the crossroads.* Hillsdale, N. J.: Lawrence Erlbaum, 1977.

6

Theoretical Divergences in the Person–Situation Debate

An Alternative Perspective

Philip K. Peake

Over the last two decades, personality psychology has struggled with a set of issues that lie at the very core of the discipline. These issues, and I wish to emphasize that the issues are many, have fallen under the general rubric of the person–situation debate. In his article, Hyland argues that the debate is primarily a methodological rather than a theoretical one, that the main contributions of the debate are limited to increased attentiveness to situational influences and the size of the class of behaviors being explained, and that the term *interaction* is repeatedly used in a fashion that is quite misleading. As with so many other discussions of this topic, I find myself less concerned with the specifics of Hyland's commentary than with his reading of the debate from which they derive. Hence, in the present discussion, I will deal with but a few of Hyland's more important points in the context of an alternative perspective of the history, sources, and nature of the person–situation debate.

1. Sources and Sides in the Person–Situation Debate

Personality psychology has a long and alluring history of struggling over the meaningful routes to understanding persons and their behavior.

Philip K. Peake • Department of Psychology, Smith College, Northampton, Massachusetts 01063.

As with any complexly organized system, there are many potential strategies that can be employed in seeking to describe and explain the human personality. Not surprisingly, any particular route will be useful for achieving a certain set of circumscribed goals, and it is critical that theorists and practitioners understand the range of convenience implicit in any selected strategy. In the context of the person–situation debate, three distinct strategies are typically distinguished as alternative routes to understanding human behavior. These approaches, personologism, situationism, and interactionism, are continually contrasted and on occasion are reduced to the respective theoretical formulations $B = f(P)$, $B = f(E)$, and $B = f(P,E)$ (Ekehammar, 1974). Hyland argues that while these positions are often set out as theoretically distinct (as represented in their respective formulas), the defining characteristics of each position are primarily methodological, not theoretical. Personologists use the correlational method, situationists use the experimental method, and the interactional perspective arose primarily from a utilization of the interaction term in an ANOVA design. However, in the writing of all three perspectives, the theoretical importance of persons, situations, and interactions is recognized. In theory, then, the participants in the debate all accept an interactional model, but the positions diverge in that the methodologies they employ lead to an emphasis on either the person, the situation, or their interaction.

Such a characterization and analysis of the person–situation debate, while consistent with most other reviews of the topic, fails to capture either the nature of the differences between personologism and situationism or the history from which those differences arose. If the difference between personology and situationism was primarily one of methodological preference, the history of the person–situation debate would have witnessed a continued contrast of correlational research in support of the person component with experimental research documenting the situational component. While such research could be sought out, the debate over persons and situations was not primarily waged by contrasting evidence from divergent research paradigms and therefore, I would argue, was not simply a methodological debate.

Hyland is entirely correct in asserting that personologists have long recognized the importance of situations and interactions. Similarly, the so-called situationists, in pointing to the importance of situations in the regulation of behavior, have never denied the fundamental fact that it is the *person* who behaves and that his or her contribution to that behavior cannot be overlooked. The recognition that it is not possible to characterize personology or situationism with simplistic formulations such as

$B = f(P)$ or $B = f(E)$, does not, however, imply that there are not theoretical differences between these approaches. It may simply imply that those distinctions are too subtle to be captured by such gross formulations.

2. Personology, Situationism, and the Person–Situation Debate

In order to grasp the nature of the theoretical distinctions between personology and situationism, and the ties of both to the various forms of interactionism, it is necessary to review the context from which the person–situation debate arose. Such an analysis will show that the fundamental differences between personology and situationism are only tangentially related to the relative contributions of either persons or situations to behavior. Moreover, this review will reveal that the label "situationism" is a simplistic misnomer that has continually distracted attention from the fundamental arguments of approaches so labeled.

To begin, let us adopt Ekehammar's view that "the term 'personologism' is a label for those views advocating stable intraorganismic constructs, such as 'traits,' 'psychic structures,' or 'internal dispositions' as the main determinants of behavior" (1974, p. 1026). Actually, personology is much more than this. As set out by Murray, "the branch of psychology which principally concerns itself with the study of human lives and the factors that influence their course, which investigates individual differences and types of personality, may be termed 'personology' instead of the psychology of personality, a clumsy and tautological expression" (1940, p. 4). Personality psychology, in general, owes a great debt to the early personologists for maintaining an emphasis on the study of the person variables during a period when academic psychology was strongly influenced by the positivistic approach which frowned on the use of such hypothetical constructs. It is also noteworthy that the early personologist did not merely acknowledge situational impact but explicitly defined systems (e.g., Murray's "press") to attempt to understand how situations interact with person variables in the production of behavior. I have adopted Ekehammar's constrained definition, recognizing its limitations, while noting that in a broad sense it captures the sense of where a great deal of personological research had gone at the outset of the person–situation debate. While the personological system was far-reaching and eclectic in its inception, the portions that took hold and were pursued vigorously were those concerning what Murray referred to as the "variables of personality" or traits. Hyland argues that this springs from a reliance on the correlational method, but I would speculate, perhaps more generously, that the emphasis derived from a belief that a full and proper understanding of these traits and their interrelations was the fundamental prerequisite for a more complete exploration of the personological system, and from the fact that the burgeoning technologies made available by psychometrics made traits appear exceedingly accessible for study. Regardless of the particular reasons for this emphasis, it is clear that one of the defining features of the personological approach is a reliance on global response dispositions as the primary units in the analysis of personality. Implicit in this adoption is the belief that these traits are sufficiently stable over time and generalized across situations to be useful in both the understanding and prediction of behavior.

It is not difficult to agree about the primary defining characteristics of personology, but situationism appears to be a rather more slippery beast. In my own experience, this distinction is best illustrated in that those commonly labeled as personologists are typically the first to admit it. On the other hand, one is hard pressed to locate a self-proclaimed situationist even by seeking out those who are most frequently labeled as such. This phenomenon may have something to do with the fact that situationism is not a self-selected label. It is, as Hyland puts it, "the name given—usually by its opponents (Allport, 1966; Bowers, 1973)—to a group of rather varied theories." By extension, it is appropriate to maintain that "situationist" is the label given to any of a small group of rather varied theorists, since it is the commentary of a rather few individuals that is repeatedly referenced along with the label situationism.

The commentaries to which I refer are a series of penetrating critiques of traditional personality theory and assessment that emerged in the 1960s, best represented in the work of Mischel (1968), Peterson (1968), and Vernon (1964). The term *situationism,* of course, had been applied to other approaches prior to any of these publications, the most popular being those of the radical behaviorist mode, and it is still not uncommon to see these critiques lumped with other forms of situationism (e.g., Bowers, 1973). While there are clear links between the thinking of these theorists and, for instance, the radical behaviorists, to catalogue them with theorists such as Skinner and then to criticize them via radical behaviorism misses completely the message they were trying to convey.

The situationist commentaries of the 1960s were not written by radical behaviorists or by psychologists strictly wedded to experimental methodology. Each of the major critiques was written by an academically oriented clinician who had become concerned with the prevalent methodologies of personality and clinical psychology, particularly those aspects concerned with the assessment and prediction of behavior. According to these critiques, numerous investigations on various aspects of the clinical enterprise raised serious questions about the utility of the units that had been popularly adopted to characterize individuals—namely, global response dispositions or traits—for achieving many of the pressing needs of the field. It should be emphasized that these critics were not denouncing the use of trait assessments for all purposes. Mischel (1968), for instance, noted that the trait approach had proved useful for a variety of purposes such as gross screening decisions. However, the trait approach evidenced severe limitations in the many areas in which psychologists were increasingly attempting to use it, and these limitations required explication.

In making the case that the traditionally conceived and assessed personality trait may not be "the most acceptable unit for the investigation in the psychology of personality" (Allport, 1937), these critics, most particularly Mischel (1968), questioned the assumption that behavior within a particular trait domain is sufficiently general across situations to warrant using trait-based methods as a primary assessment strategy. In this vein, they cited many studies demonstrating what Hartshorne and May (1928) had quite early labeled the "doctrine of specificity"—considerable variability in trait-related behavior from situation to situation. Second, they reviewed numerous studies that demonstrated the somewhat limited utility of trait-oriented assessments to predict behavior well in specific situations—a fundamental goal of both the theorist and the clinician. Lastly, they reviewed numerous experimental studies that demonstrated how predictive utility was often increased by using more context-specific, and typically less costly, methods of assessment.

There are two misconceptions that frequently arise in discussions of these critiques. First, the commentaries on the trait approach were based, almost exclusively, on a focused examination of correlational research. The criticisms do not show a methodological favoritism for experimental paradigms but are based on an evaluation of the same correlational research that is supposed to lead to an interpretive bias in favor of stability and person effects. Far from implementing the methodological bias implied, these critiques attempted to demonstrate that the trait approach simply had not worked well using the criteria on which it is supposed to work best. The debate between situationists and personologists did not derive from a reliance on differing methodologies, as Hyland contends, but from divergent readings of research from the correlational paradigm.

The second, and more common, misreading of these critiques (and the one from which much of the impetus for the person–situation debate arose) is that their aim was somehow to substitute situations for persons as the primary object of study for personality psychology. Relatedly, it is often claimed that they advocated the situation as the primary causal factor for most behavior. While the situationists explicitly called for increased attention to situations, they did not argue that such a focus should be maintained by ignoring the contribution of the person. The situationist argument was not an either/or proposition, and at its core it had little to do with the relative contributions of either persons or situations to behavior. Acknowledging that the behavior of individuals tends to vary from situation to situation does not imply that the person contributes less to behavior than the situation and need not lead to any inferences about the causal primacy of situations. Such a recognition, in fact, has little to do with the contribution of situations to behavior since it does not assume that the effect of a particular situation will be similar, or even directionally stable, across persons. Ironically, the so-called doctrine of situational specificity is little more than a descriptive recognition of what became popularly referred to as the interaction of persons and situations.

In recognizing the cross-situational variability of behavior, what was questioned was not whether situations are more important than persons, but whether the units that psychologists were using to characterize persons could adequately capture the important aspects of individuals that would lead to useful predictions about their behavior in the contexts in which they lived. Therein lies the primary theoretical difference between the two approaches. Personologists had assumed that there was sufficient generality to behavior in trait domains so that nontrivial statements could be made about individuals by assessing the persons' typical or average level. Situationists questioned that assumption and suggested that the assessment should be redirected in two distinct ways. First, if one insists on relying on typical behavior, then the domains spanned in the assessment should be decreased in breadth in a manner that allows assessments to be more context-specific. Second, and more important, they suggested that rather than worrying about characterizations of the person in general, assessment should be focused on those properties of persons that are likely to interact with the properties of situations in contributing to behavior. Since those original commentaries, researchers commonly labeled as situationists have gone on to identify some of those properties (Bandura, 1977; Mischel, 1973), and they include such things as competencies, outcome expectancies, self-efficacy expectancies, values, and goals. The theoretical relevance of these types of variables for an interactional analysis of behavior is illustrated by the fact that Mischel's (1973) discussion of encoding strategies anticipates Hyland's discussion of internal situations by a full decade. The important point for the current discussion is that the traits favored by the personologists and the process-oriented, context-bound variables favored by situationists derive from distinct theoretical differences over the most meaningful ways to characterize and understand the person. That simplistic characterizations of these positions in the form $B = f(P)$ or $B = f(E)$ fail to capture these differences is more a reflection of the inadequacy of the characterizations than the lack of theoretical differences in the positions.

3. Interactionism and the Person–Situation Issue

In the context of the person–situation debate, interactionism was presented as a potential alternative to both situationism and personology. Although an interactional approach had received considerable theoretical attention previously, it was the introduction of the interaction term in an ANOVA design using *S–R* inventories and the like that supplied the approach with its initial empirical application. The statistical interaction approach demanded considerable attention because it apparently offered a resolution to the person–situation debate upon which both situationists and personologists could agree.

Several points should be highlighted concerning these early interactional studies. First, in attempting to resolve the person–situation problem they explicitly and repeatedly juxtaposed the previously discussed simplistic models (e.g., $B = f(E)$ and $B = f(P)$) and then attempted to show that neither situation nor person was the more important factor. Instead, some form of interaction of person and situation typically carried the most weight in the ANOVA results. This approach fell into disfavor for a variety of reasons, most notably because either of the three factors could be made to account for a disproportionate amount of variance depending on the selection criteria employed by the researcher. Thus, the results seemed to reveal more about the design choices of the researcher than the behavior of individuals.

From the perspective advanced here, these ANOVA designs were intriguing but fell far short of the mark in that they bypassed the more important divergences between personologists and situationists in order to analyze the more apparent. Typically, these studies used *S–R* inventories to assess some global disposition (e.g., anxiety, dominance), in a variety of contexts and across several response modes. Person effects were restricted to the amount of variance accounted for by the globally assessed trait. Thus, although these studies attempted to heed the suggestion to attend to situations more seriously, the assessment of the person remained focused on omnibus assessments of the person in general.

The move from ANOVA forms of interaction to the theoretically more satisfactory forms of transactionism and reciprocal interactionism was indirectly a function of this latter shortcoming. Soon after these studies were introduced, it became apparent that ANOVA interactions were difficult to predict, elusive on replication, and not intrinsically interesting once established. The pressing question from an interactional perspective should not have been whether persons and situations interact, but how they interact. This shift of attention from description to process imperceptibly shifts the question back to the point of divergence in the situationist critiques. By focusing on the psychological processes that are involved in an interactional sequence, the issue of what types of units will best suit such a task comes to the fore. As noted earlier, the situationists had essentially argued that in order to begin to understand such process-oriented questions we must begin to assess properties of the individual that are likely to interact with aspects of situations, rather than relying on characterizations of the individual's average performance in a domain. Far from resolving this issue, the interaction studies, in a lengthy and indirect manner, rediscovered it. Moreover, the complexities of the problem emerged doubly; in addition to grappling with the issue of how best to characterize the individual, the interaction studies led to the equally complex issue of how to characterize situations.

The nature of these problems is illustrated in Hyland's interactional model, where "the unit" of analysis is a "situation–hypothetical construct–behavior" interactional system. First, it seems a bit preemptive to assert that it is "quite wrong" to divide such a system into its constituent parts because no part functions independently. Whether such a decision is warranted most surely depends on the particular questions asked and the scope of the analysis required (see Bandura, 1983, for a related discussion). More important, specification of a model in these terms leaves open the critical questions of how the situation, the hypothetical construct, or the behavior should be conceptualized or operationalized. Substituting the term *hypothetical construct* for the term *person variable* does not change the fact that theory-driven choices must still be made concerning the types of hypothetical constructs that will best serve such a model. Those choices, of course, will depend on the purposes at hand. If one's goal is to describe interactions in the general sense of "this type of person tends to behave this way in these situations," then a more global, trait-oriented assessment may be useful. If one wishes to understand why specific interactions occur, then more process-oriented hypothetical constructs are essential.

4. Conclusion

I have argued that the primary differences between personology and situationism, broadly conceived, are theoretical, not methodological, and have little to do with the relative contributions of situations or persons (hypothetical constructs) to behavior. Rather, the differences are realized in the units that are preferred in attempting to characterize the person for divergent purposes. The so-called situationist critiques have long been characterized as a broadside attack on the concept of personality. Such is the case only if one restricts the domain of personality to global characterizations of individuals on nomothetically defined dimensions. What was questioned by the situationists was not the concept of personality, but the viability of a trait-oriented approach for many (if not most) purposes in clinical and personality psychology. Certainly any approach, be it personological, situational, unidirectional, bidirectional, transactional, or otherwise, will be useful for some set of circumscribed goals and purposes. One of the most challenging tasks ahead is to explore, define, and, most important, recognize the uses and limitations—the range of convenience—of each of the alternative strategies we employ in attempting to understand the human personality.

5. References

Allport, G. W. *Personality: A psychological interpretation*. New York: Holt, 1937.

Allport, G. W. Traits revisited. *American Psychologist*, 1966, *21*, 1–16.

Bandura, A. Self-efficacy: Toward a unifying theory of behavioral change. *Psychological Review*, 1977, *84*, 191–215.

Bandura, A. Temporal dynamics and decomposition of reciprocal determinism: A reply to Phillips and Orton. *Psychological Review*, 1983, *90*, 166–170.

Bowers, K. S. Situationism in psychology: An analysis and critique. *Psychological Review*, 1973, *80*, 307–336.

Ekehammar, B. Interactionism in personality from a historical perspective. *Psychological Bulletin*, 1974, *81*, 1026–1048.

Hartshorne, H., & May, M. A. *Studies in the nature of character*. (Vol. 1.) *Studies in Deceit*. New York: Macmillan, 1928.

Mischel, W. *Personality and assessment*. New York: Wiley, 1968.
Mischel, W. Toward a cognitive social learning reconceptualization of personality. *Psychological Review*, 1973, *80*, 252–283.
Murray, H. A. *Explorations in personality*. New York: Oxford University Press, 1938.
Peterson, D. R. *The clinical study of social behavior*. New York: Appleton-Century-Crofts, 1968.
Vernon, P. E. *Personality assessment: A critical survey*. New York: Wiley, 1964.

6

Persons, Situations, Interactions, and the Future of Personality

Lawrence A. Pervin

Professor Hyland addresses two issues in his paper—what the person–situation debate is about and, more significantly, where we go from here. Basic to the author's position is the suggestion that what is in fact largely a methodological debate has been interpreted as one based on theoretical differences. I believe that this interpretation involves a somewhat artificial distinction and ignores much of the past and current history of the debate.

Hyland correctly notes that much of the current concern with the person–situation issue dates to the 1960s, in particular publication of Mischel's (1968) book. What was important about this book was that it presented a clear exposition of the evidence against the presumed assumptions of traditional personality theory and that it presented a clear alternative in the form of social learning theory. We shall return to the latter point later, but for now it is important to note that Mischel was not just suggesting methodological changes but rather was presenting a conceptual alternative. As I have noted elsewhere, the history of the issue suggests a profound difference in the way psychologists view phenomena (Pervin, 1978a). The person–situation debate was formulated in almost identical terms in the 1930s. For example, Woodworth (1937) asked: How consistent is the individual in his personality traits? What is consistency? He suggested that no individual shows perfect consistency in behavior since behavior depends on the situation as much as on the individual. He concluded that "consistency should mean like behavior in like situations. And we have to ask, what are like situations

Lawrence A. Pervin • Department of Psychology, Rutgers University, New Brunswick, New Jersey 08903.

to a given individual? . . . We have to ask how he sees the situation and what he is trying to do or get out of the situation" (Woodworth, 1937, p. 104). To take another example, Allport (1937) noted the controversy between generalists (traits) and specifists (situations) and emphasized the need to look at underlying individual consistencies in teleonomic trends. Thus, at least as far back as the 1930s psychologists were concerned with the person–situation issue and, as is currently the case, with the issue of consistency. Given the long-standing history of the issue, one would suspect that its roots lie deeper than methodological differences.

While conceptually distinct, my sense is that there is a fairly close relationship between theory and methodology in personality research. The kinds of variables that are of interest to one theorist are different from those of interest to another theorist, and different procedures are used in the investigation of these phenomena. Similarly, the observation of certain phenomena leads to particular kinds of conceptualizations and to the utilization of particular tools for further empirical investigation. To view the person–situation debate in theoretical or methodological terms alone is to ignore the intimate relation between the two. It also is to ignore what may be the need for both theoretical and methodological innovation if we are to go beyond the past and current impasse. At the same time, it is likely that different conceptualizations and research procedures may be useful for different purposes. An analogy may be made here to Cronbach's (1960) discussion of the concepts of bandwidth and fidelity in relation to assessment. Just as certain assessment devices may have greater or lesser value in relation to complexity and fidelity of information, certain theories may have greater or lesser value in relation to understanding broad patterns of behavior as opposed to specific pieces of behavior. Such values may overlap with but need not be identical to an emphasis on individual differences as opposed to personality processes common to all individuals.

In the field of personality we must be careful about establishing artificial dichotomies as well as failing to make important distinctions. In relation to the former, Allport's (1937) discussion of idiographic and nomothetic approaches, as well as the ensuing debate, illustrates how confusion and useless controversy can arise. As I have suggested elsewhere (Pervin, 1984), there is no essential conflict between the utilization of idiographic methods in the pursuit of nomothetic principles. Whatever the nature of their other philosophical and theoretical differences, such distinguished psychologists as Allport, Freud, Pavlov, Piaget, and Skinner all made use of the study of individual subjects with the intent of formulating general principles of psychological functioning. On the other hand, the principles emphasized by these and other theorists were different and cannot be ignored. Thus, I believe that one can distinguish between person variables and situation variables and that such a distinction is intrinsic to alternative conceptualizations. Hyland correctly notes that trait and psychodynamic theorists attend to the importance of the situation—a point often ignored by critics of traditional personality theory. However, it is also the case that internal variables are emphasized, and the reasons for this are not just methodological. Similarly, although it is true that a situation must always be tied to an organism to have a behavioral effect, there may be legitimacy and utility in defining certain variables independent of the experiencing organism. This was the case in the early social learning emphasis on evoking and maintaining conditions in situations. It continues to be the case in ecology and parts of social psychology. Thus, criticism of the emphasis by Argyle, Furnham, and Graham (1981) on situation variables which may have parallel person variables would appear to be ill-founded. Despite historic debate over how to define and measure stimuli, situations, and environments (Pervin, 1978b), objective definitions are possible and the utilization of such definitions involves theoretical as well as methodological commitments.

If one attends to current research relevant to the person–situation debate, one continues to see a relation among theory, methodology, and interpretation of data. For example, recently particular attention has focused on the issue of consistency of behavior. In a review of the literature, Mischel and Peake (1982) argue that there is little evidence of cross-situational consistency in behavior and that this can be understood in terms of the ability of the organism to make discriminations among situations. Note that a person variable is being emphasized here, at least a shift in emphasis from earlier emphases on eliciting and maintaining conditions in the environment. On the other hand, Epstein (1983) finds the evidence in support of cross-situational consistency impressive and argues:

> There is enough cross-situational consistency in everyday life so that useful statements about individual behavior can be made without having to specify the eliciting situation. This, of course, is the way a trait is usually defined, and the findings demonstrate the utility of such a concept. (Epstein, 1979, p. 1122)

What is interesting is that Epstein and Mischel agree that there is evidence both for situational specificity and for cross-situation generality. However, they disagree about the evidence for each, the magnitude of each, and the associated theoretical variables to be emphasized. It is clear that the differences between Epstein and Mischel are interpretive and theoretical as well as methodological.

Interactionism served a useful function in the history of the person–situation debate (Magnusson & Endler, 1977). However, the author quite rightly points to the many interpretations of interactionism and the fact that today almost everyone is an interactionist. Obviously, if this were all there was to it, the debate would be settled by now. Hyland comes to a transactional or systems point of view, one with which I am in complete agreement (Pervin, 1978c, 1983). However, neither transactionalism nor cybernetics or systems theory tells us what to look for or how to look for it. The value of systems theory at this point is that it focuses attention on certain principles and perhaps offers a way out of previous ways of construing the issue. In particular, systems theory suggests that we be more concerned with processes and patterns as opposed to discrete events. In addition, the concept of equifinality has important implications for our interpretation of the meaning of behavior. Let us consider both these points in further detail.

The problem with the concept of traits is that it is a static concept and does not provide for an understanding of dynamic processes. This is part of the reason why trait theorists get caught up in arguing for the sameness of behavior or the structural identity of behavior across situations rather than coming to grips with the issues of diversity and pattern. This is a point made by Mischel and Peake (1982) with which I am in complete agreement. Rather than arguing about whether the person is consistent or inconsistent, stable or varying, we should recognize patterns of functioning that involve both stability and change. Person variables could still be emphasized here and it is unfortunate in this regard that personality theory has, to a considerable extent, come to be identified almost exclusively with trait theory.

A second problem with trait theory is that it relates to behavior at the overt or manifest level. It may be that overt behavior is a poor level at which to consider organizing principles, a point which is fundamental to the consistency issue where behavior is taken as the test of alternative theoretical positions. The point here is similar to that made by Murray: "According to my prejudice, trait psychology is over-concerned with recurrences, with consistency, with what is clearly manifested (the surface of personality), with what is conscious, ordered, and rational" (1938, p. 715). Murray's preference was for the concept of need, which might be a momentary process and which might be present within the organism without becoming manifest directly or overtly. The word *directly* is perhaps of particular significance, since it suggests the potential for multiple expressions depending on factors such as the perception of environmental contingencies or internal conditions. Thus, there may be structural organization underlying overt behavioral diversity. There are parallels here, I believe, with the general systems theory principle of equifinality.

As I noted above, general systems theory is useful in providing a way of viewing phenomena, but it does not answer fundamental questions. In particular, systems theory does not in itself tell us which units to consider. If traits are not the desired person units, where else can we turn? Some, including Mischel, have suggested an emphasis on cognitive variables. Clearly, personality, or at least a major part of personality, has gone social and gone cognitive (Cantor, 1981). However, in gaining its head, personality may be in danger of losing its soul. Cognitive processes are important and play a critical role in human behavior. In emphasizing generalizations or categorizations and discriminations, cognitive psychologists can account for both generality and specificity in functioning. However, what has happened to date is that cognitive variables are emphasized not only to the exclusion of other variables but as a substitute for them! Only now are there beginning to be signs of an awakening interest in affect and motivation, both in their own right and in terms of their influence on cognitive processes.

Elsewhere I have tried to suggest a way of viewing personality that considers both stability and change, which emphasizes motivation and affect as well as cognition, and which emanates from a dynamic systems perspective (Pervin, 1983). In particular, a distinction is made between goals and plans, with goals representing the motives organizing behavior and plans representing the specific behaviors enacted toward the achievement of goals. The organization of a goal system provides for stability and, more significantly, coherence in functioning. Changes in the system, temporary as well as permanent and caused by internal factors as well as environmental changes, provide for fluidity and flexibility. In addition, changes in plans or overt behaviors also contribute to variability in functioning. Since the same goal can lead to different behaviors and the same behavior (plan) can be associated with different goals, behavioral observations alone, in particular discrete behavioral observations, are seen as inadequate for appreciating the underlying organization of personality functioning.

The concept of goals is emphasized by theorists of varying orientations including psychoanalytic (Gedo, 1979), social learning (Bandura & Cervone, 1983; Mischel, 1973), and artificial intelligence (Schank & Abelson, 1977). It is offered here not as a solution to the problem but rather as an illustration of an effort to conceptualize human behavior in terms that go beyond the person–situation debate. The debate is unlikely to be settled in favor of one or the other. In and of itself it probably is a trivial issue. As the author notes, evidence can be found in support of either point of view and research can be manipulated to demonstrate the power of each. Rather, the real significance of the person–situation debate may be in calling attention to the critical issue of understanding patterns of stability and change.

1. References

Allport, F. H. Teleonomic description in the study of personality. *Character and Personality*, 1937, *5*, 202–214.

Allport, G. W. *Personalty*. New York: Holt, 1937.

Argyle, M., Furnham, A., & Graham, J. A. *Social situations*. London: Cambridge University Press, 1981.

Bandura, A., & Cervone, D. Self-evaluative and self-efficacy mechanisms governing the motivational effects of goal systems. *Journal of Personality and Social Psychology*, 1983, *45*, 1017–1028.

Cantor, N. A cognitive-social approach to personality. In N. Cantor & J. F. Kihlstrom (Eds.), *Personality, cognition, and social interaction*. Hillsdale, N. J.: Lawrence Erlbaum, 1981.

Cronbach, L. J. *Essentials of psychological testing*. New York: Harper & Row, 1960.

Epstein, S. The stability of behavior: I. On predicting most of the people much of the time. *Journal of Personality and Social Psychology*, 1979, *37*, 1097–1126.

Epstein, S. The stability of confusion: A reply to Mischel and Peake. *Psychological Review*, 1983, *90*, 179–184.

Gedo, J. E. *Beyond interpretation: Toward a revised theory for psychoanalysis*. New York: International Universities Press, 1979.

Magnusson, D., & Endler, N. S. (Eds.). *Personality at the crossroads: Current issues in interactional psychology*. Hillsdale, N. J.: Lawrence Erlbaum, 1977.

Mischel, W. *Personality and assessment*. New York: Wiley, 1968.

Mischel, W. Toward a cognitive social learning conceptualization of personality. *Psychological Review*, 1973, *80*, 252–283.

Mischel, W., & Peake, P. K. Beyond *déjà vu* in the search for cross-situational consistency. *Psychological Review*, 1982, *89*, 730–755.

Murray, H. A. *Exploration in personality*. New York: Oxford, 1938.

Pervin, L. A. *Current controversies and issues in personality*. New York: Wiley, 1978. (a)

Pervin, L. A. Definitions, measurements, and classifications of stimuli, situations, and environments. *Human Ecology*, 1978, *6*, 71–105. (b)

Pervin, L. A. Alternative approaches to the conceptualization of individual–environment interaction. In L. A. Pervin & M. Lewis (Eds.), *Perspectives in interactional psychology*. New York: Plenum Press, 1978. (c)

Pervin, L. A. The stasis and flow of behavior. In M. M. Page (Ed.), *Personality: Current theory and research*. Lincoln: University of Nebraska Press, 1983.

Pervin, L. A. Idiographic approaches to personality. In N. Endler & J. McV. Hunt (Eds.), *Personality and the behavioral disorders* (2nd Ed.). New York: Wiley, 1984.

Schank, R. C., & Abelson, R. P. *Scripts, plans, goals, and understanding*. Hillsdale, N. J.: Lawrence Erlbaum, 1977.

Woodworth, R. S. *Psychology*. New York: Holt, 1937.

6

Interactionism and Achievement Theory

Joel O. Raynor

I have not found the *contemporary* person × situation ($p \times s$) debate very enlightening and therefore am pleased with Hyland's efforts to clarify the theoretical perspective from which one can view it. I had thought that the *historical* $p \times s$ controversy and its resolution in Lewin's (1935, 1938, 1943) programmatic equation, $B = f(P, E)$, where P represents characteristic(s) of the person and E the *perceived* environment, had long ago resolved the issue, that both person and perceived environment are needed as constructs to understand the determinants of behavior, and that the theoretical task for psychology involves the specification of what these variables are for a particular behavior and how they combine (interact) at a given point in time to determine that behavior (cf. Atkinson, 1964; Atkinson & Birch, 1978a; Raynor & Entin, 1982). Lewin (1943) and Hull (1943) were both concerned with the proper role of the environment as a determinant of action in psychological theory. My thinking has been that the Hullians had won major battles, but the Lewinians had won the war. I thought a consensus had evolved that the *perceived* rather than the objective environment in interaction with person characteristics as specified in a mathematical function rule defined the task of theory construction in psychology, and that individual differences determine the particular nature of the perceived environment in such a function rule. Why should a *contemporary* $p \times s$ debate resurrect these issues? Although I was premature concerning the existence of such a consensus, Hyland's analysis confirms to me that it must ultimately emerge.

Joel O. Raynor • Department of Psychology, State University of New York at Buffalo, Buffalo, New York 14226.

In this commentary I would like to extend and elaborate several issues raised by Hyland's analysis: (1) all variables in psychology are person variables, (2) it is necessary to specify the unit of behavioral analysis, (3) time must become a variable, and (4) cyclical interaction must be considered. Each of these points is reflected in theoretical efforts that have evolved in the study of achievement motivation (Atkinson, 1957, 1964, Chapter 10; Atkinson & Feather, 1966; Atkinson & Raynor, 1974, 1978; Lewin, Dembo, Festinger, & Sears, 1944; Raynor & Entin, 1982) which anticipate and/or coincide with suggestions made in Hyland's analysis.

1. All Variables in Psychology Are Person Variables

The debate between Lewin (1943) and Hull (1943) basically revolved around the role of the external environment in psychological theory. For Lewin, the external environment could have an effect on behavior if and only if it was presented as a fact in the lifespace of the individual (i.e., the perceived environment), and he referred to the "foreign hull" in a pun to reflect the Hullian emphasis on the objective, external stimulus as the cause of a response. In fact, Hullian theory ultimately dealt with the "effective stimulus," which I take to be functionally equivalent to what Lewin meant by the perceived environment or a fact in the lifespace. The issue is best represented in treatment of a discriminative learning situation involving objective definition of different colors in terms of wavelength (red vs. green light). A color-blind individual will not respond differentially to red and green. A diagnostic (behavioral) test must first be used to determine the capacity of the individual to discriminate between colors—that is, to discover the effective or perceived stimulus situation. The psychological variable is not wavelength but perceived color, and perceived color is a person variable rather than an environment variable. For psychology, the environment or situation can only be represented conceptually as the "perceived environment." Attempts to use objective criteria to determine the definition of the *situation* falter on this basic need to determine the "effective situation" for a particular individual at a particular point in time. This point seems as convincing in the contemporary $p \times s$ debate as it is in the historical $p \times s$ controversy.

However, Lewin was less successful in dealing with the person component of his programmatic equation $B = f(P, E)$. In principle, a person variable could represent either an enduring or momentary characteristic of the individual (trait vs. state), but his theory never systematically distinguished between the two. Evolving theory of achievement motivation (cf. Atkinson, 1957) introduced the stable person characteristic of motive (trait) as distinct from aroused motivation (state), and in the more general theory of the dynamics of action (Atkinson & Birch, 1970) the concept of "inertial tendency to act" recaptures the earlier Lewinian concept of "tension." But it has never been clear within this tradition whether diagnostic tests of person variables are restricted to objective assessment of them, as in administration of an ability test, or require subjective assessment, as in measurement of the perceived ability of an individual. Thus, although I agree with Hyland that "all variables in psychology are person variables," I disagree that the issue can or should be divorced from the so-called methodological issue concerning the kinds of operational definitions that are coordinated with definition of theoretical constructs. It seems to me that the issue as to whether objectively or subjectively assessed person variables will ultimately provide a better explanation for a particular behavior in psychology is an empirical one.

I suspect that the need for resurrection of the $p \times s$ debate in contemporary form in part stemmed from the failure of the Lewinian tradition to provide convincing arguments for assessment of what is, and what is not, a "fact in the lifespace," an issue which cannot be divorced from the methodological issue of operational definitions. The Lewinian conceptual analysis was sound, but the research tradition in which it was embedded was much less so.

2. Specification of the Unit of Behavioral Analysis

It is my view that scientific advance is best achieved by going from the specific to the general. Research-oriented theory of personality such as that concerned with achievement motivation has been for the most part very careful to define the *domain of behaviors* to which the theoretical statement is appropriate. Thus the concepts of the "success-oriented" and "failure-threatened" personalities (Atkinson, 1978) have evolved in the context of theory which specifies the definition of "achievement-oriented behavior" (McClelland, 1961) and in the context of research that distinguishes the *conditions* under which success-oriented and failure-threatened individuals are expected to act according to theoretical predictions (Atkinson & Raynor, 1978). Again, one must consider the operational definitions of theoretical terms and the research in which they evolved in order to understand the specification of the function rule relating individual differences in personality to perceived environmental influences. Although it is true that most theories of personality in principle considered such environmental influences on behavior, the research in which they are embedded often indicates that such concern is more apparent than real, whereas, for example, theory of achievement motivation can be seen as a continually evolving effort to delineate the perceived situational determinants that arouse personality constructs such as motives (cf. Raynor & Entin, 1982).

Specification of the *unit of behavioral analysis* has become a concern in theory of achievement motivation (Raynor, 1969, 1978a; Raynor & Entin, 1982). An evolving "step-path theory of action" (Raynor & Entin, 1983) considers a path consisting of a series of steps rather than isolated behavioral acts as the appropriate unit for the study of behavior under certain specified conditions. This reemphasis on the earlier Lewinian (1936) analysis of steps-in-a-path makes the important point neglected in the earlier treatment, that a so-called subgoal can have value by itself as well as value due to its instrumental relationship to a final or ultimate goal of the path. As Hyland notes, the path of becoming a psychologist creates a quite different psychological situation for the individual than the path of becoming an archeologist when faced with the immediate step of taking an introductory psychology course. Such step-path theory of action defines the behavioral unit of analysis as the path rather than the immediate next step and focuses attention on sequences of activity rather than activity in isolation. We have recently become concerned with behavior *along* such paths when these steps are functionally related so that immediate success or failure bears on the opportunity to continue along it (cf. Raynor, 1982), whereas earlier emphasis was limited to determinants of action in the first step of such paths (cf. Raynor, 1969, 1978b). A related point was made earlier by Atkinson and Birch (1970, 1978b) by their emphasis on the stream of activity and change of activity as the important focus of conceptual analysis. Taken together, these developments illustrate Hyland's concern for specification of appropriate units of behavioral analysis. They have evolved over the past decade in response to continued conceptual analysis and programmatic research in the specific domain of achievement-oriented activity. Both the earlier Lewinian and Hullian analyses were limited in their general conceptual focus to one-step behavioral situations where one "goal object" or "reinforcing event" signaled the cessation of activity, and neither provided for adequate conceptual analysis of sequences of activity over time. This is particularly apparent in their respective analyses of conflict in the context of the goal-gradient.

Hyland's analysis suggests further that we consider some average of behavior over steps in a path rather than behavior in any particular step of a path in order to increase the sample size for that behavior. However, we must be careful in doing so not to neglect possible changes in behavior predicted as a function of stage of striving. Theoretical analysis (Raynor & Entin, 1982) suggests that one might expect substantial differences which might be glossed over with such an averaging to increase the behavioral sample.

3. Time Must Become a Variable

Atkinson and Birch's (1970) theory of the dynamics of action assumes that time is a significant factor in the determinants of action, and they use differential calculus to represent the rate of change in the strength of action and negaction tendencies. This is precisely the kind of theoretical analysis called for by Hyland. However, as yet there is little evidence that Atkinson and Birch's conceptual analysis has had much impact outside of the area of achievement motivation. More recently, concern has shifted to identification of the past, present, and future as distinct time-linked sources of value which must be taken into account in order to understand the determinants of action (Raynor, 1982). The concept of "stages of striving" over time is derived, in which differential emphasis is placed on future-oriented (early-stage striving) and past-oriented (late-stage striving) determinants of action. Different kinds of psychological processes may be involved in different stages of striving; future-oriented theory such as expectancy × value theory, past-oriented theory such as cognitive consistency theory, and present-oriented theory such as that concerned with information seeking to evaluate competences deal with different time-linked sources of value. Such theories may complement each other rather than provide different predictions concerning the same psychological process when time-linked sources of value are systematically taken into account.

4. Cyclical Interaction

Historically, failure to differentiate between historical and ahistorical analyses of behavior lead to logical confusion concerning the role of psychological theory in predicting behavior. Lewin (1943) emphasized the principle of the contemporaneous determinants of action, whereas early behavioral analyses of Hull (1931, 1932, 1937) used historical (learning theory) principles to predict contemporaneous behavior without use of a function rule. Hull (1943) converted to use of such a principle of action, and henceforth there has been general agreement that the effects of the past, or the historical evolution or development of a variable, can influence action only to the extent that its contemporaneous nature and strength is specified within a function rule to indicate how it combines with other such (contemporaneous) influences. This is the meaning of the concept of interaction that seems most appropriate for conceptual analysis in psychological theory, and most contemporary theory in the field of motivational psychology (cf. Atkinson & Birch, 1978b) now takes this form. Historically, the separation of the historical and ahistorical analysis of a variable coincided with the distinction between learning and performance and the emergence of the field of motivational psychology as uniquely different from fields such as learning or developmental psychology with their emphasis on historical processes.

The clarity gained by separation of the history of a variable from its functional relationship (in conjunction with other variables) to behavior can now be seen to have been won at the price of an artificial dichotomy which now must yield to some integrating principle such as that represented by the concept of cyclical interaction over time. Strictly adhered to, the emphasis on the ahistorical excludes reference to the history of a variable, and many psychologists are not convinced that we understand such a variable unless we can trace its historical antecedents. Conversely, strict adherence to a historical analysis fails to specify the functional significance of a variable at a given point in time—how does this variable combine (interact) with other variables to determine behavior? The concept of cyclical interaction holds the promise of combining the historical and ahistorical approaches without a regressive confusion between them.

In practice, resultant valence (force) theory (Lewin *et al.*, 1944) and theory of achievement motivation (Atkinson & Feather, 1966) have utilized a quasi-cyclical analysis in the prediction of shifts in level of aspiration. A particular function rule is used to specify the determinants of initial level of aspiration (tendency = motive × expectancy × incentive); success-failure feedback in striving to attain that level of aspiration is then used to determine changes in expectancy (and hence incentive, since the two are assumed to be inversely related); new values of expectancy and incentive then replace the old values in reapplication of the function rule to predict subsequent change in level of aspiration. Such an analysis combines both the ahistorical approach (use of the function rule) and the historical approach (changes in expectancy and incentive over time), but does not consider the possible *emergent properties* of such a feedback or cyclical process.

This kind of analysis has been extended to consideration of motivational or behavioral syndromes—compulsive striving, uptight striving, apathetic striving—and more generally, to the notion of a psychological career (cf. Raynor, 1978b, 1982). The analysis of psychological careers in different stages of striving seems to be a specific example of the concept of cyclical interaction in that it views feedback from previous action in a systematic way over time to produce a predictable and specifiable end-state having properties not apparent at the outset. This approach also provides a framework for viewing the feedback TOTE unit (Miller, Galanter, & Pribram, 1960) as a specific case of a more general principle. The TOTE (test, operate, test, exit) represents what has been called a "partial contingent path" (Raynor & Entin, 1982) in which a successful test allows movement to the next step of a path, whereas an unsuccessful test allows for repetition of behavior until a successful test is obtained. However, this is but one of a number of possible functional relationships between steps in a path. In a contingent path, success allows for movement to the next step but failure rules out further striving along that path, whereas in a noncontingent path both success and failure allow for taking subsequent steps. But by itself the TOTE unit is a static concept in that it does not allow for possible changes in the psychological situation faced by the individual after a particular behavior, which may change the test criterion, the value of any of the variables which determine a subsequent action or which may lead to the emergence of an additional determinant of action. The concept of cyclical interaction seems an appropriate means of conceptualizing behavior when considered as a series of steps in a path over time.

Hylands's analysis suggests that the segmented, linear kind of conception that has been used when applying ahistorical principles (such as a function rule specifying the interaction of variables at a given point in time) must somehow be modified systematically to include feedback to indicate how variables change over time and/or how new determinants of behavior might emerge. I agree. Without recognizing it as "cyclical interaction," I have developed a method of presenting the development and functioning of a psychological career. A psychological career is defined as "an opportunity for self-identity" consisting of steps in a path involving life activity (cf. Raynor, 1982). I had been tempted to refer to it as an instance of "person–situation–self" interaction in that the emergence of a self-image defined by the outcomes of prior activity along a particular path is a predictable consequence of prior success or

Person

0 +1 +2 +3 +4

$R = f(A, B, C)$

Figure 1

failure under certain specified conditions. After reading Hyland's paper it becomes apparent that what is now needed is a direct mathematical means of representing such cyclical interaction. I have adopted a quasi-cyclical analysis similar to that described above for understanding level of aspiration. In Figure 1 the person in the here-and-now (represented by 0) is shown as facing a series of anticipated steps (represented by positive numbers) and behavior (*R*) is a function of variables *A, B,* and *C* as assessed at that time. In Figure 2 the person has moved along the path and the outcomes of action at the previous steps (represented by negative numbers) can change the values of variables *A, B,* and/or *C*. This movement has led to the emergence of the past as a possible determinant of immediate behavior (e.g., variable *D*). In addition, the individual may now see himself in terms of the anticipated future outcomes, retrospected past outcomes, and/or assessment of attributes through the immediate next step (i.e., the self-system has emerged along this path, represented as variable *E*). Thus both changes in strength of variables (*A, B, C*) over time and the emergence of new variables (*D, E*) over time can be represented in the new psychological situation as determinants of behavior. The new function rule may be similar to or different from the old one, but diagnostic assessment of variables in the new psychological situation is required for its use, or systematic principles predicting the nature and extent of change in variables *A, B,* and *C* must be specified. In Figure 3 the person has taken all of the steps in the path, no new anticipated steps exist, and hence the anticipated future no longer serves as a possible determinant of immediate behavior. Thus the function rule no longer includes variable *A* as a determinant of action, but the historical analysis of variables *D* and *E* continues.

Representation of processes such as cyclical interaction seems to me to be misleading if it merely implies a *repetition* of a process without movement in time (i.e., in this case, movement along a path so that early, middle, and late stages of striving are represented by Figures 1,

Person

-2 -1 0 +1 +2

$R = f(A, B, C, D, E)$

Figure 2

Person

$R = f(B, C, D, E)$

-4 -3 -2 -1 0

Figure 3

2, and 3, respectively). Such representation must allow for changes in the (new) psychological situation faced by the person. The concept of interaction could now refer to the combination (function) rule for the various variables at a given point in time, the effects of prior action on the nature and strength of variables at the next point in time, and/or the possible emergence of new variables at some future point in time. As long as we are clear as to which meaning of *interaction* we are concerned with, we avoid confusion. However, I prefer to restrict its meaning to the particular function rule indicating how variables combine to determine action at a particular point in time, and to use other terms to refer to these other processes.

5. Concluding Comment

Hyland's paper provides an excellent theoretical perspective on the $p \times s$ debate and valuable suggestions for new directions. My comments are meant to illustrate how one particular program of theory and research has dealt with some of the issues he raises, with the hope that this both gives substance to his comments and alerts the reader to work that has continued to be sensitive to issues raised both by the historical and contemporary versions of this basic issue in psychological theory and research. As Atkinson (1981) has noted, the study of personality in the context of an advanced motivational psychology has provided a theoretical rationale for integrating both the concepts of individual differences in personality and the effects of the perceived situation on the determinants of action.

6. References

Atkinson, J. W. Motivational determinants of risk-taking behavior. *Psychological Review*, 1957, *64*, 359–372.

Atkinson, J. W. *An introduction to motivation*. Princeton, N. J.: Van Nostrand, 1964.

Atkinson, J. W. The mainsprings of achievement-oriented activity. In J. W. Atkinson & J. O. Raynor (Eds.), *Personality, motivation, and achievement*. Washington, D. C.: Hemisphere, 1978.

Atkinson, J. W. Studying personality in the context of an advanced motivational psychology. *American Psychologist*, 1981, *36*, 117–128.

Atkinson, J. W., & Birch, D. *The dynamics of action.* New York: Wiley, 1970.
Atkinson, J. W., & Birch, D. *An introduction to motivation* (2nd Ed.). New York: Van Nostrand, 1978. (a)
Atkinson, J. W., & Birch, D. The dynamics of achievement-oriented activity. In J. W. Atkinson & J. O. Raynor (Eds.), *Personality, motivation, and achievement.* Washington, D. C.: Hemisphere, 1978. (b)
Atkinson, J. W., & Feather, N. T. (Eds.). *A theory of achievement motivation.* New York: Wiley, 1966.
Atkinson, J. W., & Raynor, J. O. (Eds.). *Motivation and achievement.* Washington, D. C.: Hemisphere, 1974.
Atkinson, J. W., & Raynor, J. O. (Eds.). *Personality, motivation, and achievement.* Washington, D. C.: Hemisphere, 1978.
Hull, C. L. Goal attraction and directing ideas conceived as habit phenomena. *Psychological Review,* 1931, *38,* 487–506.
Hull, C. L. The goal gradiant and maze learning. *Psychological Review,* 1932, *39,* 25–43.
Hull, C. L. Mind, mechanism, and adaptive behavior. *Psychological Review,* 1937, *44,* 1–32.
Hull, C. L. *Principles of behavior.* New York: Appleton-Century-Crofts, 1943.
Lewin, K. *A dynamic theory of personality.* New York: McGraw-Hill, 1935.
Lewin, K. *Principles of topological psychology.* New York: McGraw-Hill, 1936.
Lewin, K. *The conceptual representation and measurement of psychological forces.* Durham, N. C.: Duke University Press, 1938.
Lewin, K. Defining the "field at a given time." *Psychological Review,* 1943, *50,* 292–310.
Lewin, K., Dembo, T., Festinger, L., & Sears, P. S. Level of aspiration. In J. McV. Hunt (Ed.), *Personality and the behavior disorders* (Vol. 1). New York: Ronald Press, 1944, pp. 333–378.
McClelland, D. C. *The achieving society.* Princeton: Van Nostrand, 1961.
Miller, G. A., Galanter, E., & Pribram, K. H. *Plans and the structure of behavior.* New York: Holt, 1960.
Raynor J. O. Future orientation and motivation of immediate activity: An elaboration of the theory of achievement motivation. *Psychological Review,* 1969, *76,* 606–610.
Raynor, J. O. Future orientation in the study of achievement motivation. In J. W. Atkinson & J. O. Raynor (Eds.), *Personality, motivation, and achievement.* Washington, D. C.: Hemisphere, 1978. (a)
Raynor, J. O. Motivation and career striving. In J. W. Atkinson & J. O. Raynor (Eds.), *Personality, motivation, and achievement.* Washington, D. C.: Hemisphere, 1978. (b)
Raynor, J. O. A theory of personality functioning and change. In J. O. Raynor & E. E. Entin (Eds.), *Motivation, career striving, and aging.* Washington, D. C.: Hemisphere, 1982.
Raynor, J. O., & Entin, E. E. *Motivation, career striving, and aging.* Washington, D. C.: Hemisphere, 1982.
Raynor, J. O., & Entin, E. E. The function of future orientation as a determinant of human behavior in step-path theory of action. *International Journal of Psychology,* 1983, *18,* 463–487.

6

Interactionism and Control Theory

William T. Powers

Hyland has identified the person × situation debate as a methodological, not a theoretical issue. He has shown that the proposed resolution of this issue, interactionism in any of several forms, is not really a resolution but an overdue statement of the nature of the problem that still remains for psychologists to solve. And he has suggested that a real solution might be found in the directions indicated by control theory. In my opinion, he has identified the crux of the real issue, which is the difference between an old cause–effect paradigm based on a lineal form of analysis and a new paradigm based on what we might call, in the present context, principles of concurrent interactionism. This fundamental difference deserves further discussion.

Control theory can be understood correctly only if it is first understood that while behavior proceeds there are two concurrent kinds of relationships involved. One is the relationship in which behavior depends on external events according to a function that characterizes the behaving organism. The actions of an organism are at least in part a function of what is happening to it.

The other relationship, which exists at the same time as the first and operates concurrently with it, characterizes the way in which the impingement of external events on the organism depends in part on the actions of the organism. What is happening to an organism depends at least in part on what the organism is doing to the outside world.

There are two classes of variables that must satisfy both of these functional relationships at the same time: input variables, the external variables having immediate sensory or physiological importance to the organism, and output variables, the immediate effects of the organism's motor activities outside the organism.

William T. Powers • 1138 Whitfield Road, Northbrook, Illinois 60062.

It is impossible to separate these two concurrent functions under the old paradigm. Any correct expression of the relationship between "situation" and "behavior" must show the same variable on both sides of the equal sign: neither way of writing the relationship in lineal form, $B = f(S)$ or $S = f(B)$, with or without intervening variables, is correct. Instead we must say $B = f(S,B)$ or $S = f(B,S)$. This form is typical of all system equations describing closed-loop relationships, whether they describe machines or organisms. Hyland showed an advanced understanding of control theory when he pointed out that this closed-loop relationship is not a cycle that can be characterized correctly by a sequential analysis.

In order to understand the characteristics of the organism separately from properties of its environment, we must find ways of separating the variables; no approach that fails to separate them can successfully characterize either the organism or the aspect of the external situation that is important. That, I venture, is the real reason why experimental approaches to personality have resulted in the abysmally low correlations that Hyland mentions. All of the approaches discussed by Hyland share the same defect; they are based on the wrong underlying model and therefore use the wrong mathematics for identifying and analyzing experimental evidence.

There is only one mathematically correct way to treat variables linked by more than one relationship at a time, and that is to treat the relationships as simultaneous equations. Control theory consists largely of ways of setting up appropriate differential equations describing each relationship individually and then solving them simultaneously for the variables of interest. Although these methods were developed for use with artificial control systems, they are quite general and have nothing to do specifically with machines; they apply wherever closed circles of relationships are found.

There is a reason, however, why my theoretical work deals specifically with the properties of control systems and not simply with simultaneous differential equations. It is that the kinds of relationships found in control systems have particularly important properties. Control systems are capable of controlling any external variable that they can both sense and affect. They do this, furthermore, relative to an internal specification of the desired state of the external world. They are, in short, purposive. They are capable of varying their actions in any way required to bring an external variable to a preselected state and keep it there, despite disturbances of all kinds, predictable or unpredictable, detectable or invisible. There are limits, of course, relating to the magnitudes of disturbances and the speed with which they change.

Hyland cites Bandura as saying that the relationship between organism and environment is symmetrical: either can be viewed as the agent or as the object of control. That is not correct. Control theory spells out in detail what is required of any active system if it is to be described as "controlling" something else. Whatever is controlled must be stabilized by the actions of the controlling system against disturbances. This, indeed, is what distinguishes a control system from a system that behaves in a way totally determined by external forces. If there is interference with the object of control, the actions of a control system will immediately change to counteract the interference (within the limits of possible performance). Thus a control system counteracts disturbances of the consequences of its behavior, downstream in the apparent cause–effect chain. This occurs, of course, because there is really a closed loop, not a simple sequence of causes and effects.

The environment in which a control system exists normally lacks the organization that would be needed in order to control anything. In statements such as "temperature controls the rate of a chemical reaction," the word "control" is used loosely; what is meant is "affects." If someone were to drop a catalyst into the container of reagents, the temperature would do nothing to restore the original reaction rate; it does not actively control that rate. A control system, however, could *use* temperature as a *means* of controlling reaction rate, provided it could sense the reaction rate. The addition of a catalyst might speed up the reaction; the control system would then lower the temperature until the rate returned to its original value (and, if well designed, could even prevent most of the change that the catalyst would otherwise have caused).

Thus the relationship between a control system and its environment is by no means symmetrical as Bandura implied. An organism has properties that are not normally found in the passive physical world. The organization of organisms as hierarchies of control systems sets them apart from other organizations of matter; they can do things that other material objects simply cannot do. When psychology became a science, and during most of its first hundred years of existence, nobody realized that there was any way to organize matter that would result in behavior oriented around purposes. Both theory and practice became organized around the premise that only lineal cause and effect needed to be taken into account. This premise was not recognized as a theory; most scientists took it as a fundamental principle that had to be upheld even when all the evidence seemed to argue against it. This premise has determined not only the methodology of the life sciences, but the nature of the observations that determine what will and will not be considered "data." Hyland mentioned the "theory-laden" nature of all observation. The theory of lineal cause and effect is the heaviest burden of all.

Under the old paradigm, actions of an organism that correlate with independent events are understood as having been caused by those events. In personality theory, the internal properties of the organism show up in terms of the manner of the causation, shaping the form of the "response" but not the basic flow of causation. Control theory, too, recognizes this relationship, which is too clear to argue away. But under control theory the model is not causal in the same sense. If actions regularly accompany external events, this is not taken as indicating that events determine actions. Instead, the assumption is that the external event disturbed one or more variables that the organism is maintaining under control. The actions are understood as having the purpose of nullifying the disturbing effects on the controlled variables. Notice that this assumption is falsifiable; if careful study fails to reveal any variable that is actually being controlled in this way, the control hypothesis must be dropped. There is no corresponding test, under the old paradigm, of the assumption that events are causing behaviors. The control model leads to identification of the specific control organization that is acting, but only if it does in fact exist.

In the areas of behavior in which most psychologists are interested, the details of motor activity are of little interest, and there would be few occasions for using the rigorous differential-equation approach of control theory. But the existence theorem provided by the rigorous approaches can make it possible for personality theorists and cognitive psychologists and others interested in more global aspects of behavior to interpret their observations differently, and even to make new kinds of observations. Once purpose is accepted as a natural and nonmystical aspect of living systems, questions relating to purposes can be asked, and using methods suggested by control theory, potential answers can be tested. More rigor can be achieved over the whole spectrum of psychology. At one end, observations based on physical measures can be organized to take the controlling properties of organisms into account, and at the other end, vague and fanciful treatments of mental phenomena, goals, and purposes can be given more substance and brought under experimental scrutiny in a more productive way.

Professor Hyland has used my work in a way that any theoretician would take as the highest of compliments: with understanding of important subtleties, with an integrative result, and correctly. I am a toolmaker, not a psychologist. My work is useful only if it is used; otherwise it is an empty exercise. I think that Hyland is making excellent use of it, and I hope that others will also take the time to learn what control theory is really about before trying to apply it.

6

Objectives and Questions in Personality Research

Reply to Commentators

Michael E. Hyland

Of the four commentators, Peake and Pervin have published previously within the context of the person × situation debate. Neither Raynor nor Powers has published in that context, but both of these authors are associated with particular types of interactional theory. Briefly, Peake suggests an alternative interpretation of the development of the person × situation debate which emphasizes the difference between trait- and process-based personality theory. Pervin's contribution concerns the relation between theory and methodology, and Pervin goes on to suggest a goal-oriented approach to understanding personality. Raynor describes recent advances in achievement motivation theory, showing how this goal-oriented theory incorporates many of the theoretical features of interactionism, including cyclical interactions. Finally, Powers shows how control theory can contribute to an understanding of cyclical interaction. Each of the four commentators contributes in a different way to a further understanding of the theoretical issues involved, and since they are writing from different perspectives, their assessments are sometimes quite different. Both Peake and Pervin are critical of some of the points made in my paper, and their critical contribution is assessed in the present remark.

Michael E. Hyland • Department of Psychology, Plymouth Polytechnic, Drake Circus, Plymouth, Devon PL4 8AA, England.

1. Theoretical Basis of the Debate

I suggest that the difference between personologism and situationism is largely methodological rather than theoretical. Both Peake and Pervin argue that those positions reflect theoretical differences, but they do so for different reasons. The basis for my argument is that personologism and situationism as characterized by the formulae $B = f(P)$ and $B = f(E)$ have not been advocated as theoretical positions; instead, personologism and situationism reflect the use of different methodologies.

Pervin points out that methodology and theory are inextricably linked and that to extract one from the other creates a false dichotomy. I entirely agree with Pervin that the selection of methodology reflects underlying theoretical assumptions. There is a wealth of argument by philosophers of science to that effect (Hanson, 1958; Popper, 1963); indeed, it forms part of the discussion relating to the theory-ladenness of observables. My paper certainly failed to discuss what the underlying theoretical assumptions of experimental versus correlational designs might be, but whatever they are they are certainly not the crude theoretical positions of $B = f(P)$ and $B = f(S)$. They are not those crude theoretical positions, because, as I argue, those positions never existed.

Peake asserts that theoretical positions corresponding to personologism and situationism can be distinguished but that they are not accurately characterized by the simple $B = f(P)$ and $B = f(E)$ formulae: "That simplistic characterizations of these positions in the form $B = f(P)$ or $B = f(E)$ fail to capture these differences is more a reflection of the inadequacy of the characterizations than the lack of theoretical differences in the positions."

In characterizing situationism, Peake distinguishes radical behaviorism from those personality theories (e.g., Mischel, 1973; Vernon, 1964) described by their critics as situationist. According to Peake, theoretical differences arise between the personality-situationist position and personologism with regard to the sort of person variable selected to explain behavior: Traditional personality theories employ traits as person variables whereas their critics advocate process based person variables:

> In recognizing the cross-situational variability of behavior, what was questioned was not whether situations are more important than persons, but whether the units that psychologists were using to characterize persons could adequately capture the important aspects of individuals that would lead to useful predictions about their behavior in the contexts in which they lived. (p. 334)

Peake is, of course, quite right, though one doubts whether everyone taking part in the debate had this level of insight. Certainly it is not evident in many of the ANOVA studies featuring in the debate, which despite referencing Mischel's theoretical contribution failed to do more than examine the parameters affecting the amount of person, situation, and interaction variance.

At least part of the difference between Peake and myself arises over different uses of the terms *personologism* and *situationism*. There are four quite separate types of psychology which attract one of these labels: Trait-based personality theory; personality theory critical of trait theory (e.g., Mischel's contribution), which advocates processes as person variables; radical behaviorism; experimental psychologies which include person variables (process theories) but do not normally consider individual differences (e.g., much of cognitive psychology).

I classify Mischel's theoretical contribution by his own label of *interactionist;* Peake uses the term *situationist* because this was the label used (unfairly) by Mischel's critics during the debate. Peake uses the word *personologism* in the sense of trait theory alone. My use reflects Murray's original definition (1938, p. 4), which includes process theories—such as Murray's. To compound this terminological difference, I have followed Bowers (1973, 1978) in including under the situationist heading those experimental psychologies which exclude consideration of individual differences, a category which Peake does not consider but which broadens the discussion beyond personality theory. The methodological difference between personologism and situationism is implicit in my, but not Peake's, use of those terms, just as the theoretical difference follows from Peake's use but not mine. Perhaps the words *situationism* and *personologism* should be avoided.

Leaving aside questions of terminology, it is evident that Peake's trait versus process distinction provides an accurate characterization of an important theoretical difference between trait theorists and their situationist-interactionist critics. The selection of either sort of personality variable is not arbitrary but reflects what Peake, in the quotation above, calls "useful predictions of . . . behavior." A neglected difference between the older trait theories (and possibly some recent trait theories) and their critics may arise simply from different meanings of the phrase "useful prediction." A prediction is useful only if one knows the use to which that prediction is to be put; useful predictions do not occur in a vacuum.

There are two main areas in which personality tests are used. They are used within a clinical context and they are used in occupational settings. In the latter instance a personality test is commonly used as a general selection device the purpose of which is to ensure that performance in a selected population is at least somewhat better than a random sample. In such circumstances correlations of only 0.3 and the low predictive power of the test may not cause concern. While such a general test might lead to individual injustices, it would be useful in the sense that it improves performance in the selected population—even though it may not be useful in predicting the behavior of any one individual at any one time: The test has *organizational* utility. If, however, we insist that any test should be able to predict an individual's behavior in a given context (so, for example, individual injustices are not done), then low specificity of prediction jeopardizes the test's usefulness. As Peake points out, the recent criticism of trait theory came from academically oriented clinicians, and it would seem reasonable to infer that their objectives would have been at an individual rather than an organizational level.

Unfortunately, tests which are only capable of making predictions at an organizational level are sometimes used, quite erroneously, as though they could make accurate predictions at an individual level (in Britain, academic selection tests at age 11 are criticized for just this reason). Moreover, tests which are designed to provide general predictions may then be construed within the context of a theory which "explains behavior." The phrase "explain behavior" is ambiguous as the specificity of what is being explained or predicted is not stated, and it is easy to imagine that these general theories should be capable of specific prediction of behavior.

Interestingly, one response of trait theorists to critics has been to set a lower level of specificity for the data to be explained. For example, Epstein (1979, 1980, 1983) suggests that behavior is found to be consistent if it is averaged over time for any one individual. That is, the average of a series of behaviors can be predicted even if specific behaviors can not (but see Raynor's comment, pp. 348–349). It may be that relatively nonspecific predictions were the covert goal of the early trait theorists—and that it needed the person × situation debate to make these covert assumptions visible.

Traits and processes may reflect different objectives to which those person variables are put, but that does not necessarily imply that the two sorts of person variables must be considered equally valid. If both trait and process variables can be used to explain individual differences (and the latter *are* used for just this purpose, as Raynor remarks) then a commitment to personality variables only capable of nonspecific prediction would not seem the best recipe for scientific advance. Furthermore, the trait concept can only be used to explain individual differences, whereas process theories are used to explain both individual differences and situational differences. As confidence in the ontological validity of a theoretical entity is increased by multiple and different "sightings"

(Hyland, 1981; Underwood, 1975), the use of theoretical concepts which have the wider application would seem to hold greater promise.

2. The Value of the Person × Situation Debate

Some similarities between Peake, Pervin, and myself should not obscure a fundamental difference in orientation. Both Peake and Pervin see the person × situation debate as having significant theoretical value. In particular, they both suggest that the debate has focused attention on alternatives to the trait approach of personality description. Peake refers to the work of Mischel (1973) and Bandura (1977) to illustrate these alternative person variable descriptions. Pervin refers to his own motivational theory (Pervin, 1982).

The reason I do not share this positive orientation to the debate is that almost any theoretical idea which has arisen can also be found in the psychological literature quite independantly of the person × situation issue. For example, the advocacy of motivational rather than trait theories of personality is realized in the *Dynamics of action* research program to which Raynor has contributed extensively (Atkinson, 1981). In his comment Raynor distinguishes state from trait theories of personality, a distinction equivalent to Peake's process versus trait distinction. Raynor shows how motivational theory (a state or process theory) "anticipates and/or coincides with" theoretical suggestions emerging from the person × situation debate. Although I share Peake's and Pervin's aversion to traits as a person variable, there is little evidence that trait theorists are abandoning their preferred concept in the light of the person × situation debate any more than they abandoned it the light of earlier theoretical arguments (McClelland, 1951; Murray, 1938). That, perhaps, is more a comment on the convenience of trait theory than on the value of the debate.

Another reason for my less positive orientation to the person × situation debate arises from the use of the term *interaction* as some kind of solution to the problem. An interaction—apart from having many meanings—is not a solution but the statement of a problem, a point which is emphasized by Powers. Peake refers to Bandura's recent interpretation of reciprocal determinism (Bandura, 1983), in which Bandura asserts, contrary to Pervin's transactionalism, that an interacting system must be decomposed into a series of separate unidirectional cause-and-effect relations. Powers shows here why Bandura is wrong and Pervin right: Analysis of simultaneous cause interactions *is* possible. Control theory, based on the use of simultaneous equations, was developed in an engineering context, (see also Powers, 1973, 1978) to deal specifically with simultaneous, not sequential, causal relations in an interaction. Powers argues that there is a fundamental difference between unidirectional process theories and those based on cyclical causality.

Although both process theories and consideration of interactions or transactions have been developed outside the context of the person × situation debate, the debate still has value when viewed in a more general historical context. First, it has forced trait theorists into a reevaluation of their position. Second, it has focused attention on the complexity of the concept of an interaction. Finally, and perhaps most important, it has drawn attention to a very neglected aspect of psychological investigation, the questions which are asked; on this point Peake and I are in complete agreement.

3. Questions

It is generally true to say that the value of an answer depends on the question asked (Bateson, 1979). In the context of the person × situation debate, questions of the form, How much variance do persons, situations, and interactions contribute to behavior? are thought to have little theoretical interest and contribute to so-called pseudoissues. Recently, the variance question has been reformulated (Epstein, 1979, 1980, 1983) in terms of stability coefficients from aggregated scores. However, Day, Marshall, Hamilton, and Christy (1983) show that stability coefficients are a function of the number of behavioral observations taken in any particular aggregation. As in the ANOVA studies, the answer depends on the parameters chosen. There is an inescapable feeling of *déjà vu* (see also Mischel & Peake, 1982) that questions relating to stability coefficients represent, yet again, a pseudoissue.

In rejecting such pseudoissues, many authors suggest that the more sensible question is, How do the person and situation interact to cause behavior? Although this question is certainly not new and also implies a unidirectional form of interactionism, as a question it would seem far too general actually to help direct research. A better and more precise form of question is to ask about the type of person variable and situation variable description in relation to certain explanatory objectives. The need to ask this last question has (to my knowledge) only emerged from the person × situation debate, and it is an issue which is sadly neglected in theoretical psychology.

To an extent, questions about description depend on assumptions concerning objectivity. Pervin suggests here that it is possible to arrive at an objective description of situations independently of person variables. Nowadays, the notion of objectivity (which is a characteristic of logical positivism) is generally rejected by philosophers of science (Chalmers, 1978), and instead it is agreed that descriptions of observation reflect theoretical assumptions. If this "theory-ladenness of observables" argument is applied to psychology, then any situational description, whether it is a "stimulus" or a "press," reflects theoretical assumptions about the etiology of behavior. To extend the argument further, situational description follows from assumptions inherent in the person variable concept (Hyland, 1981). In other words, it is a mistake to consider or classify situational description independently of person variables, for the former affects the description of the latter. A similar point is made by Raynor in relation to Lewin's theory. It follows, therefore, that theoretical assumptions embodied in the person variables—or hypothetical constructs, as I still like to call them—constitute the basis for psychological explanation. It is questions about these fundamental theoretical assumptions, rather than any other questions, which might be best used to direct further research.

4. References

Atkinson, J. W. Studying personality in the context of an advanced motivational psychology. *American Psychologist*, 1981, *36*, 117–128.

Bandura, A. Self-efficacy: Toward a unifying theory of behavioral change. *Psychological Review*, 1977, *84*, 191–215.

Bandura, A. Temporal dynamics and decomposition of reciprocal determinism: A reply to Philips and Orton. *Psychological Review*, 1983, *90*, 166–170.

Bateson, G. *Mind and nature*. London: Wildwood House, 1979.

Bowers, K. S. Situationism in psychology: An analysis and a critique. *Psychological Review*, 1973, *80*, 307–336.

Bowers, K. S. There's more to Iago than meets the eye: A clinical account of personal consistency. In D. Magnusson & N. S. Endler (Eds.), *Personality at the crossroads*. Hillsdale, N.J.: Lawrence Erlbaum, 1977.

Chalmers, A. F. *What is this thing called science?* Milton Keynes, U.K.: Open University Press, 1978.

Day, H. D., Marshal, D., Hamilton, B. & Christy, J. Some cautionary notes regarding the use of aggregated scores as a measure of behavioral stability. *Journal of Research in Personality*, 1983, *17*, 97–109.

Epstein, S. The stability of behavior: I. On predicting most of the people much of the time. *Journal of Personality and Social Psychology*, 1979, *37*, 1097–1126.

Epstein, S. The stability of behavior: II. Implications for psychological research. *American Psychologist*, 1980, *35*, 790–806.

Epstein, S. The stability of confusion: A reply to Mischel and Peake. *Psychological Review*, 1983, *90*, 179–184.

Hanson, N. R. *Patterns of discovery: An inquiry into the conceptual foundations of science.* London: Cambridge University Press, 1958.
Hyland, M. E. *Introduction to theoretical psychology.* London: Macmillan, 1981.
McClelland, D. C. *Personality.* New York: Sloane, 1951.
Mischel, W. Toward a cognitive social learning conceptualization of personality. *Psychological Review,* 1973, *80,* 252–283.
Mischel, W., & Peake, P. K. Beyond déjà vu in the search for cross-situational consistency. *Psychological Review,* 1982, *89,* 730–755.
Murray, H. A. *Explorations in personality.* New York: Oxford University Press, 1938.
Pervin, L. A. The stasis and flow of behavior: Toward a theory of goals. In M. M. Page (Ed.), *Nebraska symposium on motivation.* Nebraska: University of Nebraska Press, 1982.
Popper, K. R. *Conjectures and refutations.* London: Routledge, 1963.
Powers, W. T. *Behavior: The control of perception.* Chicago: Aldine, 1973.
Powers, W. T. Quantitative analysis of purposive systems. *Psychological Review,* 1978, *85,* 417–535.
Underwood, B. J. Individual differences as a crucible in theory construction. *American Psychologist,* 1975, *30,* 128–134.
Vernon, P. E. *Personality assessment: A critical survey.* New York: Wiley, 1964.

Author Index

Italic numbers indicate pages where complete reference citations are given.

Subject Index